Burris Numerical System - Expressing numbers as a function of space and time.

VOLUME 2

MY LEFT OUT RESEARCH

Written by Lloyd Dudley Burris

http://time-travel.institute

Volume 2 is all my left out research computer programs written in basic that my publisher would not let me fit into my first book. This book is not a book you read it is a book for studying my computer programs as they relate to my work on my Burris Numerical System. Which is a system for representing numbers in both time and space. So, that the entire number itself which was coded by a BNS computer program can be written down and carried around in your pocket no matter how large the number was.

When the BNS number that was written down and put into your pocket is put back into the BNS computer program then the entire number no matter how large it was can be extracted and used as seen fit.

This is not true data compression but an algorithm that simulates space-time itself. It simulates how space-time stores information. So, with that said there is nothing more here to say. Enjoy the computer programs. You need QB64 written by microsoft here is a link to download it from my web-sight. If you can't get it from me than go out on the internet and google it and download it.

http://time-travel.institute/qb64.zip

Good luck,

Lloyd Burris

```
"C:\Users\Reactor1967\vcns\work\102103B.BAS"
DECLARE SUB control (v#, r#, d#, sr#)
REM control does not work properly it does not hit every distance needed.
REM What this program is missing is a roll. Sometimes d# sames the same
REM while the speed of r# changes and sometimes the speed of r# stays the
REM same while d# changes. I call this the roll. Thats what this might need.
v# = 4
r# = 1
d# = 0
sr# = 0
CLS
RANDOMIZE TIMER
DO
z = INT(RND * 2) + 0
IF z = 0 THEN v# = v# + (v# - r# + 1)
IF z = 1 THEN v# = v# + (v# - r# + 2)
PRINT , , z; v#; r#; v# - r#; d#; sr#; "-------------"
a$ = INKEY$
IF a$ = "s" THEN STOP
CALL control(v#, r#, d#, sr#)
LOOP

SUB control (v#, r#, d#, sr#)
sr# = 0
d# = 0
DO
vr2# = (d# * 2) + 1
vr3# = (d# * 2) + 2
vr# = d# + sr#
REM PRINT , , d# + sr#; d#; sr#; (d# * 2) + 1; (d# * 2) + 2
IF (v# - r#) = vr# THEN EXIT DO
a$ = INKEY$
INPUT a$
IF a$ = "s" THEN STOP
d# = d# + 1
IF d# - sr# >= 3 THEN sr# = sr# + 4
LOOP
r# = r# + sr#
END SUB
```

```
"C:\Users\Reactor1967\vcns\work\102103A.BAS"
REM d# = v# - r# before coding data:sr# = speed of r# at current d#
REM and speed of r# for v# - r# after coding as listed below next 2 lines.
REM (d# * 2) + 1 is (v# - r#) after coding
REM (d# * 2) + 2 is (v# - r#) after coding
d# = 0
sr# = 0
CLS
DO
PRINT , , d# + sr#; d#; sr#; (d# * 2) + 1; (d# * 2) + 2
a$ = INKEY$
INPUT a$
IF a$ = "s" THEN STOP
d# = d# + 1
IF d# - sr# >= 3 THEN sr# = sr# + 4
LOOP




"C:\Users\Reactor1967\vcns\work\101903A.BAS"
REM Here I directly controlled the speed of r# to keep of with the speed
REM of v#. Also this is decodable.
REM You might be able to use d# as your test for decode. If d2# is the same
REM as d1# then your speed of r# decreases. If d2# is not the same as d1# then
REM your speed of r# stays the same. Only d# or sr# changes but not but at
REM the same time.
CLS
v# = 1
r# = 1
sr# = 0
z = 0
RANDOMIZE TIMER
DO
d# = v# - r#
IF test = -1 THEN z = 1 ELSE z = 0
REM z = INT(RND * 2) + 0
IF z = 0 THEN v# = v# + (v# - r#) + 1
IF z = 1 THEN v# = v# + (v# - r#) + 2
m# = (v# / 4) - INT(v# / 4): m# = m# * 10: m# = INT(m#)
m2# = ((v# - r#) / 4) - INT((v# - r#) / 4): m2# = m2# * 10: m2# = INT(m2#)
PRINT , z; m#; v#; r#; v# - r#; m2#; d#; d# - sr#; sr#
IF m2# = 5 THEN STOP
a$ = INKEY$
```

```
REM INPUT a$
IF a$ = "s" THEN STOP
IF a$ = "d" THEN EXIT DO
test = d# - sr# >= 3
IF test = -1 THEN sr# = sr# + 4
r# = r# + sr#
LOOP
PRINT "--------------------------------------------------------------------"
DO
z = (v# / 4) - INT(v# / 4): z = z * 10: z = INT(z)
test1 = (z = 2) OR (z = 7)
test2 = (z = 0) OR (z = 5)
IF test1 = -1 THEN z = 1
IF test2 = -1 THEN z = 0
d# = (v# - r#)
IF (d# / 2) - INT(d# / 2) = .5 THEN d# = d# - 1
d# = d# / 2
d# = d# + 1
IF z = 0 THEN d# = d# - 1
IF z = 1 THEN d# = d# - 2
PRINT , z; v#; r#; v# - r#; d#; d# - sr#; sr#
IF v# <= 1 THEN STOP
a$ = INKEY$
REM INPUT a$
IF a$ = "s" THEN STOP
dist# = v# - r#
IF (dist# / 2) - INT(dist# / 2) = .5 THEN dist# = dist# - 1
dist# = dist# / 2
dist# = dist# + 1
v# = v# - dist#
d# = v# - r#
r# = r# - sr#
IF z = 1 THEN sr# = sr# - 4
LOOP
```

"C:\Users\Reactor1967\vcns\work\101803A.BAS"
REM Here im just controlling the speed of r#. This seems like a good place
REM to be able to find a way to do a control code and find a place to shove
REM in some random data. Anyway here the point is to find a way to control
REM sr# and d# at the same time.
REM program not finished it has error that has to be traced.
v# = 1
r# = 1
d# = 0

```
sr# = 0
z = 1
CLS
RANDOMIZE TIMER
DO
REM z = INT(RND * 2) + 0
d# = v# - r#
IF (d# - sr#) = 2 THEN z = INT(RND * 2) + 0
IF z = 0 THEN v# = v# + (v# - r# + 1)
IF z = 1 THEN v# = v# + (v# - r# + 2)
m# = (v# / 4) - INT(v# / 4): m# = m# * 10: m# = INT(m#)
PRINT , z; m#; v#; r#; v# - r#; d#; d# - sr#; sr#
a$ = INKEY$
REM INPUT a$
 IF a$ = "s" THEN STOP
IF a$ = "d" THEN EXIT DO
test = d# - sr# >= 3
IF d# - sr# >= 3 THEN sr# = sr# + 4
r# = r# + sr#
IF test = -1 THEN z = 1 ELSE z = 0
LOOP
DO
m# = (v# / 4) - INT(v# / 4): m# = m# * 10: m# = INT(m#)
PRINT , z; m#; v#; r#; v# - r#; d#; d# - sr#; sr#
IF v# <= 1 THEN STOP
a$ = INKEY$
INPUT a$
IF a$ = "s" THEN STOP
z = (v# / 4) - INT(v# / 4): z = z * 10: z = INT(z)
test1 = (z = 2) OR (z = 7)
test2 = (z = 0) OR (z = 5)
IF test1 = -1 THEN z = 1
IF test2 = -1 THEN z = 2
dist# = (v# - r#)
IF (dist# / 2) - INT(dist# / 2) = .5 THEN dist# = dist# - 1
dist# = dist# / 2
dist# = dist# + 1
d# = dist#
IF z = 0 THEN d# = d# - 1
IF z = 1 THEN d# = d# - 2
test = d# - sr# = 2
r# = r# - sr#
IF test = -1 THEN GOTO skip:
IF z = 1 THEN sr# = sr# - 4
skip:
v# = v# - dist#
```

```
"C:\Users\Reactor1967\vcns\work\101703A.BAS"
REM Im studying the speed of v# - N and the speed of r#. What I learned is
REM for v# and r# to stay up with each other there has to be a correctional
REM system. What ever happens on one side also has to happen on the other.
REM N is determined by 1 or both things.
REM (1. the speed of r# such that v# / 4 = n)
REM (2. ((v# - r#) / 4) - int((v# - r#) / 4) = n)
REM The thing is to correct both v# and r# you need to know N at v2 and v1.
REM so the speed of r# here will be very important. I find that the speed
REM of r# should be divisiable by the base.
REM 0 to a 1 is d# = d# + 1 & sr# = sr# + 4
REM 1 to a 0 is d# = d# + 3 & sr# = sr# + 0
REM 0 to a 0 or 1 to a 1 d# = d# + 4 & sr# = sr# + 4
v# = 1: REM for prog1
r# = 1: REM for prog1
REM v# = 26: REM for prog2
REM r# = 9: REM for prog2
REM d# = 8: REM for prog2
REM sr# = 4: REM for prog2
REM z = 1: for prog2
CLS
RANDOMIZE TIMER
REM GOTO prog2:
prog1:
DO
REM z = 1
REM z = 0
REM IF z = 0 THEN z = 1 ELSE z = 0
z = INT(RND * 2) + 0
d# = v# - r#
IF z = 0 THEN v# = v# + (v# - r# + 1)
IF z = 1 THEN v# = v# + (v# - r# + 2)
PRINT , z; v#; r#; v# - r#; d#; d# - x1#; r# - sr#; (r# - sr#) - x2#
x1# = d#: x2# = (r# - sr#)
sr# = r#
a$ = INKEY$
REM INPUT a$
IF a$ = "s" THEN STOP
d# = INT((v# - r#) / 2)
DO
d# = d# + 1
s# = (v# - r#) - d#
```

```
IF (s# / 4) - INT(s# / 4) = 0 THEN EXIT DO
a$ = INKEY$: IF a$ = "s" THEN STOP
LOOP
r# = v# - d#
LOOP
STOP
prog2:
REM 0 to a 1 is d# = d# + 1 & sr# = r# + 4
REM 1 to a 0 is d# = d# + 3 & sr# = r# + 0
REM 0 to a 0 or 1 to a 1 d# = d# + 4 & sr# = r# + 4
DO
sz = z
z = INT(RND * 2) + 0
test1 = (z = 0) AND (sz = 0)
test2 = (z = 0) AND (sz = 1)
test3 = (z = 1) AND (sz = 1)
test4 = (z = 1) AND (sz = 0)
IF test1 = -1 THEN sr# = sr# + 4
IF test1 = -1 THEN d# = d# + 4
IF test2 = -1 THEN d# = d# + 3
IF test2 = -1 THEN sr# = sr# + 0
IF test3 = -1 THEN d# = d# + 4
IF test3 = -1 THEN sr# = sr# + 4
IF test4 = -1 THEN d# = d# + 1
IF test4 = -1 THEN sr# = sr# + 4
r# = r# + sr#
test = (v# - r#) = d#
REM IF test = 0 THEN STOP
IF z = 0 THEN v# = v# + (v# - r# + 1)
IF z = 1 THEN v# = v# + (v# - r# + 2)
PRINT , , v#; r#; v# - r#; d#; sr#
INPUT a$
IF a$ = "s" THEN STOP
LOOP


"C:\Users\Reactor1967\vcns\work\101603A.BAS"
REM Keeping sysmetery by controlling the speed of r# and also controlling
REM the how big or small v# - r# gets.
v# = 1029
r# = 1
RANDOMIZE TIMER
CLS
redo:
DO
```

```
z = INT(RND * 2) + 0
s# = INT((v# - r#) / 2)
DO
test = (s# / 4) - INT(s# / 4) = 0
IF test = -1 THEN EXIT DO
s# = s# + 1
a$ = INKEY$
IF a$ = "s" THEN STOP
LOOP
r# = r# + s#
d# = v# - r#
IF z = 0 THEN v# = v# + (v# - r# + 1)
IF z = 1 THEN v# = v# + (v# - r# + 2)
m# = (v# / 4) - INT(v# / 4): m# = m# * 10: m# = INT(m#)
PRINT , , z; m#; v#; r#; v# - r#; d#; d# - (r# - sr#); r# - sr#
sr# = r#
a$ = INKEY$
REM INPUT a$
IF a$ = "s" THEN STOP
LOOP UNTIL (v# - r#) <= 50
DO
z = INT(RND * 2) + 0
d# = INT((v# - r#) / 2)
DO
s# = (v# - r#) - d#
test = (s# / 4) - INT(s# / 4) = 0
IF test = -1 THEN EXIT DO
d# = d# + 1
a$ = INKEY$
IF a$ = "s" THEN STOP
LOOP
r# = v# - d#
d# = v# - r#
IF z = 0 THEN v# = v# + (v# - r# + 1)
IF z = 1 THEN v# = v# + (v# - r# + 2)
m# = (v# / 4) - INT(v# / 4): m# = m# * 10: m# = INT(m#)
PRINT , , z; m#; v#; r#; v# - r#; d#; d# - (r# - sr#); r# - sr#
sr# = r#
a$ = INKEY$
REM INPUT a$
IF a$ = "s" THEN STOP
LOOP UNTIL (v# - r#) > 1000
GOTO redo:
```

```
"C:\Users\Reactor1967\vcns\work\101503B.BAS"
REM this board is not correct. It seemed to work but does not.
REM this program only codes certain patterns or binary correctly. Some
REM implemention of a cordinate increment decrement system needs to be
REM put in place but from looking at this thing I don,t know thats possible
REM here. You can code all zero's at 8 successfully.
DIM can(10000)
datacount = 1
can(datacount) = 27: datacount = datacount + 1
a1# = 27
a2# = 4
a3# = 31
a4# = 32
a5# = 0
carry# = 27
cord# = 0
CLS
RANDOMIZE TIMER
DO
test1 = (carry# = a1#)
IF test1 = -1 THEN z = INT(RND * 2) + 0
REM z = 1
REM IF a2# = 8 THEN z = 0 this fix no good only codes 0's
test2 = (carry# = a1#) AND (z = 0)
test3 = (carry# = a1#) AND (z = 1)
IF test2 = -1 THEN carry# = a3#
IF test3 = -1 THEN carry# = a4#
REM ------------------------------------------------------------------------
REM  error checking and testing routine disabled
IF test2 = -1 THEN can(datacount) = carry#
IF test2 = -1 THEN datacount = datacount + 1
IF test3 = -1 THEN can(datacount) = carry#
IF test3 = -1 THEN datacount = datacount + 1
REM  IF datacount = 8000 THEN can(datacount - 1) = 1967: REM Trying to
test error catching to make sure it will catch an error.
REM ------------------------------------------------------------------------
REM COORDINATE CODING SYSTEM
ctest1 = (test1 = -1) AND (a2# <= 3): REM cord = cord + 0
ctest2 = (test1 = -1) AND (a2# = 4) AND (z = 0): REM cord = cord + 0
ctest3 = (test1 = -1) AND (a2# = 4) AND (z = 1): REM cord = cord + 1
ctest4 = (test1 = -1) AND (a2# >= 5) AND (a2# <= 7): REM cord = cord + 1
ctest5 = (test1 = -1) AND (a2# = 8): REM cord = cord + 1
IF ctest1 = -1 THEN cord# = cord# + 0
IF ctest2 = -1 THEN cord# = cord# + 0
IF ctest3 = -1 THEN cord# = cord# + 1
```

```
IF ctest4 = -1 THEN cord# = cord# + 1
IF ctest5 = -1 THEN cord# = cord# + 1
REM if test1 = -1 then cord# = a2#
REM ------------------------------------------------------------------------
m# = (carry# / 4) - INT(carry# / 4): m# = m# * 10: m# = INT(m#)
PRINT datacount, , a1#; a2#; a3#; a4#; a5#; carry#; z; m#; cord#
REM PRINT a1#; a2#; a3#; a4#; a5#; carry#; cord#
REM PRINT , , a1#; a2#; a3#; a4#; a5#
a$ = INKEY$
INPUT a$
IF a$ = "s" THEN STOP
IF a$ = "d" THEN EXIT DO
IF datacount = 10001 THEN EXIT DO
a1# = a1# + 1
a2# = a2# + 1
IF a2# = 9 THEN a5# = a5# + 1
IF a2# = 9 THEN a2# = 1
test1 = (a2# = 1)
test2 = (a2# = 2)
test3 = (a2# = 3)
test4 = (a2# = 4)
test5 = (a2# = 5)
test6 = (a2# = 6)
test7 = (a2# = 7)
test8 = (a2# = 8)
IF test1 = -1 THEN a3# = a3# - 6
IF test1 = -1 THEN a4# = a4# + 2
IF test2 = -1 THEN a3# = a3# + 2
IF test2 = -1 THEN a4# = a4# + 2
IF test3 = -1 THEN a3# = a3# + 2
IF test3 = -1 THEN a4# = a4# + 2
IF test4 = -1 THEN a3# = a3# + 2
IF test4 = -1 THEN a4# = a4# + 2
IF test5 = -1 THEN a3# = a3# + 2
IF test5 = -1 THEN a4# = a4# + 2
IF test6 = -1 THEN a3# = a3# + 2
IF test6 = -1 THEN a4# = a4# + 2
IF test7 = -1 THEN a3# = a3# + 2
IF test7 = -1 THEN a4# = a4# + 2
IF test8 = -1 THEN a3# = a3# + 2
IF test8 = -1 THEN a4# = a4# - 6
LOOP
PRINT "--------------------------------------------------------------------"
datacount = datacount - 2
DO
test = (cord# = a5#) AND ((carry# = a3#) OR (carry# = a4#))
```

```
IF test = -1 THEN carry# = a1#
IF test = -1 THEN cord# = cord# - 1
IF cord# < 0 THEN cord# = 0
            REM some system needs to be in place here to
            REM correct decode errors.(Cord# correction here.)
PRINT datacount, , a1#; a2#; a3#; a4#; a5#; carry#; cord#
REM PRINT a1#; a2#; a3#; a4#; a5#; carry#; cord#
REM -----------------------------------------------------------------
vcns = (test = -1) AND (carry# <> (can(datacount)))
IF vcns = -1 THEN PRINT , , "DATA DECODE ERROR!!!"; can(datacount)
IF vcns = -1 THEN INPUT z$
IF vcns = -1 THEN STOP
IF test = -1 THEN datacount = datacount - 1
REM -----------------------------------------------------------------
IF a1# <= 27 THEN EXIT DO
a$ = INKEY$
REM INPUT a$
IF a$ = "s" THEN STOP
a1# = a1# - 1
a2# = a2# - 1
IF a2# = 0 THEN a5# = a5# - 1
IF a2# = 0 THEN a2# = 8
test1 = (a2# = 1)
test2 = (a2# = 2)
test3 = (a2# = 3)
test4 = (a2# = 4)
test5 = (a2# = 5)
test6 = (a2# = 6)
test7 = (a2# = 7)
test8 = (a2# = 8)
IF test1 = -1 THEN a3# = a3# + 6
IF test1 = -1 THEN a4# = a4# - 2
IF test2 = -1 THEN a3# = a3# - 2
IF test2 = -1 THEN a4# = a4# - 2
IF test3 = -1 THEN a3# = a3# - 2
IF test3 = -1 THEN a4# = a4# - 2
IF test4 = -1 THEN a3# = a3# - 2
IF test4 = -1 THEN a4# = a4# - 2
IF test5 = -1 THEN a3# = a3# - 2
IF test5 = -1 THEN a4# = a4# - 2
IF test6 = -1 THEN a3# = a3# - 2
IF test6 = -1 THEN a4# = a4# - 2
IF test7 = -1 THEN a3# = a3# - 2
IF test7 = -1 THEN a4# = a4# - 2
IF test8 = -1 THEN a3# = a3# - 2
IF test8 = -1 THEN a4# = a4# + 6
```

```
LOOP
PRINT "Data decoded correctly."




"C:\Users\Reactor1967\vcns\work\101503A.BAS"
REM test notes:In base three its not as simple as base to cord incrments by
REM 2 sometimes instead of 1. So, it might be wise to use a binary system
REM and keep track of cord since when you incode or decode cord always
REM increments or decrements by 1.
vr# = 1
CLS
y# = 0
cord# = 0
carry1# = 1
carry2# = cord#
RANDOMIZE TIMER
DO
x# = vr#
DO
x# = x# + 1
m# = (x# / 3) - INT(x# / 3): m# = m# * 10: m# = INT(m#)
test1 = (m# = 0)
d# = INT(x# / 3)
IF test1 = -1 THEN d# = d# - 1
s# = vr# - d#
test2 = (s# / 3) - INT(s# / 3) = 0
test = (m# = 3) AND (x# - vr# > y#) AND (test2 = -1)
IF test = -1 THEN EXIT DO
a$ = INKEY$: IF a$ = "s" THEN STOP
LOOP
y# = x# - vr#
m1# = (x# / 3) - INT(x# / 3): m1# = m1# * 10: m1# = INT(m1#)
m2# = ((x# + 1) / 3) - INT((x# + 1) / 3): m2# = m2# * 10: m2# = INT(m2#)
m3# = ((x# + 2) / 3) - INT((x# + 2) / 3): m3# = m3# * 10: m3# = INT(m3#)
z = INT(RND * 3) + 0
test0 = (vr# = carry1#)
test1 = (vr# = carry1#) AND (z = 0)
test2 = (vr# = carry1#) AND (z = 1)
test3 = (vr# = carry1#) AND (z = 2)
IF test1 = -1 THEN carry1# = x#
IF test2 = -1 THEN carry1# = (x# + 1)
IF test3 = -1 THEN carry1# = (x# + 2)
IF test0 = -1 THEN carry2# = cord#
REM PRINT vr#; y#; x#; m1#; x# + 1; m2#; x# + 2; m3#; cord#; carry1#;
carry2#
```

```
PRINT vr#; x#; m1#; x# + 1; m2#; x# + 2; m3#; cord#; carry1#; carry2#; y#
REM IF carry2# - scarry2# > 1 THEN STOP
IF y# < sy# THEN cord# = cord# + 1
sy# = y#
IF test0 = -1 THEN scarry2# = carry2#
IF y# >= 21 THEN y# = 0
a$ = INKEY$
REM INPUT a$
IF a$ = "s" THEN STOP
vr# = vr# + 1
LOOP




"C:\Users\Reactor1967\vcns\work\101403A.BAS"
REM This program was to help me lear how to find d# in base 3
REM for 0 and 1 just divide d# = (v# - r#) / base
REM for 2 divide d# = ((v# - r#) / base) + 1
v# = 1
r# = 1
d# = 0
CLS
RANDOMIZE TIMER
DO
z = INT(RND * 3) + 0
r# = v# - d#
IF z = 0 THEN v# = v# + ((v# - r#) * 2) + 1
IF z = 1 THEN v# = v# + ((v# - r#) * 2) + 2
IF z = 2 THEN v# = v# + ((v# - r#) * 2) + 3
m# = ((v# - r#) / 3) - INT((v# - r#) / 3): m# = m# * 10: m# = INT(m#)
m2# = (v# / 3) - INT(v# / 3): m2# = m2# * 10: m2# = INT(m2#)
IF z = 0 THEN PRINT z; m2#; v#; r#; v# - r#; m#; d#; d# - (r# - sr#); r# - sr#;
INT((v# - r#) / 3); d# - sd#
IF z = 1 THEN PRINT z; m2#; v#; r#; v# - r#; m#; d#; d# - (r# - sr#); r# - sr#;
INT((v# - r#) / 3); d# - sd#
IF z = 2 THEN PRINT z; m2#; v#; r#; v# - r#; m#; d#; d# - (r# - sr#); r# - sr#;
INT((v# - r#) / 3) - 1; d# - sd#
sr# = r#
sd# = d#
a$ = INKEY$
IF a$ = "s" THEN STOP
DO
d# = d# + 1
s# = (v# - r#) - d#
IF (s# / 3) - INT(s# / 3) = 0 THEN EXIT DO
```

```
a$ = INKEY$
IF a$ = "s" THEN STOP
LOOP
LOOP




"C:\Users\Reactor1967\vcns\work\101103B.BAS"
CLS
REM cord# stands for coordinate on code decode chart something like that
REM would have to be used for this to be useful.
cord# = 0
vr# = 27
DO
x# = vr#
z = 0
redo:
DO
x# = x# + 1
test1 = ((x# / 2) - INT(x# / 2) = 0) AND (z = 1)
test2 = ((x# / 2) - INT(x# / 2) = .5) AND (z = 0)
IF test1 = -1 THEN EXIT DO
IF test2 = -1 THEN EXIT DO
a$ = INKEY$
IF a$ = "s" THEN STOP
LOOP
dist# = x#
IF (dist# / 2) - INT(dist# / 2) = .5 THEN dist# = dist# - 1
dist# = dist# / 2
dist# = dist# + 1
IF z = 0 THEN dist# = dist# - 1
IF z = 1 THEN dist# = dist# - 2
s# = vr# - dist#
test = (s# / 4) - INT(s# / 4) = 0
a$ = INKEY$: IF a$ = "s" THEN STOP
IF test = 0 THEN GOTO redo:
IF ((dist# * 2) + z + 1) - vr# = 1 THEN cord# = cord# + 1
PRINT , , vr#; ((dist# * 2) + z + 1) - vr#; (dist# * 2) + z + 1;
z = 1
x# = vr#
redo2:
DO
x# = x# + 1
test1 = ((x# / 2) - INT(x# / 2) = 0) AND (z = 1)
test2 = ((x# / 2) - INT(x# / 2) = .5) AND (z = 0)
IF test1 = -1 THEN EXIT DO
```

```
IF test2 = -1 THEN EXIT DO
a$ = INKEY$
IF a$ = "s" THEN STOP
LOOP
dist# = x#
IF (dist# / 2) - INT(dist# / 2) = .5 THEN dist# = dist# - 1
dist# = dist# / 2
dist# = dist# + 1
IF z = 0 THEN dist# = dist# - 1
IF z = 1 THEN dist# = dist# - 2
s# = vr# - dist#
test = (s# / 4) - INT(s# / 4) = 0
a$ = INKEY$: IF a$ = "s" THEN STOP
IF test = 0 THEN GOTO redo2:
PRINT (dist# * 2) + z + 1; cord#
x# = x# + 1
a$ = INKEY$
REM INPUT a$
IF a$ = "s" THEN STOP
vr# = vr# + 1
LOOP




"C:\Users\Reactor1967\vcns\work\101103A.BAS"
v# = 1
r# = 1
CLS
DO
z = INT(RND * 2) + 0
x# = v# - r#
redo:
DO
x# = x# + 1
test1 = ((x# / 2) - INT(x# / 2) = 0) AND (z = 1)
test2 = ((x# / 2) - INT(x# / 2) = .5) AND (z = 0)
IF test1 = -1 THEN EXIT DO
IF test2 = -1 THEN EXIT DO
a$ = INKEY$
IF a$ = "s" THEN STOP
LOOP
dist# = x#
IF (dist# / 2) - INT(dist# / 2) = .5 THEN dist# = dist# - 1
dist# = dist# / 2
```

```
dist# = dist# + 1
IF z = 0 THEN dist# = dist# - 1
IF z = 1 THEN dist# = dist# - 2
r2# = v# - dist#
y# = r2# - r1#
a$ = INKEY$: IF a$ = "s" THEN STOP
IF (y# / 4) - INT(y# / 4) <> 0 THEN GOTO redo:
r# = v# - dist#
d# = v# - r#
IF z = 0 THEN v# = v# + (v# - r# + 1)
IF z = 1 THEN v# = v# + (v# - r# + 2)
m# = (v# / 4) - INT(v# / 4): m# = m# * 10: m# = INT(m#)
PRINT , z; m#; v#; r#; v# - r#; d# - (r# - sr#); d#; r# - sr#
sr# = r#
a$ = INKEY$
IF a$ = "s" THEN STOP
LOOP




"C:\Users\Reactor1967\vcns\work\101003E.BAS"
REM It seems either v# - r# will represent sysmetery or
REM v# / base will represent sysmetery one or the other.
REM -----------------------------------------------------------
REM Implement changes below in next program.
REM Trying something new. Try to always know c# So that if your
REM Maybe looking in a previous program wil help figure x's out.
REM going from a 0 to a 0 c# = c# + x1#
REM going from a 0 to a 1 c# = c# + x2#
REM going from a 1 to a 0 c# = c# + x3#
REM going from a 1 to a 1 c# = c# + x4#
REM So that either the speed of r# can be found by v# / base
REM or so that when (v2# - r1#) - c# = (v1# - r1#) you know what
REM to subtract from c# which would be x?#.
v# = 121
r# = 99
z = 1
CLS
RANDOMIZE TIMER
s# = 0
DO
sz = z
z = INT(RND * 2) + 0
test = (sz = z)
IF test = -1 THEN c# = 2
IF test = 0 THEN c# = 3
```

```
d# = (v# - r#) + c#
IF (d# / 2) - INT(d# / 2) = .5 THEN d# = d# - 1
d# = d# / 2
d# = d# + 1
IF z = 0 THEN d# = d# - 1
IF z = 1 THEN d# = d# - 2
r# = v# - d#
IF z = 0 THEN v# = v# + (v# - r# + 1)
IF z = 1 THEN v# = v# + (v# - r# + 2)
m# = ((v# - r#) / 4) - INT((v# - r#) / 4): m# = m# * 10: m# = INT(m#)
m2# = (v# / 4) - INT(v# / 4): m2# = m2# * 10: m2# = INT(m2#)
PRINT z; m#; m2#; v#; r#; v# - r#; c#; d#; r# - sr#; (r# - sr#) - rsr#
rsr# = (r# - sr#)
sr# = r#
a$ = INKEY$
REM INPUT a$
IF a$ = "s" THEN STOP
LOOP
```

"C:\Users\Reactor1967\vcns\work\101003D.BAS"
```
REM this program explores the c# variable
REM if e# - (d# - e#) is negative then v# is going down in value toward
REM 0. If e# - (d# - e#) is postive then v# is going up in value away
REM from zero. If v# is negative then reverse this.
REM Next time now we know how c# works lets learn how to use it.
REM write a program to keep c# constant as possible.
d# = 2
e# = 0
CLS
DO
PRINT , , d#; e#; d# - e#; " "; e# - (d# - e#)
a$ = INKEY$
INPUT a$'
IF a$ = "s" THEN STOP
e# = e# + 1
test = e# > d#
IF test = -1 THEN d# = d# + 1
IF test = -1 THEN e# = 0
LOOP
```

"C:\Users\Reactor1967\vcns\work\101003C.BAS"
```
REM IDEAL HERE IS TO TELL n BY (V# - R#) / BASE
```

```
REM So, subtract your (v2# - r2#) - c# to find (v1# - r1#).
REM then find your N for v1#. Looking at N1 and N2 update c# so
REM c# will reflect what is needed to repeat this process.
REM Right now this programs does not work correctly as intended.
REM I believe writing a program that controls the difference
REM between your speed or v# minus your speed of r# is the trick.
REM if you can control the difference between the two speeds in a
REM controlable way you got it.
REM c# = c# +- (x#) +- difference in N
v# = 1
r# = 1
c# = 1
z = 1
CLS
RANDOMIZE TIMER
DO
sz = z
z = INT(RND * 2) + 0
IF c# < 100 THEN flag = 1
IF c# > 1000 THEN flag = 0
IF flag = 1 THEN c# = c# + 4
IF flag = 0 THEN c# = c# - 4
test1 = (sz = 0) AND (z = 1)
test2 = (sz = 1) AND (z = 0)
IF test1 = -1 THEN c# = c# + 1
IF test2 = -1 THEN c# = c# - 1
y# = (v# - r#) + c#
IF (y# / 2) - INT(y# / 2) = .5 THEN y# = y# - 1
y# = y# / 2
y# = y# + 1
IF z = 0 THEN y# = y# - 1
IF z = 1 THEN y# = y# - 2
IF z = 0 THEN c# = c# - 1
IF z = 1 THEN c# = c# - 2
r# = v# - y#
i# = v# - r#
IF z = 0 THEN v# = v# + (v# - r# + 1)
IF z = 1 THEN v# = v# + (v# - r# + 2)
m# = ((v# - r#) / 4) - INT((v# - r#) / 4): m# = m# * 10: m# = INT(m#)
PRINT z; m#; v#; r#; v# - r#; (v# - r#) - svr#; i#; r# - sr#; i# - (r# - sr#)
sr# = r#
svr# = (v# - r#)
REM IF c# >= 1051 THEN STOP
a$ = INKEY$
REM INPUT a$
IF a$ = "s" THEN STOP
```

LOOP
REM decoding

"C:\Users\Reactor1967\vcns\work\101003B.BAS"
REM NEW MATH EQUATION (SPEED OF R2#) - ((C# / BASE) - n) =
SPEED OF R1#
REM correction (speed of r1# - (c1# / 2) round down to the even number.
REM correction the speed of r# is changed by the previous c#/2
REM Here I have really advanced c# so that no multiplication is needed
REM with a method I used way back with fractions. Fractions are great
REM sometimes.
REM Now that it is easier to figure c# try to find a way to decode c# with
REM v# and r#. Once you got c# you have it made just see my other equations.
REM WHAT YOU NEED TO MAKE THIS WORK IS TO USE N1 & N2 TO
TELL YOU
REM WHAT YOU NEED TO DO WITH YOUR MATH EQUATIONS. THEN
WHEN YOU DECODE
REM N2 AND N1 WILL TELL YOU WHAT YOU NEED TO DO WITH
YOUR MATH EQUATIONS
REM TO DECODE. THATS WHAT YOU DO TO MAKE THIS WORK.
REM develope a chart for c that tells you what c you goto for N. be able
REM to use that chart for decode and you got it. You can run a program
REM like this to look at your posibilities for c to c from N to N
v# = 1
r# = 1
c# = 1
z = 1
CLS
RANDOMIZE TIMER
DO
sz = z
z = INT(RND * 2) + 0
x1# = (c# / 4) - INT(c# / 4): x1# = x1# * 10: x1# = INT(x1#)
IF x1# = 2 THEN c# = c# + 1: REM equals 5 then c# = c# - 1
IF x1# = 5 THEN c# = c# + 2: REM equals 0 then c# = c# - 2
IF x1# = 7 THEN c# = c# + 3: REM equals 5 then c# = c# - 3
IF x1# = 0 THEN c# = c# + 4: REM equals 0 then c# = c# - 4
test1 = (sz = 0) AND (z = 1)
test2 = (sz = 1) AND (z = 0)
IF test1 = -1 THEN c# = c# + 1
IF test2 = -1 THEN c# = c# - 1
y# = (v# - r#) + c#
IF (y# / 2) - INT(y# / 2) = .5 THEN y# = y# - 1

```
y# = y# / 2
y# = y# + 1
IF z = 0 THEN y# = y# - 1
IF z = 1 THEN y# = y# - 2
r# = v# - y#
i# = v# - r#
IF z = 0 THEN v# = v# + (v# - r# + 1)
IF z = 1 THEN v# = v# + (v# - r# + 2)
m# = (v# / 4) - INT(v# / 4): m# = m# * 10: m# = INT(m#)
PRINT z; m#; v#; r#; v# - r#; c#; c# - sc#; i#; i# - si#; r# - sr#; (r# - sr#) - rsr#
REM IF c# >= 1051 THEN STOP
si# = i#
rsr# = (r# - sr#)
sc# = c#
sr# = r#
a$ = INKEY$
REM INPUT a$
IF a$ = "s" THEN STOP
IF c# = 9 THEN c# = 1
LOOP
REM decoding




"C:\Users\Reactor1967\vcns\work\101003A.BAS"
REM c# = (v# - r#) - svr#
REM What I have learned. You reall do not what your c# to increase then
REM revert to a lower number. C# needs to increase and decrease
REM in a controled fashion. Because with c# you can calculate your
REM previous speed of r# which is very important if not critical to
REM this program ever working.
v# = 1
r# = 1
CLS
DO
z = INT(RND * 2) + 0
d# = v# - r#
IF z = 0 THEN v# = v# + (v# - r# + 1)
IF z = 1 THEN v# = v# + (v# - r# + 2)
PRINT , , z; v#; r#; v# - r#; (v# - r#) - svr#; d#; r# - sr#
svr# = (v# - r#)
sr# = r#
dist# = INT((v# - r#) / 2)
DO
dist# = dist# + 1
s# = (v# - r#) - dist#
```

```
test = (s# / 4) - INT(s# / 4) = 0
IF test = -1 THEN EXIT DO
a$ = INKEY$: IF a$ = "s" THEN STOP
LOOP
r# = v# - dist#
LOOP
```


```
"C:\Users\Reactor1967\vcns\work\100903B.BAS"
v# = 1
r# = 1
c# = 1
RANDOMIZE TIMER
CLS
DO
z = INT(RND * 2) + 0
IF z = 0 THEN x# = (v# - r#) + c#
IF z = 1 THEN x# = (v# - r#) + c# + 1
IF (x# / 2) - INT(x# / 2) = .5 THEN x# = x# - 1
x# = x# / 2
x# = x# + 1
IF z = 0 THEN x# = x# - 1
IF z = 1 THEN x# = x# - 2
r# = v# - x#
IF z = 0 THEN v# = v# + (v# - r#) + 1
IF z = 1 THEN v# = v# + (v# - r#) + 2
m# = ((v# - r#) / 4) - INT((v# - r#) / 4): m# = m# * 10: m# = INT(m#)
PRINT , , z; m#; v#; r#; v# - r#; c#
a$ = INKEY$
REM INPUT a$
IF a$ = "s" THEN STOP
IF z = 0 THEN c# = (c# * 2) - 1
IF z = 1 THEN c# = (c# * 2)
test = c# > 8
IF test = -1 THEN c# = c# - 8
LOOP
```


```
"C:\Users\Reactor1967\vcns\work\100903A.BAS"
REM New discovery. Speed of r# only increases when c# > 8 here.
REM NEW MATH EQUATION SPEED OF R# = (D# - C# - N)
REM D# = (V# - R#) BEFORE CODING. N = numerical value being coded.
REM c# = (v2# - r2#) - (v1# - r1#):r# = reference point.
REM NEW MATH EQUATION C# = (SPEED OF V#) - (SPEED OF R#) + n
```

```
REM im working with a new variable called c#. Its what you add to v# - r#
REM to get your next v# - r#. You have to know each v# - r# before you code
REM it. But, now you have to learn to know what v# - r# to decode too when
REM you get your next n when decoding.
REM This program correctly calculated c# and used it.
REM now I have successfully worked with c# I can try to come up with a
REM different way of calculating c# so I can decode. Maybe creating a
REM chart or new math equations for c#. Who knows.
v# = 1
r# = 1
c# = 1
z = 1
CLS
RANDOMIZE TIMER
DO
sz = z
z = INT(RND * 2) + 0
c# = c# * 2
IF c# > 8 THEN c# = c# - 8
test1 = (sz = 0) AND (z = 1)
test2 = (sz = 1) AND (z = 0)
IF test1 = -1 THEN c# = c# + 1
IF test2 = -1 THEN c# = c# - 1
y# = (v# - r#) + c#
IF (y# / 2) - INT(y# / 2) = .5 THEN y# = y# - 1
y# = y# / 2
y# = y# + 1
IF z = 0 THEN y# = y# - 1
IF z = 1 THEN y# = y# - 2
r# = v# - y#
i# = v# - r#
IF z = 0 THEN v# = v# + (v# - r# + 1)
IF z = 1 THEN v# = v# + (v# - r# + 2)
m# = (v# / 4) - INT(v# / 4): m# = m# * 10: m# = INT(m#)
PRINT z; m#; v#; r#; v# - r#; c#; i#; r# - sr#
sr# = r#
a$ = INKEY$
INPUT a$
IF a$ = "s" THEN STOP
IF c# = 9 THEN c# = 1
LOOP
```

"C:\Users\Reactor1967\vcns\work\100803E.BAS"

```
REM NEW MATH EQUATION C# = C# * 2 FOR 0 OR C# = (C# * 2) + 1
FOR 1
REM IF C# IS GREATER THEN YOUR LARGEST C# THEN C# = C# -
(LARGEST C#)
REM ALSO I BELIEVE THE SPEED OF R# IS (C# / BASE).
REM this program shows what v# - r# goes to each v# - r# for N.
REM this is why todays programs where having trouble decoding.
REM The first choice is not always the correct choice when decoding.
REM that has to be fixed before a decode is possible. You have to
REM have to know specificly what to do before decoding there can
REM be no doubt.
REM FIX 1:If you know what v# you started with you could chart the
REM v's onto this so by looking at v# you could tell what v# - r#
REM to goto. You would have to have specific ranges of v# to make
REM this work.
REM Maybe not. Just having a even r or a odd r could be all the difference
REM to tell how for to go back. This might can be worked out some strange
REM way.
REM this program is wrong on how to do c# refer to 100803d.bas
CLS
dist# = 1
DO
x# = dist#
m1# = (dist# / 4) - INT(dist# / 4): m1# = m1# * 10: m1# = INT(m1#)
IF m1# = 0 THEN z = 1
IF m1# = 2 THEN z = 0
IF m1# = 5 THEN z = 1
IF m1# = 7 THEN z = 0
IF (x# / 2) - INT(x# / 2) = .5 THEN x# = x# - 1
x# = x# / 2
x# = x# + 1
IF z = 0 THEN x# = x# - 1
IF z = 1 THEN x# = x# - 2
DO
x# = x# + 1
m2# = (x# / 4) - INT(x# / 4): m2# = m2# * 10: m2# = INT(m2#)
test = (m1# = m2#)
IF test = -1 THEN EXIT DO
a$ = INKEY$: IF a$ = "s" THEN STOP
LOOP
l# = (dist# / 4) - INT(dist# / 4): l# = l# * 10: l# = INT(l#)
c# = ((x# * 2) + 1) - dist#
PRINT , , dist#; c#; (x# * 2) + 1; (x# * 2) + 2
INPUT a$
IF a$ = "s" THEN STOP
dist# = dist# + 1
```

```
"C:\Users\Reactor1967\vcns\work\100803D.BAS"
REM NEW MATH EQUATION: RATE OF CHANGE OF D = (((SPEED OF
V) - (SPEED OF R)) / BASE)
REM rate of change of v - r = (rate of change of d) + (rate of change of v)
REM must add more test during coding that can be used during
REM decoding to be successful here.
REM GOAL STRIVING OR HERE IS TO DECODE. Look at N before you
decode:
REM Look at N after you decode. From that you know how much to decrement
REM your v# - r#, or d#, or speed of r# what ever. Just anything to
REM decode.
REM It seems like reading your binary ahead of time and organizing
REM some time of using coding pattern that can be reconized with decoding
REM is the best option here.
CLS
v# = 1
r# = 1
d# = 0
RANDOMIZE TIMER
DO
z = INT(RND * 2) + 0
dist# = v# - r#
x# = (v# / 4) - INT(v# / 4): x# = x# * 10: x# = INT(x#)
DO
d# = d# + 1
x2# = (d# / 4) - INT(d# / 4): x2# = x2# * 10: x2# = INT(x2#)
test = (x# = x2#)
a$ = INKEY$: IF a$ = "s" THEN STOP
IF test = -1 THEN EXIT DO
LOOP
r# = v# - d#
sv# = v#
py# = v# - r#
IF z = 0 THEN v# = v# + (v# - r# + 1)
IF z = 1 THEN v# = v# + (v# - r# + 2)
y# = (v# - r#)
m3# = (py# / 4) - INT(py# / 4): m3# = m3# * 10: m3# = INT(m3#)
m2# = (y# / 4) - INT(y# / 4): m2# = m2# * 10: m2# = INT(m2#)
m# = (v# / 4) - INT(v# / 4): m# = m# * 10: m# = INT(m#)
PRINT z; m#; m2#; m3#; v#; r#; v# - r#; (v# - r#) - svr#; d#; d# - sd#; v# - sv#;
(v# - sv#) - vsv#; r# - sr#; (r# - sr#) - rsr#
rsr# = (r# - sr#)
```

```basic
vsv# = (v# - sv#)
svr# = v# - r#
sd# = d#
sr# = r#
a$ = INKEY$
REM INPUT a$
IF a$ = "s" THEN STOP
LOOP
```

"C:\Users\Reactor1967\vcns\work\100803C.BAS"

```basic
REM This program here is the stuff man. Now when we code we move v# - r#
REM up. Check the speed of v# - r# and check N. If everything falls into
REM place we code to that distance. When we decode Look at N by dividing
REM v# by 4 and examing its remainder. If you know its remainder then you
REM should know what the remainder of v# - r# is for that N. Just derement
REM v# - r# to you get that remainder. Repeat decode until finished.
REM binary 0 equals a odd distance
REM binary 1 equals a even distance
REM Conclusion of testing results. Sometimes the first match when
REM decrement v# - r# is not the correct one. Try to get it there.
REM Write down all your information that you need in for the criteria
REM that your variables must meet when coding and decoding and code
REM for the same method that can be reversed and used to decode with.
v# = 100
r# = 99
RANDOMIZE TIMER
CLS
DO
z2 = INT(RND * 2) + 0
dist2# = (v# - r#)
lb# = r#
redo:
dist2# = dist2# + 1
dist# = dist2#
test = (dist# / 2) - INT(dist# / 2)
IF test = .5 THEN z = 0
IF test = 0 THEN z = 1
a$ = INKEY$: IF a$ = "s" THEN STOP
IF z <> z2 THEN GOTO redo:
IF test = .5 THEN dist# = dist# - 1
dist# = dist# / 2
dist# = dist# + 1
IF z = 0 THEN dist# = dist# - 1
IF z = 1 THEN dist# = dist# - 2
```

```
r# = v# - dist#
test = ((r# - lb#) / 4) - INT((r# - lb#) / 4) = 0
a$ = INKEY$: IF a$ = "s" THEN STOP
IF test = 0 THEN GOTO redo:
dist# = (v# - r#)
IF z = 0 THEN v# = v# + (v# - r# + 1)
IF z = 1 THEN v# = v# + (v# - r# + 2)
y# = v# - r#
m3# = (dist# / 4) - INT(dist# / 4): m3# = m3# * 10: m3# = INT(m3#)
m1# = (y# / 4) - INT(y# / 4): m1# = m1# * 10: m1# = INT(m1#)
m# = (v# / 4) - INT(v# / 4): m# = m# * 10: m# = INT(m#)
PRINT z; m#; m1#; m3#; v#; r#; v# - r#; dist#; r# - sr#
sr# = r#
a$ = INKEY$
REM INPUT a$
IF a$ = "s" THEN STOP
IF a$ = "d" THEN EXIT DO
LOOP
DO
q# = (v# / 4) - INT(v# / 4): q# = q# * 10: q# = INT(q#)
IF q# = 0 THEN z = 0
IF q# = 2 THEN z = 1
IF q# = 5 THEN z = 0
IF q# = 7 THEN z = 1
y# = v# - r#
m1# = (y# / 4) - INT(y# / 4): m1# = m1# * 10: m1# = INT(m1#)
m# = (v# / 4) - INT(v# / 4): m# = m# * 10: m# = INT(m#)
PRINT z; m#; m1#; v#; r#; v# - r#; dist#; sr# - r#
INPUT a$
IF a$ = "s" THEN STOP
REM ----------------------------------------------------------
x# = v# - r#: REM storing the value of v# - r#
REM ----------------------------------------------------------
REM Decoding for previous v#
dist# = v# - r#
IF (dist# / 2) - INT(dist# / 2) = .5 THEN dist# = dist# - 1
dist# = dist# / 2
dist# = dist# + 1
v# = v# - dist#
REM ----------------------------------------------------------
REM Get N value of v#
q2# = (v# / 4) - INT(v# / 4): q2# = q2# * 10: q2# = INT(q2#)
IF q2# = 0 THEN z2 = 0
IF q2# = 2 THEN z2 = 1
IF q2# = 5 THEN z2 = 0
IF q2# = 7 THEN z2 = 1
```

```
REM -----------------------------------------------------------
REM Get your N remainder then find your remainder for V# - r#
m# = (v# / 4) - INT(v# / 4): m# = m# * 10: m# = INT(m#)
IF m# = 0 THEN rm# = 2
IF m# = 2 THEN rm# = 5
IF m# = 5 THEN rm# = 7
IF m# = 7 THEN rm# = 0
REM -----------------------------------------------------------
REM Decrement v# - r# and test for remainder. Do not change v#
count = 0
redo2:
count = count + 1
DO
x# = x# - 1
m# = (x# / 4) - INT(x# / 4): m# = m# * 10: m# = INT(m#)
IF m# = rm# THEN EXIT DO
a$ = INKEY$: IF a$ = "s" THEN STOP
LOOP
r# = v# - x#
test = (z <> z2) AND (count = 2)
a$ = INKEY$: IF a$ = "s" THEN STOP
REM IF test = -1 THEN GOTO redo2:
LOOP




"C:\Users\Reactor1967\vcns\work\100803B.BAS"
REM binary 0 equals a odd distance
REM binary 1 equals a even distance
v# = 999
r# = 1
RANDOMIZE TIMER
CLS
redo:
count = 0
DO
count = count + 1
dist# = (v# - r#)
dist# = dist# - 1
test = (dist# / 2) - INT(dist# / 2)
IF test = .5 THEN z = 0
IF test = 0 THEN z = 1
IF test = .5 THEN dist# = dist# - 1
dist# = dist# / 2
dist# = dist# + 1
IF z = 0 THEN dist# = dist# - 1
```

```
IF z = 1 THEN dist# = dist# - 2
r# = v# - dist#
IF z = 0 THEN v# = v# + (v# - r# + 1)
IF z = 1 THEN v# = v# + (v# - r# + 2)
y# = v# - r#
m1# = (y# / 4) - INT(y# / 4): m1# = m1# * 10: m1# = INT(m1#)
PRINT , , z; m1#; v#; r#; v# - r#; dist#; r# - sr#; count
sr# = r#
a$ = INKEY$
REM INPUT a$
IF a$ = "s" THEN STOP
IF v# - r# = 1 THEN EXIT DO
LOOP
DO
z = INT(RND * 2) + 0
dist# = v# - r#
IF z = 0 THEN v# = v# + (v# - r# + 1)
IF z = 1 THEN v# = v# + (v# - r# + 2)
y# = v# - r#
m1# = (y# / 4) - INT(y# / 4): m1# = m1# * 10: m1# = INT(m1#)
PRINT , , z; m1#; v#; r#; v# - r#; dist#; r# - sr#
sr# = r#
a$ = INKEY$
REM INPUT a$
IF a$ = "s" THEN STOP
LOOP UNTIL v# - r# >= 1000
INPUT a$
IF a$ = "s" THEN STOP
GOTO redo:


"C:\Users\Reactor1967\vcns\work\100803A.BAS"
REM binary 0 equals a odd distance
REM binary 1 equals a even distance
v# = 999
r# = 1
RANDOMIZE TIMER
CLS
redo:
count = 0
DO
count = count + 1
dist# = (v# - r#)
dist# = dist# - 1
test = (dist# / 2) - INT(dist# / 2)
```

```
IF test = .5 THEN z = 0
IF test = 0 THEN z = 1
IF test = .5 THEN dist# = dist# - 1
dist# = dist# / 2
dist# = dist# + 1
IF z = 0 THEN dist# = dist# - 1
IF z = 1 THEN dist# = dist# - 2
r# = v# - dist#
IF z = 0 THEN v# = v# + (v# - r# + 1)
IF z = 1 THEN v# = v# + (v# - r# + 2)
y# = v# - r#
m1# = (y# / 4) - INT(y# / 4): m1# = m1# * 10: m1# = INT(m1#)
PRINT , , z; m1#; v#; r#; v# - r#; dist#; r# - sr#; count
sr# = r#
a$ = INKEY$
REM INPUT a$
IF a$ = "s" THEN STOP
IF v# - r# = 1 THEN EXIT DO
LOOP
DO
z = INT(RND * 2) + 0
dist# = v# - r#
IF z = 0 THEN v# = v# + (v# - r# + 1)
IF z = 1 THEN v# = v# + (v# - r# + 2)
y# = v# - r#
m1# = (y# / 4) - INT(y# / 4): m1# = m1# * 10: m1# = INT(m1#)
PRINT , , z; m1#; v#; r#; v# - r#; dist#; r# - sr#
sr# = r#
a$ = INKEY$
REM INPUT a$
IF a$ = "s" THEN STOP
LOOP UNTIL v# - r# >= 1000
INPUT a$
IF a$ = "s" THEN STOP
GOTO redo:



"C:\Users\Reactor1967\vcns\work\100703D.BAS"
REM This program tells what N is by dividing (v# - r#) by 4 and looking
REM at the remainder. It goes from a large v# - r# to a small v# - r#
v# = 999
r# = 1
d# = INT((v# - r#) / 2)
CLS
DO
```

```
z = INT(RND * 2) + 0
REM IF z = 0 THEN d# = d# - 1
REM IF z = 1 THEN d# = d# - 2
IF z = 0 THEN v# = v# + (v# - r# + 1)
IF z = 1 THEN v# = v# + (v# - r# + 2)
m# = ((v# - r#) / 4) - INT((v# - r#) / 4)
m# = m# * 10
m# = INT(m#)
PRINT , , z; m#; v#; r#; v# - r#; d#; r# - sr#
sr# = r#
a$ = INKEY$
REM INPUT a$
IF a$ = "s" THEN STOP
d# = d# - 1
r# = v# - d#
LOOP UNTIL v# - r# < 4
```



```
"C:\Users\Reactor1967\vcns\work\100703C.BAS"
REM We already know that v# - r# increases or decreases by
REM rate of change of v# - r# = ((d1# - d2#) * base) +- (n1 - n2)
REM So try to increase d# predictable so that you can find N and
REM always decode for v# - r#.
REM GOAL NOT ACCOMPLISHED!!! Here D needs be known before you code
REM not afterwards.
v# = 99
r# = 99
d# = 0
REM 1 to a 1 or 0 to a 0 d# = d# + 2
REM 1 to a 0 d# = d# + 4
REM 0 to a 1 d# = d# + 3
z = 1
CLS
RANDOMIZE TIMER
DO
z2 = z
z = INT(RND * 2) + 0
x# = v# - r#
IF z = 0 THEN v# = v# + (v# - r#) + 1
IF z = 1 THEN v# = v# + (v# - r#) + 2
m# = (v# / 4) - INT(v# / 4): m# = m# * 10: m# = INT(m#)
PRINT z; m#; v#; r#; v# - r#; x#; x# - sx#; r# - sr#; INT(((((r# - sr#) / 4) -
INT((r# - sr#) / 4)) * 10)
sr# = r#
```

```
sx# = x#
a$ = INKEY$
REM INPUT a$
IF a$ = "s" THEN STOP
test1 = (z = 0) AND (z2 = 0)
test2 = (z = 0) AND (z2 = 1)
test3 = (z = 1) AND (z2 = 0)
test4 = (z = 1) AND (z2 = 1)
IF test1 = -1 THEN d# = d# + 2
IF test2 = -1 THEN d# = d# + 1
IF test3 = -1 THEN d# = d# + 1
IF test4 = -1 THEN d# = d# + 2
r# = v# - d#
LOOP



"C:\Users\Reactor1967\vcns\work\100703B.BAS"
REM We already know that v# - r# increases or decreases by
REM rate of change of v# - r# = ((d1# - d2#) * base) +- (n1 - n2)
v# = 99
r# = 99
d# = 0
REM 1 to a 1 or 0 to a 0 d# = d# + 2
REM 1 to a 0 d# = d# + 4
REM 0 to a 1 d# = d# + 3
z = 1
CLS
RANDOMIZE TIMER
DO
z2 = z
z = INT(RND * 2) + 0
x# = v# - r#
IF z = 0 THEN v# = v# + (v# - r#) + 1
IF z = 1 THEN v# = v# + (v# - r#) + 2
m# = (v# / 4) - INT(v# / 4): m# = m# * 10: m# = INT(m#)
PRINT z; m#; v#; r#; v# - r#; x#; x# - sx#; r# - sr#
sr# = r#
sx# = x#
a$ = INKEY$
REM INPUT a$
IF a$ = "s" THEN STOP
DO
d# = d# + 1
s# = (v# - r#) - d#
test = (s# / 4) - INT(s# / 4) = 0
```

```
IF test = -1 THEN EXIT DO
a$ = INKEY$: IF a$ = "s" THEN STOP
LOOP
r# = v# - d#
LOOP
```

```
"C:\Users\Reactor1967\vcns\work\100309A.BAS"
REM New discovery. Speed of r# only increases when c# > 8 here.
REM NEW MATH EQUATION SPEED OF R# = (D# - C# - N)
REM D# = (V# - R#) BEFORE CODING. N = numerical value being coded.
REM c# = (v2# - r2#) - (v1# - r1#):r# = reference point.
REM NEW MATH EQUATION C# = (SPEED OF V#) - (SPEED OF R#) + n
REM im working with a new variable called c#. Its what you add to v# - r#
REM to get your next v# - r#. You have to know each v# - r# before you code
REM it. But, now you have to learn to know what v# - r# to decode too when
REM you get your next n when decoding.
REM This program correctly calculated c# and used it.
REM now I have successfully worked with c# I can try to come up with a
REM different way of calculating c# so I can decode. Maybe creating a
REM chart or new math equations for c#. Who knows.
v# = 1
r# = 1
c# = 1
z = 1
CLS
RANDOMIZE TIMER
DO
sz = z
z = INT(RND * 2) + 0
x1# = (c# / 4) - INT(c# / 4): x1# = x1# * 10: x1# = INT(x1#)
IF x1# = 2 THEN c# = c# + 1
IF x1# = 5 THEN c# = c# + 2
IF x1# = 7 THEN c# = c# + 3
IF x1# = 0 THEN c# = c# + 4
test1 = (sz = 0) AND (z = 1)
test2 = (sz = 1) AND (z = 0)
IF test1 = -1 THEN c# = c# + 1
IF test2 = -1 THEN c# = c# - 1
y# = (v# - r#) + c#
IF (y# / 2) - INT(y# / 2) = .5 THEN y# = y# - 1
y# = y# / 2
y# = y# + 1
IF z = 0 THEN y# = y# - 1
IF z = 1 THEN y# = y# - 2
```

```
r# = v# - y#
i# = v# - r#
IF z = 0 THEN v# = v# + (v# - r# + 1)
IF z = 1 THEN v# = v# + (v# - r# + 2)
m# = (v# / 4) - INT(v# / 4): m# = m# * 10: m# = INT(m#)
PRINT z; m#; v#; r#; v# - r#; c#; i#; r# - sr#
sr# = r#
a$ = INKEY$
REM INPUT a$
IF a$ = "s" THEN STOP
IF c# = 9 THEN c# = 1
LOOP
"C:\Users\Reactor1967\vcns\work\010603I.BAS"
v# = 1
r# = 1
v2# = 1
r2# = 1
dist# = 1
CLS
RANDOMIZE TIMER
DO
z = INT(RND * 2) + 0
IF z = 0 THEN v# = v# + (v# - r# + 1)
IF z = 1 THEN v# = v# + (v# - r# + 2)
IF r# - store# = 2 THEN z2 = 0
IF r# - store# = 4 THEN z2 = 1
hg# = (v2# - r2#)
IF z2 = 0 THEN v2# = v2# + (v2# - r2# + 1)
IF z2 = 1 THEN v2# = v2# + (v2# - r2# + 2)
PRINT v#; r#; v# - r#; r# - store#; "|"; z2; v2#; r2#; v2# - r2#
r2# = r2# + hg#
store# = r#
a$ = INKEY$
REM INPUT a$
IF a$ = "s" THEN STOP
REM dist# = dist# + 1
IF (v# / 2) - INT(v# / 2) = 0 THEN r# = v# - 1
IF (v# / 2) - INT(v# / 2) = .5 THEN r# = v# - 2
LOOP



"C:\Users\Reactor1967\vcns\work\010603H.BAS"
v# = 1
r# = 1
v2# = 1
```

```
r2# = 1
dist# = 0
CLS
RANDOMIZE TIMER
DO
REM Lesson learned. Not all distances are good canidates so we have to
REM learn to weed the bad ones out. Now, the trick is being able to
REM mathmaticly code and decode to the goods ones.
REM either a chart or an equation is needed here.
REM or we could go back to letting R# float between even and odd
REM and be able to know what the dist# is but earlier I could not
REM get that to work.
REM I will try to generate a chart such as
REM d1 = d2 or d3
REM d2 = d4 or d5
REM d3 = d6 or d7
REM then try to find an equation for generating distances and see
REM where that leads.
REM trying to use distance for a control side
z = INT(RND * 2) + 0
REM INPUT z
test = (v# / 2) - INT(v# / 2)
test1 = (test = 0) AND (z = 0)
test2 = (test = 0) AND (z = 1)
test3 = (test = .5) AND (z = 0)
test4 = (test = .5) AND (z = 1)
IF test1 = -1 THEN dist# = dist# + 2
IF test2 = -1 THEN dist# = dist# + 1
IF test3 = -1 THEN dist# = dist# + 1
IF test4 = -1 THEN dist# = dist# + 2
mr# = 0
redo:
mr# = mr# + 1
x# = dist#
IF z = 0 THEN x# = x# - 1
IF z = 1 THEN x# = x# - 2
x# = INT(x# / 2)
r# = v# - x#
IF (r# / 2) - INT(r# / 2) = 0 THEN dist# = dist# + 2
IF (r# / 2) - INT(r# / 2) = 0 THEN GOTO redo:
IF z = 0 THEN v# = v# + (v# - r# + 1)
IF z = 1 THEN v# = v# + (v# - r# + 2)
IF mr# = 1 THEN z2 = 0
IF mr# = 2 THEN z2 = 1
x2# = dist#
IF (x2# / 2) - INT(x2# / 2) = .5 THEN x2# = x2# - 1
```

```
x2# = x2# / 2
r2# = v2# - x2#
IF (r2# / 2) - INT(r2# / 2) = 0 THEN r2# = r2# - 1
IF z2 = 0 THEN v2# = v2# + (v2# - r2# + 1)
IF z2 = 1 THEN v2# = v2# + (v2# - r2# + 2)
PRINT , z; v#; r#; dist#; "|"; z2; v2#; r2#; v2# - r2#
store# = dist#
a$ = INKEY$
REM INPUT a$
IF a$ = "s" THEN STOP
LOOP




"C:\Users\Reactor1967\vcns\work\010603G.BAS"
REM Lesson Learned
REM try to have specific increases in v# - r#
REM example say a 0 to a 0 (v# - r#) increases by two
REM example say a 0 1 a 1 (v# - r#) increases by three
REM example say a 1 to a 0 (v# - r#) increases by 1
REM example say a 1 to a 1 (v# - r#) increases by 2
REM when decoding do the reverse so you can always find your
REM previous r#
v# = 1
r# = 1
CLS
RANDOMIZE TIMER
DO
z = INT(RND * 2) + 0
IF z = 0 THEN v# = v# + (v# - r# + 1)
IF z = 1 THEN v# = v# + (v# - r# + 2)
PRINT , , z; v#; r#; v# - r#
a$ = INKEY$
IF a$ = "s" THEN STOP
dist# = (v# - r#)
IF (dist# / 2) - INT(dist# / 2) = .5 THEN dist# = dist# - 1
dist# = dist# / 2
IF (dist# / 2) - INT(dist# / 2) = .5 THEN dist# = dist# - 1
r# = r# + dist#
LOOP




"C:\Users\Reactor1967\vcns\work\010603F.BAS"
v# = 1
r# = 1
```

```
dist# = 0
CLS
RANDOMIZE TIMER
DO
z = INT(RND * 2) + 0
IF z = 0 THEN v# = v# + ((v# - r#) * 1) + 1
IF z = 1 THEN v# = v# + ((v# - r#) * 1) + 2
PRINT z; v#; r#; v# - r#; "|"; (v# - r# + 1); v# - r# + 2; v# - r# + 3; v# - r# + 4;
v# - r# + 5
store# = (v# - r#)
a$ = INKEY$
REM INPUT a$
IF a$ = "s" THEN STOP
dist# = dist# + 1
test = ((v# - dist#) / 2) - INT((v# - dist#) / 2) = .5
IF test = 0 THEN dist# = dist# + 1
r# = v# - dist#
LOOP


"C:\Users\Reactor1967\vcns\work\010603E.BAS"
REM CONTROLLING THE DISTANCE AFTER THE CODE!!!
v# = 1
r# = 1
dist# = 0
RANDOMIZE TIMER
CLS
DO
z = INT(RND * 2) + 0
IF z = 0 THEN dist# = dist# + 1
IF z = 1 THEN dist# = dist# + 2
x# = dist#
IF (x# / 2) - INT(x# / 2) = .5 THEN x# = x# - 1
x# = INT(x# / 2)
x# = x# + 1
IF z = 0 THEN x# = x# - 1
IF z = 1 THEN x# = x# - 2
r# = v# - x#
IF z = 0 THEN v# = v# + (v# - r# + 1)
IF z = 1 THEN v# = v# + (v# - r# + 2)
PRINT , , z; v#; r#; dist#, v# - dist#
a$ = INKEY$
IF a$ = "s" THEN STOP
LOOP
```

```
"C:\Users\Reactor1967\vcns\work\010603D.BAS"
v# = 18
r# = 5
dist# = 6
CLS
RANDOMIZE TIMER
DO
REM Lesson learned. Maybe I should control the distance after the code
REM not control the dist# before the code. Lloyd Burris
r# = v# - dist#
z = INT(RND * 2) + 0
test1 = (z = 0) AND (((v# + dist# + 1) / 2) - INT((v# + dist# + 1) / 2) = 0)
test2 = (z = 0) AND (((v# + dist# + 2) / 2) - INT((v# + dist# + 2) / 2) = 0)
test3 = (z = 1) AND (((v# + dist# + 1) / 2) - INT((v# + dist# + 1) / 2) = .5)
test4 = (z = 1) AND (((v# + dist# + 2) / 2) - INT((v# + dist# + 1) / 2) = .5)
IF test1 = -1 THEN v# = v# + dist# + 1
IF test2 = -1 THEN v# = v# + dist# + 2
IF test3 = -1 THEN v# = v# + dist# + 1
IF test4 = -1 THEN v# = v# + dist# + 2
PRINT , , z; v#; r#; dist#; v# - r#
IF z = 0 THEN dist# = dist# + 1
IF z = 1 THEN dist# = dist# + 2
a$ = INKEY$
IF a$ = "s" THEN STOP
LOOP
```

```
"C:\Users\Reactor1967\vcns\work\010603C.BAS"
v# = 1
r# = 1
dist# = (v# - r#)
RANDOMIZE TIMER
CLS
DO
r# = v# - dist# - 1
IF (v# / 2) - INT(v# / 2) = 0 THEN dist# = dist# + 1
IF (v# / 2) - INT(v# / 2) = .5 THEN dist# = dist# + 2
z = INT(RND * 2) + 0
d1# = v# + dist# + 1
d2# = v# + dist# + 2
test1 = (z = 0) AND ((d1# / 2) - INT(d1# / 2) = 0)
test2 = (z = 0) AND ((d2# / 2) - INT(d2# / 2) = 0)
test3 = (z = 1) AND ((d1# / 2) - INT(d1# / 2) = .5)
```

```
test4 = (z = 1) AND ((d2# / 2) - INT(d2# / 2) = .5)
IF test1 = -1 THEN v# = d1#
IF test2 = -1 THEN v# = d2#
IF test3 = -1 THEN v# = d1#
IF test4 = -1 THEN v# = d2#
PRINT , , z; v#; r#; dist#; v# - r#
IF test1 = -1 THEN dist# = dist# + 1
IF test2 = -1 THEN dist# = dist# + 2
IF test3 = -1 THEN dist# = dist# + 1
IF test4 = -1 THEN dist# = dist# + 2
a$ = INKEY$
IF a$ = "s" THEN STOP
LOOP




"C:\Users\Reactor1967\vcns\work\010603B.BAS"
v# = 2
r# = 1
CLS
DO
dist# = (v# - r#)
z = INT(RND * 2) + 0
test1 = (z = 0) AND ((v# + (((v# - r#) * 1) + 1) / 2) - INT(v# + (((v# - r#) * 1) +
1) / 2) = 0)
test2 = (z = 0) AND ((v# + (((v# - r#) * 1) + 2) / 2) - INT(v# + (((v# - r#) * 1) +
2) / 2) = 0)
test3 = (z = 1) AND ((v# + (((v# - r#) * 1) + 1) / 2) - INT(v# + (((v# - r#) * 1) +
1) / 2) = .5)
test4 = (z = 1) AND ((v# + (((v# - r#) * 1) + 2) / 2) - INT(v# + (((v# - r#) * 1) +
2) / 2) = .5)
IF test1 = -1 THEN v# = v# + (v# - r# + 1)
IF test2 = -1 THEN v# = v# + (v# - r# + 2)
IF test3 = -1 THEN v# = v# + (v# - r# + 1)
IF test4 = -1 THEN v# = v# + (v# - r# + 2)
PRINT , , z; v#; r#; dist#; v# - r#
r# = r# + dist#
a$ = INKEY$
IF a$ = "s" THEN STOP
LOOP




"C:\Users\Reactor1967\vcns\work\010603.BAS"
v# = 1
r# = 1
```

```
t# = 1
RANDOMIZE TIMER
CLS
DO
dist# = v# - r#
t# = t# + 1
z = INT(RND * 2) + 0
IF z = 0 THEN v# = v# + dist# + 1
IF z = 1 THEN v# = v# + dist# + 2
PRINT , , z; v#; r#; dist#; v# - r#
r# = r# + dist#
IF (r# / 2) - INT(r# / 2) = 0 THEN r# = r# - 1
a$ = INKEY$
INPUT a$
IF a$ = "s" THEN STOP
LOOP




"C:\Users\Reactor1967\vcns\work\010503C.BAS"
v# = 1
r# = 1
spv# = 0
spr# = 0
CLS
RANDOMIZE TIMER
t# = 0
DO
t# = t# + 1
z = INT(RND * 2) + 0
spv# = (v# - r#)
IF z = 0 THEN spv# = spv# + 1
IF z = 1 THEN spv# = spv# + 2
spr# = spv#
IF (spr# / 2) - INT(spr# / 2) = .5 THEN spr# = spr# - 1
v# = v# + spv#
PRINT t#; z; v#; r#; spv#; spr#; v# - r#; r# + spr#; store1# - v#; store2# - r#
PRINT t#; z; v#; r#; spv#; spr#; r# + spr#; store1#; store2#
store1# = v#
store2# = r#
r# = r# + spr#
a$ = INKEY$
aIF a$ = "s" THEN STOP
LOOP
```

```
"C:\Users\Reactor1967\vcns\work\010503B.BAS"
v# = 1
r# = 1
CLS
RANDOMIZE TIMER
DO
dist# = (v# - r#)
rd# = dist#
IF (dist# / 2) - INT(dist# / 2) = .5 THEN dist# = dist# - 1
z = INT(RND * 2) + 0
IF z = 0 THEN v# = v# + (v# - r# + 1)
IF z = 1 THEN v# = v# + (v# - r# + 2)
REM COLOR 2, 0
PRINT z; "|"; v#; r#; "|"; v# - r#; "|"; rd#; "|"; dist#; "|"; (v# - r#) - rd#
r# = r# + dist#
a$ = INKEY$
IF a$ = "s" THEN STOP
LOOP




"C:\Users\Reactor1967\vcns\work\010503A.BAS"
v# = 1
r# = 1
r2# = r#
add# = 0
dist# = 0
CLS
RANDOMIZE TIMER
DO
z = INT(RND * 2) + 0
dist# = dist# + add#
IF z = 0 THEN add# = 1
IF z = 1 THEN add# = 2
IF z = 0 THEN v# = v# + (v# - r# + 1)
IF z = 1 THEN v# = v# + (v# - r# + 2)
REM PRINT (v# / 2); z; v#; r#; v# - r#; dist#; (v# - r#) - dist#; ((r# + 1) / 2);
((r# + 2) / 2)
PRINT , z; v#; r#; v# - r#; dist#; (v# - r#) - dist#
r# = r# + dist#
a$ = INKEY$
REM INPUT a$
IF a$ = "s" THEN STOP
```

```
IF a$ = "d" THEN EXIT DO
LOOP




"C:\Users\Reactor1967\vcns\work\010403C.BAS"
v# = 1
r# = 1
r2# = r#
add# = 0
dist# = 0
CLS
RANDOMIZE TIMER
DO
z = INT(RND * 2) + 0
dist# = dist# + add#
IF z = 0 THEN add# = 1
IF z = 1 THEN add# = 2
IF z = 0 THEN v# = v# + (v# - r# + 1)
IF z = 1 THEN v# = v# + (v# - r# + 2)
REM PRINT (v# / 2); z; v#; r#; v# - r#; dist#; (v# - r#) - dist#; ((r# + 1) / 2);
((r# + 2) / 2)
PRINT , , z; v#; r#; v# - r#; dist#; (v# - r#) - dist#
r# = r# + dist#
a$ = INKEY$
REM INPUT a$
IF a$ = "s" THEN STOP
IF a$ = "d" THEN EXIT DO
LOOP




"C:\Users\Reactor1967\vcns\work\010403B.BAS"
REM DISCOVERY, YOUR PRESENT DISTANCE + PREVIOUS
DISTANCE SUBTRACTED FROM
REM THE PREVIOUS VECTOR EQUAL THE PREVIOUS REFERENCE
POINT OR R!!!!
v# = 1
r# = 1
dist# = 0
CLS
RANDOMIZE TIMER
DO
z = INT(RND * 2) + 0
IF z = 0 THEN v# = v# + (v# - r# + 1)
IF z = 1 THEN v# = v# + (v# - r# + 2)
```

```
PRINT , z; v#; r#; dist#; (v# - r#) - dist#; v# - r#
a$ = INKEY$
IF a$ = "s" THEN STOP
IF a$ = "d" THEN EXIT DO
r# = r# + dist#
IF z = 0 THEN dist# = dist# + 1
IF z = 1 THEN dist# = dist# + 2
LOOP
PRINT "--------------------------------------------"
DO
x1# = (v# / 2) - INT(v# / 2)
IF ((r# + 1) / 2) - INT((r# + 1) / 2) = x1# THEN z = 0
IF ((r# + 2) / 2) - INT((r# + 2) / 2) = x1# THEN z = 1
x2# = dist#
x1# = (v# - r#)
IF (x1# / 2) - INT(x1# / 2) = .5 THEN x1# = x1# - 1
x1# = x1# / 2
x1# = x1# + 1
v# = v# - x1#
IF z = 0 THEN dist# = dist# - 1
IF z = 1 THEN dist# = dist# - 2
x1# = x2# + dist#
r# = v# - x1#
PRINT , z; v#; r#; dist#; (v# - r#) - dist#; v# - r#
INPUT a$
IF a$ = "s" THEN STOP
LOOP




"C:\Users\Reactor1967\vcns\work\010403A.BAS"
v# = 1
r# = 1
dist# = 0
CLS
RANDOMIZE TIMER
DO
z = INT(RND * 2) + 0
IF z = 0 THEN v# = v# + (v# - r# + 1)
IF z = 1 THEN v# = v# + (v# - r# + 2)
PRINT , , z; v#; r#; dist#; (v# - r#) - dist#
a$ = INKEY$
IF a$ = "s" THEN STOP
IF a$ = "d" THEN EXIT DO
r# = r# + dist#
IF z = 0 THEN dist# = dist# + 1
```

```
IF z = 1 THEN dist# = dist# + 2
LOOP
REM Decode(It does not fully work yet).
DO
x# = (v# / 2) - INT(v# / 2)
IF ((r# + 1) / 2) - INT((r# + 1) / 2) = x# THEN z = 0
IF ((r# + 2) / 2) - INT((r# + 2) / 2) = x# THEN z = 1
x# = (v# - r#)
IF (x# / 2) - INT(x# / 2) = .5 THEN x# = x# - 1
x# = x# / 2
x# = x# + 1
v# = v# - x#
IF z = 0 THEN dist# = dist# - 1
IF z = 1 THEN dist# = dist# - 2
r# = r# - dist#
PRINT , , z; v#; r#; dist#
INPUT a$
IF a$ = "s" THEN STOP
LOOP
```

```
"C:\Users\Reactor1967\vcns\work\010307B.BAS"
REM this is a distance calculator for figuring distance before
REM coding a vector and distance after coding a vector
REM you can start off with anything and add anything
REM If should closely go by the vector equations you are using
REM v# = v# + ((v# - r#) * base-1) + n
REM v# = v# - ((r# - v#) * base-1) + n
REM the plus and minus signs are interchangable depending on what
REM you are trying to do. So is the v# and r#
dist# = 2
CLS
DO
PRINT , , dist#; (dist# + 2) + dist#; (dist# + 4) + dist#
dist# = dist# + 2
a$ = INKEY$
INPUT a$
IF a$ = "s" THEN STOP
LOOP
```

```
"C:\Users\Reactor1967\vcns\work\010303F.BAS"
v# = 1
```

```
r# = 1
dist# = 0
CLS
RANDOMIZE TIMER
DO
z = INT(RND * 2) + 0
IF z = 0 THEN v# = v# + (v# - r# + 2)
IF z = 1 THEN v# = v# + (v# - r# + 4)
COLOR 2, 0
PRINT , z; v#; r#; dist#; (v# - r#) - dist#
r# = r# + dist#
IF z = 0 THEN dist# = dist# + 2
IF z = 1 THEN dist# = dist# + 4
a$ = INKEY$
REM INPUT a$
IF a$ = "s" THEN STOP
LOOP



"C:\Users\Reactor1967\vcns\work\010303E.BAS"
v# = 100
r# = v#
r# = v# - dist#
dist# = 0
CLS
RANDOMIZE TIMER
DO
z = INT(RND * 2) + 0
dx# = dist#
IF z = 0 THEN dist# = dist# + 2
IF z = 1 THEN dist# = dist# + 4
REM r# = r# + dist#
r# = v# - dist#
IF z = 0 THEN v# = v# + ((v# - r#) * 1) + 2
IF z = 1 THEN v# = v# + ((v# - r#) * 1) + 4
COLOR 2, 0
REM PRINT (v# / 4); z; v#; r#; dist#; (r# + 2) / 4; (r# + 4) / 4
PRINT , , z; v#; r#; dist#
a$ = INKEY$
REM INPUT a$
IF a$ = "s" THEN STOP
REM EXIT DO
LOOP
dist# = (v# - r#)
IF (dist# / 2) - INT(dist# / 2) = .5 THEN dist# = dist# - 1
```

```
dist# = dist# / 2
dist# = dist# + 2
v# = v# - dist#
PRINT v#
```

```
"C:\Users\Reactor1967\vcns\work\010303D.BAS"
v# = 1
r# = 1
dist# = 0
CLS
RANDOMIZE TIMER
DO
z = INT(RND * 2) + 0
IF z = 0 THEN v# = v# + ((v# - r#) * 1) + 2
IF z = 1 THEN v# = v# + ((v# - r#) * 1) + 4
x# = r#
COLOR 2, 0
r# = r# + dist#
IF z = 0 THEN dist# = dist# + 2
IF z = 1 THEN dist# = dist# + 4
PRINT v# / 4; z; v#; x#; dist#; (x# + 2) / 4; (x# + 4) / 4
a$ = INKEY$
IF a$ = "s" THEN STOP
LOOP
```

```
"C:\Users\Reactor1967\vcns\work\010303C.BAS"
v# = 1
r# = 1
dist# = 0
CLS
RANDOMIZE TIMER
DO
z = INT(RND * 2) + 0
IF z = 0 THEN v# = v# + ((v# - r#) * 1) + 2
IF z = 1 THEN v# = v# + ((v# - r#) * 1) + 4
COLOR 2, 0
PRINT v# / 4; z; v#; r#; dist#; (r# + 2) / 4; (r# + 4) / 4
r# = r# + dist#
IF z = 0 THEN dist# = dist# + 2
IF z = 1 THEN dist# = dist# + 4
a$ = INKEY$
IF a$ = "s" THEN STOP
```

```
LOOP



"C:\Users\Reactor1967\vcns\work\010303B.BAS"
v# = 1
r# = 1
dist# = 0
CLS
RANDOMIZE TIMER
DO
z = INT(RND * 2) + 0
IF z = 0 THEN v# = v# + ((v# - r#) * 1) + 2
IF z = 1 THEN v# = v# + ((v# - r#) * 1) + 4
COLOR 2, 0
PRINT , z, v#; r#; dist#
r# = r# + dist#
IF z = 0 THEN dist# = dist# + 2
IF z = 1 THEN dist# = dist# + 4
a$ = INKEY$
IF a$ = "s" THEN STOP
LOOP



"C:\Users\Reactor1967\vcns\work\010303A.BAS"
v# = 1
r# = 1
dist# = 0
RANDOMIZE TIMER
CLS
DO
z = INT(RND * 2) + 0
IF z = 0 THEN v# = v# + (v# - r# + 1)
IF z = 1 THEN v# = v# + (v# - r# + 2)
IF z = 0 THEN dist# = dist# + 1
IF z = 1 THEN dist# = dist# + 2
r1# = r1# + dist#
r# = r1#
IF (r# / 2) - INT(r# / 2) = 0 THEN r# = r# - 1
PRINT , z#; v#; r#; r1#; dist#
a$ = INKEY$
IF a$ = "s" THEN STOP
LOOP
```

```
"C:\Users\Reactor1967\vcns\work\010203D.BAS"
dist# = 0
v# = 1
r# = 1
CLS
RANDOMIZE TIMER
DO
REM Trying to find previous R mathmatically
z = INT(RND * 2) + 0
IF z = 0 THEN v# = v# + (v# - r# + 1)
IF z = 1 THEN v# = v# + (v# - r# + 2)
PRINT , z; v#; r#; dist#
store# = r#
a$ = INKEY$
IF a$ = "d" THEN EXIT DO
IF a$ = "s" THEN STOP
IF z = 0 THEN dist# = dist# + 1
IF z = 1 THEN dist# = dist# + 2
r# = (v# - dist#)
IF (r# / 2) - INT(r# / 2) = 0 THEN r# = r# - 1
LOOP
```

```
"C:\Users\Reactor1967\vcns\work\010203C.BAS"
v# = 2
r# = 2
dist# = 0
CLS
DO
z = INT(RND * 2) + 0
IF z = 0 THEN v# = v# + (v# - r# + 2)
IF z = 1 THEN v# = v# + (v# - r# + 4)
COLOR 2, 0
r# = r# + dist#
IF z = 0 THEN dist# = dist# + 2
IF z = 1 THEN dist# = dist# + 4
PRINT , z; v#; r#; dist#
a$ = INKEY$
REM INPUT a$
IF a$ = "s" THEN STOP
LOOP
```

```
"C:\Users\Reactor1967\vcns\work\010203B.BAS"
v# = 2
r# = 2
dist# = 0
CLS
DO
z = INT(RND * 2) + 0
IF z = 0 THEN v# = v# + (v# - r# + 2)
IF z = 1 THEN v# = v# + (v# - r# + 4)
r# = r# + dist#
IF z = 0 THEN dist# = dist# + 2
IF z = 1 THEN dist# = dist# + 4
COLOR 2, 0
PRINT , z; v#; r#; dist#
a$ = INKEY$
REM INPUT a$
IF a$ = "s" THEN STOP
LOOP
```

```
"C:\Users\Reactor1967\vcns\work\010203A.BAS"
v# = 2
r# = 0
dist# = 0
CLS
DO
z = INT(RND * 2) + 0
dist# = v# - r#
IF z = 0 THEN v# = v# + (v# - r# + 2)
IF z = 1 THEN v# = v# + (v# - r# + 4)
COLOR 2, 0
PRINT z; v#; r#; dist#
r# = r# + dist#
a$ = INKEY$
REM INPUT a$
IF a$ = "s" THEN STOP
LOOP
```

```
"C:\Users\Reactor1967\vcns\work\010203.BAS"
v# = 2
r# = 0
dist# = 0
CLS
```

```
DO
z = INT(RND * 2) + 0
dist# = v# - r#
IF z = 0 THEN v# = v# + (v# - r# + 2)
IF z = 1 THEN v# = v# + (v# - r# + 4)
COLOR 2, 0
PRINT z; v#; r#; dist#; (r# + 2) / 4; (r# + 4) / 4, v# / 4
r# = r# + dist#
a$ = INKEY$
REM INPUT a$
IF a$ = "s" THEN STOP
LOOP




"C:\Users\Reactor1967\vcns\work\8902.BAS"
REM Lesson learned here. Study how far v# increased when encoded
REM use that value to compute a value to increase R by.
REM When decoding study how far v# decreaded in value when decoded.
REM use that value to compute the value for R to decrease by.
REM See if can put this together to for a plan to increase R and
REM decrease R so can code data using any base of vcns.
high1# = 0
low1# = 999999999999999#
high2# = 0
low2# = 999999999999999#
v# = 1
r# = 1
chart$ = "chart"
CLS
RANDOMIZE TIMER
DO
IF (v# + ((v# - r#) * 2) + 3) >= (r# + 100) THEN z = 2 ELSE z = INT(RND *
2) + 0
sv# = v#
IF z = 0 THEN v# = v# + ((v# - r#) * 2) + 1
IF z = 1 THEN v# = v# + ((v# - r#) * 2) + 2
IF z = 2 THEN v# = v# + ((v# - r#) * 2) + 3
COLOR 2, 0
PRINT , , r# - sr#; r#; z; v#; v# - sv#
PRINT , , low1#, high1#
PRINT , , low2#, high2#
test2 = ((r# - sr#) = 0)
test3 = (test2 = -1) AND ((v# - sv#) > high1#)
test4 = (test2 = -1) AND ((v# - sv#) < low1#)
IF test3 = -1 THEN high1# = (v# - sv#)
```

```
IF test4 = -1 THEN low1# = (v# - sv#)
test5 = ((r# - sr#) = 100)
test6 = (test5 = -1) AND ((v# - sv#) > high2#)
test7 = (test5 = -1) AND ((v# - sv#) < low2#)
IF test6 = -1 THEN high2# = (v# - sv#)
IF test7 = -1 THEN low2# = (v# - sv#)
REM OPEN chart$ FOR APPEND AS #1
REM WRITE #1, r# - sr#, v# - sv#
REM CLOSE #1
sr# = r#
IF z = 2 THEN r# = (r# + 100)
test = (z = 2) AND (v# - r#) >= 100
IF test = -1 THEN r# = r# + 102
a$ = INKEY$
IF a$ = "s" THEN STOP
LOOP




"C:\Users\Reactor1967\vcns\work\7902.BAS"
REM
*****************************************************************
*****
REM (Look at the specific value of the Vector
REM you are about to code then code using.
REM reference point from there. When decoding just
REM look at the value of the vector then you know
REM your R that you are going to decode with.)
REM
*****************************************************************
******
v# = 2
r# = 1
RANDOMIZE TIMER
CLS
count = 0
DO
count = count + 1
z = INT(RND * 2) + 0
vs# = v#
IF z = 0 THEN v# = v# + (v# - r# + 1)
IF z = 1 THEN v# = v# + (v# - r# + 2)
COLOR 2, 0
PRINT , , z; v# - vs#; v#; r#; r# - rs#
IF v# >= 99 THEN STOP
a$ = INKEY$
```

```
REM INPUT a$
IF a$ = "s" THEN STOP
IF a$ = "d" THEN GOTO decode:
dist# = v# - r#
rs# = r#
r# = r# + dist#
IF (r# / 2) - INT(r# / 2) = 0 THEN r# = r# - 1
LOOP
decode:



"C:\Users\Reactor1967\vcns\work\7602.BAS"
v# = 1
r# = 1
CLS
RANDOMIZE TIMER
DO
v# = v# + (v# - r# + 1)
COLOR 2, 0
PRINT , , v#; r#
IF v# - r# > 100 THEN r# = r# + 100
z = INT(RND * 2) + 0
IF z = 0 THEN v# = v# + 4
IF z = 1 THEN v# = v# + 5
COLOR 2, 0
PRINT , , v#; r#
IF v# - r# > 100 THEN r# = r# + 100
v# = v# + (v# - r# + 2)
COLOR 2, 0
PRINT , , v#; r#
IF v# - r# > 100 THEN r# = r# + 100
z = INT(RND * 2) + 0
IF z = 0 THEN v# = v# + 3
IF z = 1 THEN v# = v# + 4
COLOR 2, 0
PRINT , , v#; r#
IF v# - r# > 100 THEN r# = r# + 100
INPUT a$
IF a$ = "s" THEN STOP
LOOP



"C:\Users\Reactor1967\vcns\work\924CHART.BAS"
v# = 1
```

```
r# = 1
r2# = r#
vcarry# = 1
rcarry# = 1
CLS
RANDOMIZE TIMER
DO
test = v# + (v# - r#) >= (r# + 100)
v1# = v# + (v# - r# + 1)
v2# = v# + (v# - r# + 2)
PRINT v#; v1#; v2#; r#; r2#; vcarry#; rcarry#
IF test = -1 THEN r2# = r# + 100
test2 = (vcarry# = v#) AND (rcarry# = r#)
IF test2 = -1 THEN z = INT(RND * 2) + 0
test3 = (test2 = -1) AND (z = 0)
test4 = (test2 = -1) AND (z = 1)
IF test3 = -1 THEN vcarry# = v1#
IF test4 = -1 THEN vcarry# = v2#
IF test3 = -1 THEN rcarry# = r2#
IF test4 = -1 THEN rcarry# = r2#
v# = v# + 1
IF v# >= (r# + 100) THEN r# = r# + 100
a$ = INKEY$
IF a$ = "s" THEN STOP
LOOP




"C:\Users\Reactor1967\vcns\work\4-WAY.BAS"
r# = 1
v# = 1
r2# = 1
CLS
DO
z = INT(RND * 2) + 0
IF z = 1 THEN r2# = r# + (v# - r# + 2)
IF z = 0 THEN r2# = r# + (v# - r# + 1)
IF z = 0 THEN v# = v# + (v# - r# + 3)
IF z = 1 THEN v# = v# + (v# - r# + 4)
r# = r2#
COLOR 2, 0
vr# = v# - vs#
vr# = vr# - 1
vr# = vr# * 2
vr# = v# - vr#
```

```
rr# = r# - rs#
rr# = rr# - 1
rr# = rr# * 2
rr# = r# - rr#
PRINT , vr# - ds# - 2; vr#; v#; z; r#; rr#; rr# - ds2#
ds# = vr#
ds2# = rr#
vs# = v#
rs# = r#
a$ = INKEY$
IF a$ = "s" THEN STOP
LOOP
```

```
"C:\Users\Reactor1967\vcns\work\3STATEB.BAS"
r1# = 1
v# = 4
DO
PRINT v#
v1# = v# + (v# - r1#) + 1
v2# = v# + (v# - r1#) + 3
v3# = v# + (v# - r1#) + 2
v4# = v# + (v# - r1#) + 4
PRINT v1# / 2; v1#
PRINT v2# / 2; v2#
PRINT v3# / 2; v3#
PRINT v4# / 2; v4#
v# = v# + 2
INPUT a$
IF a$ = "s" THEN STOP
LOOP
```

```
"C:\Users\Reactor1967\vcns\work\2STATEC.BAS"
REM 2 x 3 SYSTEM
REM IM USING A BASE TWO NON-LINEAR NUMERICAL SYSTEM
REM THAT HAS 2 POSIBBLE STATES AT ALL TIMES.
REM A BINARY 0 WITH A BINARY 0 OR BINARY 1 STATE
REM A BINARY 1 WITH A BINARY 0 OR BINARY 1 STATE
r1# = 1
v# = 4
DO
```

```
PRINT v#
v1# = v# + (v# - r1#) + 1
v2# = v# + (v# - r1#) + 3
v3# = v# + (v# - r1#) + 2
v4# = v# + (v# - r1#) + 4
PRINT v1# / 2; v1#
PRINT v2# / 2; v2#
PRINT v3# / 2; v3#
PRINT v4# / 2; v4#
v# = v# + 2
INPUT a$
IF a$ = "s" THEN STOP
LOOP
```

```
"C:\Users\Reactor1967\vcns\work\2STATEB.BAS"
r1# = 1
v# = 4
DO
PRINT v#
v1# = v# + (v# - r1#) + 1
v2# = v# + (v# - r1#) + 3
v3# = v# + (v# - r1#) + 2
v4# = v# + (v# - r1#) + 4
PRINT v1# / 2; v1#
PRINT v2# / 2; v2#
PRINT v3# / 2; v3#
PRINT v4# / 2; v4#
v# = v# + 2
INPUT a$
IF a$ = "s" THEN STOP
LOOP
```

```
"C:\Users\Reactor1967\vcns\work\WORKS.BAS"
v# = 110
r# = 103
CLS
RANDOMIZE TIMER
COLOR 2, 0
PRINT , , v#; r#
DO
```

```
z = INT(RND * 2) + 0
store2# = r#
r1# = r#
DO
IF z = 0 THEN v1# = v# + (v# - r1# + 1)
IF z = 1 THEN v1# = v# + (v# - r1# + 2)
REM test1 = ((v1# - r1#) < 11) AND (r1# - r# = 6) AND (z = 0)
REM test2 = ((v1# - r1#) < 11) AND (r1# - r# = 4) AND (z = 1)
REM IF test1 = -1 THEN EXIT DO
REM IF test2 = -1 THEN EXIT DO
IF v1# - r1# < 11 THEN EXIT DO
a$ = INKEY$
REM INPUT a$
IF a$ = "s" THEN STOP
IF a$ = "d" THEN EXIT DO
r1# = r1# + 2
LOOP
IF r1# - r# = 6 THEN z = 0
IF r1# - r# = 4 THEN z = 1
r# = r1#
store1# = v#
IF z = 0 THEN v# = v# + (v# - r# + 1)
IF z = 1 THEN v# = v# + (v# - r# + 2)
COLOR 2, 0
PRINT , , z; v#; r#; r# - store2#
REM INPUT a$
a$ = INKEY$
IF a$ = "s" THEN STOP
IF a$ = "d" THEN EXIT DO
IF a$ = "D" THEN EXIT DO
LOOP
decode:
DO
test = (v# / 2) - INT(v# / 2)
dist# = v# - r#
IF (dist# / 2) - INT(dist# / 2) = .5 THEN dist# = dist# - 1
dist# = dist# / 2
dist# = dist# + 1
v# = v# - dist#
IF test = .5 THEN r# = r# - 4
IF test = 0 THEN r# = r# - 6
IF test = .5 THEN z = 1 ELSE z = 0
COLOR 2, 0
PRINT , , z; v#; r#
IF v# = 110 THEN PRINT "Decode Finished"
IF v# = 110 THEN SYSTEM
```

```
REM INPUT a$
a$ = INKEY$
IF a$ = "s" THEN STOP
LOOP
```

```
"C:\Users\Reactor1967\vcns\work\WORKME.BAS"
v# = 2
r# = 1
RANDOMIZE TIMER
CLS
DO
z = INT(RND * 2) + 0
vs# = v#
IF z = 0 THEN v# = v# + (v# - r# + 1)
IF z = 1 THEN v# = v# + (v# - r# + 2)
COLOR 2, 0
PRINT , , z; v# - vs#; v#; r#; r# - rs#
REM a$ = INKEY$
INPUT a$
IF a$ = "s" THEN STOP
IF a$ = "d" THEN GOTO decode:
dist# = v# - r#
rs# = r#
r# = r# + dist#
IF (r# / 2) - INT(r# / 2) = 0 THEN r# = r# - 1
LOOP
decode:
```

```
"C:\Users\Reactor1967\vcns\work\WIDGET2.BAS"
a# = 1: b# = 2
CLS
redo:
DO
dist# = a# - b#
IF dist# <= 0 THEN dist# = dist# * -1
a# = a# + dist#
b# = b# + (dist# * 2)
IF b# = a# THEN b# = b# + dist#
COLOR 2, 0
PRINT , , a#; b#
INPUT a$
IF a$ = "s" THEN STOP
```

```
IF a$ = "d" THEN EXIT DO
LOOP
DO
dist# = a# - b#
IF dist# <= 0 THEN dist# = dist# * -1
b# = b# + dist#
a# = a# + (dist# * 2)
IF a# = b# THEN a# = a# + dist#
COLOR 2, 0
PRINT , , a#; b#
INPUT a$
IF a$ = "s" THEN STOP
IF a$ = "d" THEN EXIT DO
LOOP
GOTO redo:


"C:\Users\Reactor1967\vcns\work\WIDGET.BAS"
REM Since the word widget does not seem to have any meaning I
REM decided to give it one. I named my method here widget for
REM generating distances to use as (d-r) in my math equations
REM for vcns. I just hope it works now. L.B.
d1# = 1
d2# = 2
CLS
RANDOMIZE TIMER
PRINT d1#; d2#
redo:
DO
z = INT(RND * 2) + 0
store# = d1#
IF z = 0 THEN d1# = (d1# * 2)
IF z = 1 THEN d1# = (d1# * 2) + 1
dist# = d1# - store#
d2# = d2# + dist#
PRINT , , d1#; d2#
a$ = INKEY$
IF a$ = "s" THEN STOP
LOOP UNTIL d1# * 2 >= 999999999999999#
DO
test1 = (d1# / 2) - INT(d1# / 2)
test2 = (d2# / 2) - INT(d2# / 2)
IF test1 = 0 THEN z = 0
IF test2 = 0 THEN z = 1
IF z = 0 THEN dist# = d1# - (d1# / 2)
```

```
IF z = 1 THEN dist# = d2# - (d2# / 2)
d1# = d1# - dist#
d2# = d2# - dist#




"C:\Users\Reactor1967\vcns\work\WHENFLIP.BAS"
p# = 1
r# = 1
r2# = 999999999999999#
CLS
RANDOMIZE TIMER
repeat:
add# = 250000000000000#
DO
store# = p#
c = INT(RND * 2) + 0
IF c = 0 THEN p# = p# + (p# - r# + 1)
IF c = 1 THEN p# = p# + (p# - r# + 1)
subtract# = p# - store#
add# = add# - subtract#
PRINT p#; add#
a$ = INKEY$
IF a$ = "s" THEN STOP
LOOP UNTIL p# > 500000000000000#
p# = p# - 500000000000000#
INPUT a$
IF a$ = "s" THEN STOP
GOTO repeat:




"C:\Users\Reactor1967\vcns\work\VRDIST.BAS"
v# = 1
r# = 1
CLS
RANDOMIZE TIMER
DO
z = INT(RND * 3) + 0
PRINT v# / 3; (((v# - r#) * 2) + 1) / 3
PRINT v# / 3; (((v# - r#) * 2) + 2) / 3
PRINT v# / 3; (((v# - r#) * 2) + 3) / 3
IF z = 0 THEN v# = v# + ((v# - r#) * 2) + 1
IF z = 1 THEN v# = v# + ((v# - r#) * 2) + 2
IF z = 2 THEN v# = v# + ((v# - r#) * 2) + 3
COLOR 2, 0
```

```
PRINT , , z; v# / 3
a$ = ineky$
INPUT a$
IF a$ = "s" THEN STOP
LOOP




"C:\Users\Reactor1967\vcns\work\VRCHRT4.BAS"
v# = 1000
r# = 1001
count1# = 1: REM beginning num for bank.txt
count2# = 1: REM ending num for bank.txt
count3# = 1: REM record num for chart.txt
CLS
test1$ = "chart"
test2$ = "bank"
OPEN test2$ FOR OUTPUT AS #2
WRITE #2, count1#, r#, v#
CLOSE #2
DO
OPEN test2$ FOR INPUT AS #2
DO
INPUT #2, record#, reference#, vector#
stop$ = INKEY$
IF stop$ = "s" THEN STOP
LOOP UNTIL record# = count1#
CLOSE #2
count1# = count1# + 1: REM beginning record num for bank.txt
v# = vector#
rs# = reference#
dist# = 4
redo:
r# = 1001
DO
v1# = v# - (r# - v# + 1)
v2# = v# - (r# - v# + 2)
test = (r# - v1#) <= dist#
IF test = -1 THEN EXIT DO
r# = r# - 1
IF (r# / 2) - INT(r# / 2) = 0 THEN r# = r# - 1
stop$ = INKEY$
IF stop$ = "s" THEN STOP
LOOP
OPEN test2$ FOR INPUT AS #2
```

```
REM verifying
DO
INPUT #2, record#, reference#, vector#
test1 = (r# = reference#) AND (v1# = vector#)
test2 = (r# = reference#) AND (v2# = vector#)
IF test1 = -1 THEN EXIT DO
IF test2 = -1 THEN EXIT DO
stop$ = INKEY$
IF stop$ = "s" THEN STOP
LOOP UNTIL record# = count2#
CLOSE #2
REM end verifying
IF test1 = -1 THEN dist# = dist# + 1
IF test2 = -1 THEN dist# = dist# + 1
IF test1 = -1 THEN GOTO redo:
IF test2 = -1 THEN GOTO redo:
count2# = count2# + 1
IF v1# > r# THEN STOP
IF v2# > r# THEN STOP
OPEN test2$ FOR APPEND AS #2
WRITE #2, count2#, r#, v1#
count2# = count2# + 1
WRITE #2, count2#, r#, v2#
CLOSE #2
REM PRINT count2#; r#; v1#; v2#
OPEN test1$ FOR APPEND AS #1
COLOR 2, 0
PRINT , , count3#; rs#; v#; v1#; v2#; r#; r# - v1#
WRITE #1, count3#, rs#, v#, v1#, v2#, r#
CLOSE #1
count3# = count3# + 1
a$ = INKEY$
REM INPUT a$
IF a$ = "s" THEN CLOSE #1, #2
IF a$ = "s" THEN STOP
IF v1# <= 1 THEN EXIT DO
IF v2# <= 1 THEN EXIT DO
LOOP
PRINT "Task Completed"




"C:\Users\Reactor1967\vcns\work\VRCHRT3.BAS"
v# = 100
r# = 1
count1# = 1: REM beginning num for bank.txt
```

```
count2# = 1: REM ending num for bank.txt
count3# = 1: REM record num for chart.txt
CLS
test1$ = "chart"
test2$ = "bank"
OPEN test2$ FOR OUTPUT AS #2
WRITE #2, count1#, r#, v#
CLOSE #2
DO
OPEN test2$ FOR INPUT AS #2
DO
INPUT #2, record#, reference#, vector#
stop$ = INKEY$
IF stop$ = "s" THEN STOP
LOOP UNTIL record# = count1#
CLOSE #2
count1# = count1# + 1: REM beginning record num for bank.txt
v# = vector#
rs# = reference#
dist# = 4
redo:
r# = 1
DO
v1# = v# + (v# - r# + 1)
v2# = v# + (v# - r# + 2)
test = (v1# - r#) <= dist#
IF test = -1 THEN EXIT DO
r# = r# + 1
IF (r# / 2) - INT(r# / 2) = 0 THEN r# = r# + 1
stop$ = INKEY$
IF stop$ = "s" THEN STOP
LOOP
OPEN test2$ FOR INPUT AS #2
REM verifying
DO
INPUT #2, record#, reference#, vector#
test1 = (r# = reference#) AND (v1# = vector#)
test2 = (r# = reference#) AND (v2# = vector#)
IF test1 = -1 THEN EXIT DO
IF test2 = -1 THEN EXIT DO
stop$ = INKEY$
IF stop$ = "s" THEN STOP
LOOP UNTIL record# = count2#
CLOSE #2
REM end verifying
IF test1 = -1 THEN dist# = dist# + 1
```

```
IF test2 = -1 THEN dist# = dist# + 1
IF test1 = -1 THEN GOTO redo:
IF test2 = -1 THEN GOTO redo:
count2# = count2# + 1
IF r# > v1# THEN STOP
IF r# > v2# THEN STOP
OPEN test2$ FOR APPEND AS #2
WRITE #2, count2#, r#, v1#
count2# = count2# + 1
WRITE #2, count2#, r#, v2#
CLOSE #2
REM PRINT count2#; r#; v1#; v2#
OPEN test1$ FOR APPEND AS #1
COLOR 2, 0
PRINT , , count3#; rs#; v#; v1#; v2#; r#; v1# - rs#
WRITE #1, count3#, rs#, v#, v1#, v2#, r#
CLOSE #1
count3# = count3# + 1
a$ = INKEY$
REM INPUT a$
IF a$ = "s" THEN CLOSE #1, #2
IF a$ = "s" THEN STOP
IF v1# >= 1000 THEN EXIT DO
IF v2# >= 1000 THEN EXIT DO
LOOP
PRINT "Task Completed"




"C:\Users\Reactor1967\vcns\work\VRCHRT2.BAS"
v# = 100
r# = 99
RANDOMIZE TIMER
count1# = 1
count2# = 1
dist# = 5
CLS
test1$ = "chart.txt": REM Vector Chart.
test2$ = "bank.txt": REM Store used vectors and referrence points here.
OPEN test1$ FOR OUTPUT AS #1
OPEN test2$ FOR OUTPUT AS #2
WRITE #2, count2#, r#, v#
CLOSE #2
DO
OPEN test2$ FOR INPUT AS #2
DO
```

```
INPUT #2, j2#, rs2#, v2#
stop$ = INKEY$
IF stop$ = "s" THEN STOP
LOOP UNTIL j2# = count2#
v# = v2#
r# = rs2#
CLOSE #2
count2# = count2# + 1
rs# = r#
r1# = 1
DO
r1# = r1# + 1
test = (r1# / 2) - INT(r1# / 2)
IF test = 0 THEN r1# = r1# + 1
v1# = v# + (v# - r1# + 1)
v2# = v# + (v# - r1# + 2)
testa = v1# - r1# <= dist#
IF testa = 0 THEN GOTO skip:
REM verifying ----------------------
OPEN test2$ FOR INPUT AS #2
DO
INPUT #2, j2#, rs2#, v2#
testz = (v1# = v2#) AND (r1# = rs2#)
IF testz = -1 THEN EXIT DO
stop$ = INKEY$
IF stop$ = "s" THEN STOP
LOOP UNTIL j2# = count1#
CLOSE #2
REM End verifying ------------------
skip:
testa = v1# - r1# <= dist# AND testz = 0
IF testa = -1 THEN EXIT DO
a$ = INKEY$
IF a$ = "s" THEN CLOSE #1
IF a$ = "s" THEN STOP
LOOP
IF x = 0 THEN dist# = dist# + 1
IF x = 1 THEN dist# = dist# - 1
IF dist# >= 50 THEN x = 1
IF dist# <= 10 THEN x = 0
r# = r1#
COLOR 2, 0
PRINT , , count1#; rs#; v#; "="; v1#; v2#; r#; dist#; testing
WRITE #1, count1#, rs#, v#, v1#, v2#, r#
count1# = count1# + 1
OPEN test2$ FOR APPEND AS #2
```

```
WRITE #2, count1#, r#, v1#
count1# = count1# + 1
WRITE #2, count1#, r#, v2#
CLOSE #2
IF count1# >= 1001 THEN EXIT DO
REM INPUT a$
a$ = INKEY$
IF a$ = "s" THEN STOP
LOOP
CLOSE #1
PRINT "Task Completed"




"C:\Users\Reactor1967\vcns\work\VRCHRT.BAS"
DECLARE SUB verify (v1#, r1#, vbank#(), rbank#(), testz!)
v# = 100
r# = 99
RANDOMIZE TIMER
DIM vbank#(1001)
DIM rbank#(1001)
count1 = 1
vbank#(count1) = v#
rbank#(count1) = r#
count2 = 1
dist# = 5
CLS
test$ = "test.txt"
OPEN test$ FOR OUTPUT AS #1
REM WRITE #1, r#, v#, "=", r#, a#, b#
DO
v# = vbank#(count2)
r# = rbank#(count2)
count2 = count2 + 1
rs# = r#
r1# = 1
DO
r1# = r1# + 1
test = (r1# / 2) - INT(r1# / 2)
IF test = 0 THEN r1# = r1# + 1
v1# = v# + (v# - r1# + 1)
v2# = v# + (v# - r1# + 2)
CALL verify(v1#, r1#, vbank#(), rbank#(), testz)
testa = v1# - r1# <= dist# AND testz = 0
IF testa = -1 THEN EXIT DO
a$ = INKEY$
```

```
IF a$ = "s" THEN CLOSE #1
IF a$ = "s" THEN STOP
LOOP
IF x = 0 THEN dist# = dist# + 1
IF x = 1 THEN dist# = dist# - 1
IF dist# >= 50 THEN x = 1
IF dist# <= 10 THEN x = 0
r# = r1#
COLOR 2, 0
PRINT , , count1; rs#; v#; "="; v1#; v2#; r#; dist#
WRITE #1, count1, rs#, v#, "=", v1#, v2#, r#, dist#
count1 = count1 + 1
vbank#(count1) = v1#
rbank#(count1) = r#
count1 = count1 + 1
vbank#(count1) = v2#
rbank#(count1) = r#
IF count1 >= 1001 THEN EXIT DO
REM INPUT a$
a$ = INKEY$
IF a$ = "s" THEN STOP
LOOP
CLOSE #1
PRINT "Task Completed"

SUB verify (v1#, r1#, vbank#(), rbank#(), testz)
FOR count = 1 TO 1001
testz = (v1# = vbank#(count)) AND (r1# = rbank#(count))
IF testz = -1 THEN EXIT FOR
NEXT count
END SUB




"C:\Users\Reactor1967\vcns\work\VRC.BAS"
v# = 1
r# = 1
RANDOMIZE TIMER
DO
z = INT(RND * 2) + 0
IF z = 0 THEN v# = v# + (v# - r# + 1)
IF z = 1 THEN v# = v# + (v# - r# + 2)
COLOR 2, 0
PRINT , , "v"; r#; v#; z
a$ = INKEY$
IF a$ = "s" THEN STOP
```

```
IF (v# / 2) - INT(v# / 2) = 0 THEN r# = r# + (v# - r# - 1)
IF (v# / 2) - INT(v# / 2) = .5 THEN r# = r# + (v# - r# - 2)
COLOR 2, 0
PRINT , , "r"; r#; v#; z
LOOP
```

```
"C:\Users\Reactor1967\vcns\work\VCTEST.BAS"
REM                     COPYRIGHT C 2002
REM                 Lloyd Dudley Burris
REM ----------------------Initilize R#------------------------------------
CLS
DIM ref#(255)
ref#(1) = 1
pt1# = 5
rf# = 100
count = 2
a# = 0
replaced1:
DO
a# = a# + pt1#
test = (a# / 2) - INT(a# / 2)
ref#(count) = a#
IF (ref#(count) / 2) - INT(ref#(count) / 2) = 0 THEN ref#(count) = ref#(count)
+ 1
count = count + 1
a$ = INKEY$
IF a$ = "s" THEN STOP
IF count >= 254 THEN EXIT DO
LOOP UNTIL a# = rf#
pt1# = pt1# * 10
rf# = rf# * 10
IF count >= 254 THEN GOTO replaced2:
IF a# + pt1# > 999999999999999# THEN GOTO replaced2:
GOTO replaced1:
replaced2:
ref#(254) = 950000000000001#
ref#(255) = 999999999999999#
REM ----------------------Code P# up.------------------------------------
range = 237
p# = 100000000000001#
RANDOMIZE TIMER
repeat:
count = 1
p# = 1
```

```
DO
c = INT(RND * 2) + 0
IF c = 0 THEN p# = p# + (p# - ref#(count) + 1)
IF c = 1 THEN p# = p# + (p# - ref#(count) + 2)
bt = (p# / 2) - INT(p# / 2)
IF bt >= .5 THEN bt = 1
PRINT p#; ref#(count)
IF (p# + (p# - ref#(count) + 1)) > (ref#(count + 1)) THEN c = 1 ELSE c = 0
IF c = 0 THEN p# = p# + (p# - (ref#(count)) + 1)
IF c = 1 THEN p# = p# + (p# - (ref#(count)) + 2)
IF p# > (ref#(count + 4)) THEN count = count + 4
bt = (p# / 2) - INT(p# / 2)
IF bt >= .5 THEN bt = 1
PRINT p#; ref#(count)
INPUT a$
REM a$ = INKEY$
IF a$ = "s" THEN STOP
LOOP UNTIL ((p# + (p# - ref#(count))) + (p# + (p# - ref#(count)))) >=
999999999999999#
PRINT p#; ref#(count)
```



```
"C:\Users\Reactor1967\vcns\work\VCRF.BAS"
REM vector coordinate reducing factor demo
REM p# = p# + (p# - r# + 1) - reduction factor
r# = 1
p# = 1
CLS
RANDOMIZE TIMER
repeat:
count = 1
DO
c = INT(RND * 2) + 0
test# = INT(.9 * p#)
IF c = 0 THEN p# = p# + (p# - r# + 1) - INT(.95 * p#)
IF c = 1 THEN p# = p# + (p# - r# + 2) - INT(.95 * p#)
PRINT p#; test#; count
REM INPUT a$
a$ = INKEY$
IF a$ = "s" THEN SYSTEM
count = count + 1
LOOP UNTIL p# + (p# - r#) - INT(.9 * p#) > 999999999999999#
PRINT "                "
p# = p# - 500000000000000#
a$ = INKEY$
```

```
REM INPUT a$
IF a$ = "s" THEN STOP
GOTO repeat:




"C:\Users\Reactor1967\vcns\work\VCNSTE.BAS"
v# = 100
cr# = 0
CLS
RANDOMIZE TIMER
DO
z = INT(RND * 2) + 0
d# = 1
store1# = cr#
DO

DO
v1# = v# + d#
test = (v# / 2) - INT(v# / 2)
IF test = .5 THEN test = 1
IF test = z THEN EXIT DO
d# = d# + 1
stop$ = INKEY$
IF stop$ = "s" THEN STOP
LOOP
dist# = v1# - v#
dist# = (dist# * 2) - 1
cr# = v1# - dist#
test = ABS((v1# - v#) - (cr# - store1#)) = 1
IF test = -1 THEN EXIT DO
a$ = INKEY$
IF a$ = "s" THEN STOP
LOOP
PRINT "exit"
STOP
LOOP




"C:\Users\Reactor1967\vcns\work\VCNSRPT.BAS"
RANDOMIZE TIMER
p# = 1
CLS
```

```
start:
r# = p#
count = 1
DO
c = INT(RND * 2) + 0
IF c = 0 THEN p# = p# + (p# - r# + 1)
IF c = 1 THEN p# = p# + (p# - r# + 2)
PRINT p#; p# - r#; count; r#
count = count + 1
INPUT a$
REM a$ = INKEY$
IF a$ = "s" THEN STOP
LOOP UNTIL p# + (p# - r# + 2) > 99999999999999#
r# = p#
count = 1
DO
c = INT(RND * 2) + 0
IF c = 0 THEN p# = p# - (r# - p# + 1)
IF c = 1 THEN p# = p# - (r# - p# + 2)
PRINT p#; p# - r#; count; r#
count = count + 1
INPUT a$
REM a$ = INKEY$
IF a$ = "s" THEN STOP
LOOP UNTIL p# - (r# - p# + 2) < 0
GOTO start:
```

```
"C:\Users\Reactor1967\vcns\work\VCNSBETA.BAS"
REM lessons learned:
REM You goto to know which number is the flip number.
REM In this program the lowest in value and the
REM highest in value trade places that you can,t
REM tell which one to flip.
REM SOLUTION ++++++++++++++++++++++++++++
REM Always keep one number lower and one higher.
REM if the highest becomes the lowest then roll the lowest
REM until it flips under the highest and be able to roll
REM the lowest number back. Also, keep coding that one
REM on all the flips and code two ones when both flips.
COMMON SHARED p1#, p2#, rdc#, test, c, r#, flip$
DECLARE SUB codeandrecord ()
DECLARE SUB flipandrecord ()
DECLARE SUB compare ()
```

```
p1# = 1
p2# = 2
r# = 1
rdc# = 500000000000000#
RANDOMIZE TIMER
CLS
DO
CALL compare
IF p1# > p2# THEN PRINT p1#; "*"; p2#
IF p2# > p1# THEN PRINT p1#; p2#; "*"
IF test = -1 THEN PRINT "FLIPPING FLIPPING FLIPPING FLIPPING
FLIPPING-------"
IF test = -1 THEN CALL flipandrecord ELSE CALL codeandrecord
REM a$ = INKEY$
INPUT a$
IF a$ = "s" THEN STOP
LOOP

SUB codeandrecord
c = INT(RND * 2) + 0
IF p1# < p2# THEN chart = 1
IF p2# < p1# THEN chart = 2
test1 = (c = 0) AND chart = 1
test2 = (c = 1) AND chart = 1
test3 = (c = 0) AND chart = 2
test4 = (c = 1) AND chart = 2
IF test1 = -1 THEN GOTO test1:
IF test2 = -1 THEN GOTO test2:
IF test3 = -1 THEN GOTO test3:
IF test4 = -1 THEN GOTO test4:
STOP
test1:
p2# = p2# + (p2# - r# + 1)
IF flip$ = "" THEN p1# = p1# + (p1# - r# + 1)
IF flip$ = "flip" THEN p1# = p1# + (p1# - r# + 2)
flip$ = ""
EXIT SUB
test2:
p2# = p2# + (p2# - r# + 2)
IF flip$ = "" THEN p1# = p1# + (p1# - r# + 1)
IF flip$ = "flip" THEN p1# = p1# + (p1# - r# + 2)
flip$ = ""
EXIT SUB
test3:
p1# = p1# + (p1# - r# + 1)
IF flip$ = "" THEN p2# = p2# + (p2# - r# + 1)
```

```
IF flip$ = "flip" THEN p2# = p2# + (p2# - r# + 2)
flip$ = ""
EXIT SUB
test4:
p1# = p1# + (p1# - r# + 2)
IF flip$ = "" THEN p2# = p2# + (p2# - r# + 1)
IF flip$ = "flip" THEN p2# = p2# + (p2# - r# + 2)
flip$ = ""
EXIT SUB
END SUB


SUB compare
test = (p2# > rdc#) OR (p1# > rdc#)
END SUB


SUB flipandrecord
flip$ = "flip"
test1 = (p1# > rdc#) AND (p2# > rdc#)
test2 = (p1# > rdc#) AND (p2# <= rdc#)
test3 = (p1# <= rdc#) AND (p2# > rdc#)
IF test1 = -1 THEN GOTO test1B:
IF test2 = -1 THEN GOTO test2B:
IF test3 = -1 THEN GOTO test3B:
STOP
test1B:
p1# = p1# - rdc#
p2# = p2# - rdc#
REM p1# = p1# + (p1# - r# + 2)
REM p2# = p2# + (p2# - r# + 2)
EXIT SUB
test2B:
p1# = p1# - rdc#
EXIT SUB:
test3B:
p2# = p2# - rdc#
EXIT SUB
END SUB




"C:\Users\Reactor1967\vcns\work\VCNS3.BAS"
v# = 100
cr# = 0
CLS
RANDOMIZE TIMER
```

```
DO
d# = 1
store2# = cr#
z = INT(RND * 2) + 0
store1# = v#
DO
r# = v# - d#
REM IF (r# / 2) - INT(r# / 2) = 0 THEN d# = d# + 1
REM IF (r# / 2) - INT(r# / 2) = 0 THEN r# = v# - d#
IF z = 0 THEN v1# = v# + (v# - r# + 1)
IF z = 1 THEN v1# = v# + (v# - r# + 2)
dist# = v1# - v#
dist# = (dist# * 2) - 1
cr# = v1# - dist#
test = ABS((v1# - v#) - (cr# - store2#)) = 1
test2 = ABS((v1# - v#) - (cr# - store2#)) = 2
REM PRINT z; v1# - v#; cr# - store2#
IF test = -1 THEN EXIT DO
IF test2 = -1 THEN EXIT DO
d# = d# + 1
REM INPUT a$
a$ = INKEY$
IF a$ = "s" THEN STOP
LOOP
v# = v1#
COLOR 2, 0
PRINT , , v# - store1#; z; v#; v# - cr#; cr#; cr# - store2#
INPUT a$
REM a$ = INKEY$
IF a$ = "s" THEN STOP
LOOP


"C:\Users\Reactor1967\vcns\work\VCNS2.BAS"
REM GOT TO GET R TO SYNCHRONIZE WITH P
REM THERE IS A GAP BETWEEN 1 AND THE NEXT R#
REM FILL IN THAT GAP OR ELIMANATE THAT RANGE ALL
TOGETHER.
REM ----------------------Initilize R#-------------------------------------
CLS
DIM r#(255)
r#(1) = 1
REM PRINT r#(1)
pt1# = 5
```

```
rf# = 100
count = 2
a# = 0
replaced1:
DO
a# = a# + pt1#
test = (a# / 2) - INT(a# / 2)
r#(count) = a#
IF (r#(count) / 2) - INT(r#(count) / 2) = 0 THEN r#(count) = r#(count) + 1
REM PRINT r#(count); count
count = count + 1
REM INPUT a$
a$ = INKEY$
IF a$ = "s" THEN STOP
IF count >= 254 THEN EXIT DO
LOOP UNTIL a# = rf#
pt1# = pt1# * 10
rf# = rf# * 10
IF count >= 254 THEN GOTO replaced2:
IF a# + pt1# > 999999999999999# THEN GOTO replaced2:
GOTO replaced1:
replaced2:
r#(254) = 950000000000001#
r#(255) = 999999999999999#
REM ----------------------Code P# up.--------------------------------------
range = 1
p# = 1
RANDOMIZE TIMER
repeat:
DO
bt = (p# / 2) - INT(p# / 2)
IF bt >= .5 THEN bt = 1
PRINT , , p#; bt; r#(range)
REM c = INT(RND * 2) + 0
IF p# > 1000 THEN c = INT(RND * 2) + 0 ELSE c = 0
IF c = 0 THEN p# = p# + (p# - (r#(range)) + 1)
IF c = 1 THEN p# = p# + (p# - (r#(range)) + 2)
REM INPUT a$
a$ = INKEY$
IF a$ = "s" THEN STOP
REM IF range = 236 THEN EXIT DO
IF p# > (r#(range + 1)) THEN range = range + 1
LOOP UNTIL p# + (p# - (r#(range)) + 2) >= 999999999999999#
bt = (p# / 2) - INT(p# / 2)
IF bt >= .5 THEN bt = 1
PRINT , , p#; bt; r#(range)
```

```
REM INPUT a$
IF a$ = "s" THEN STOP
REM ---------------------Reduce P# down.---------------------------------
range = range + 1
count = 0
DO
bt = (p# / 2) - INT(p# / 2)
IF bt >= .5 THEN bt = 1
PRINT p#; bt; r#(range)
c = INT(RND * 2) + 0
IF c = 0 THEN p# = p# - ((r#(range)) - p# + 1)
IF c = 1 THEN p# = p# - ((r#(range)) - p# + 2)
REM p# = p# - ((r#(range)) - p# + 1)
REM IF p# < (r#(range - 1)) THEN p# = p# + 1
IF p# < (r#(range - 1)) THEN range = range - 1
count = count + 1
LOOP UNTIL p# < 150000000000000#
bt = (p# / 2) - INT(p# / 2)
IF bt >= .5 THEN bt = 1
PRINT p#; bt; r#(range)
REM INPUT a$
IF a$ = "s" THEN STOP
range = range - 1
GOTO repeat:




"C:\Users\Reactor1967\vcns\work\VCNS.BAS"
DIM r#(51)
r1# = 999999999999999#
range = 1
count = 1
DO
r#(count) = r1#
REM PRINT r1#; count
r1# = INT(r1# / 2)
c = (r1# / 2) - INT(r1# / 2)
IF c = 0 THEN r1# = r1# + 1
a$ = INKEY$
REM INPUT a$
IF a$ = "s" THEN STOP
count = count + 1
LOOP UNTIL count = 51
p# = 1
r1# = 1
m# = 500000000000000#
```

```
repeat:
count = 0
DO
count = count + 1
bt = (p# / 2) - INT(p# / 2)
IF bt >= .5 THEN bt = 1
PRINT , , p#; bt; count
REM IF p# + (p# - r1#) + p# + (p# - r1#) > m# THEN c = INT(RND * 2) + 0
IF p# + (p# - r1#) + p# + (p# - r1#) > m# THEN c = 1
IF p# + (p# - r1#) + p# + (p# - r1#) < m# THEN c = 0
IF c = 0 THEN p# = p# + (p# - r1# + 1)
IF c = 1 THEN p# = p# + (p# - r1# + 2)
a$ = INKEY$
IF a$ = "s" THEN STOP
LOOP UNTIL p# > m#
bt = (p# / 2) - INT(p# / 2)
IF bt >= .5 THEN bt = 1
PRINT , , p#; bt; count
PRINT "                              "
REM INPUT "Are you ready to decode y/n"; vcns$
IF vcns$ = "y" THEN GOTO decode:
IF vcns$ = "Y" THEN GOTO decode:
IF vcns$ = "s" THEN STOP
up = 0
range = 1
DO
store# = p#
IF p# <= r#(range + 1) THEN up = 1
IF up = 1 THEN p# = p# + 1
IF up = 1 THEN range = range + 1
p# = p# - (r#(range) - p# + 1)
up = 0
a$ = INKEY$
REM INPUT a$
IF a$ = "s" THEN STOP
REM PRINT p#; r#(range); range
LOOP UNTIL p# < 0
p# = store#
bt = (p# / 2) - INT(p# / 2)
IF bt = 0 THEN p# = p# + 1
GOTO repeat:
decode:
REM: Read first One here.
pl = (p# / 2) - INT(p# / 2)
IF pl = 0 THEN SYSTEM
IF pl >= .5 THEN p# = p# - ((p# - 1 + 2) / 2)
```

```
REM: Get data Here
pl = (p# / 2) - INT(p# / 2)
IF pl = 0 THEN c = 0
IF pl >= .5 THEN c = 1
IF pl = 0 THEN p# = p# - ((p# - 1 + 1) / 2)
IF pl >= .5 THEN p# = p# - ((p# - 1 + 2) / 2)
REM: read zeros and wait for flip:
DO
pl = (p# / 2) - INT(p# / 2)
IF pl = 0 THEN p# = p# - ((p# - 1 + 1) / 2)
LOOP UNTIL pl >= .5
REM  -----------------------------------------------------------------
range = 1
DO
store = range
t# = p# - (r#(range) - p# + 1)
IF t# > 0 THEN EXIT DO
range = range + 1
a$ = INKEY$
IF a$ = "s" THEN STOP
PRINT p#; r#(range)
LOOP




"C:\Users\Reactor1967\vcns\work\VCFRAME9.BAS"
DECLARE SUB codedown (p#, r#(), range!, c!)
DECLARE SUB inflate (p#, r#(), range!)
DECLARE SUB rangef (p#, r#(), range!)
DECLARE SUB deflate (p#)
REM                     COPYRIGHT C 2002
REM              Lloyd Dudley Burris
REM ----------------------Initilize R#-------------------------------------
CLS
DIM r#(255)
r#(1) = 1
REM PRINT r#(1)
pt1# = 5
rf# = 100
count = 2
a# = 0
point1:
DO
a# = a# + pt1#
test = (a# / 2) - INT(a# / 2)
r#(count) = a#
```

```
IF (r#(count) / 2) - INT(r#(count) / 2) = 0 THEN r#(count) = r#(count) + 1
count = count + 1
IF a$ = "s" THEN STOP
IF count >= 254 THEN EXIT DO
LOOP UNTIL a# = rf#
pt1# = pt1# * 10
rf# = rf# * 10
IF count >= 254 THEN GOTO point2:
IF a# + pt1# > 999999999999999# THEN GOTO point2:
GOTO point1:
point2:
r#(254) = 950000000000001#
r#(255) = 999999999999999#
REM -----------------------------End Init r#()-----------------------------
RANDOMIZE TIMER
p# = 451234567890123#
repeat:
DO
GOSUB printer:
c = 0
IF c = 0 THEN p# = p# + (p# - r#(range) + 1)
IF c = 1 THEN p# = p# + (p# - r#(range) + 2)
CALL rangef(p#, r#(), range)
test = p# > (r#(range) + 25000000000001#)
REM IF test = -1 THEN EXIT DO
ex = (test = -1) AND (range = 253)
IF test = -1 THEN GOSUB printer:
IF test = -1 THEN flip# = (r#(range) - p#) * -1
IF test = -1 THEN p# = (p# - r#(range) + r#(range + 1) - 25000000000001#)
IF ex = -1 THEN EXIT DO
a$ = INKEY$
IF a$ = "s" THEN STOP
LOOP
GOSUB printer:
CALL deflate(p#)
INPUT a$
IF a$ = "s" THEN STOP
GOTO repeat:
printer:
CALL rangef(p#, r#(), range)
bt = (p# / 2) - INT(p# / 2)
IF bt >= .5 THEN bt = 1
PRINT p#; range; bt
RETURN
decode:
PRINT "Coming Soon"
```

```
SUB codedown (p#, r#(), range, c)
test# = ABS((r#(range) - 1) - p#)
test2# = (r#(range + 2) - 1)
p# = test# - 1 - test2#
REM PRINT test#; test2#
IF c = 0 THEN p# = test2# - 2
IF c = 1 THEN p# = test2# - 1
REM PRINT test2#; c
END SUB

SUB codeup (p#, r#(), range, c)
c = p# / 2 - INT(p# / 2)
IF c >= .5 THEN c = 1
IF c = 0 THEN p# = p# + 1
IF c = 1 THEN p# = p# + 2
test# = (r#(range + 1) - 1) - p#
p# = (r#(range - 1) - 1) + test#

REM test# = ABS((r#(range) - 1) - p#)
REM p# = (r#(range + 2) - 1) - test#
REM IF c = 0 THEN test2# = test2# - 1
REM IF c = 1 THEN test2# = test2# - 2
REM PRINT test2#; c
END SUB

SUB deflate (p#)
DO
p# = p# - 50000000000000#
LOOP UNTIL p# < 500000000000000#
END SUB

SUB inflate (p#, r#(), range)
DO
p# = p# + 50000000000000#
REM LOOP UNTIL p# > 550000000000000#
LOOP UNTIL p# >= r#(246) + 25000000000000#
END SUB

SUB rangef (p#, r#(), range)
range = 255
DO
test = r#(range) <= p#
IF test = -1 THEN EXIT DO
range = range - 1
a$ = INKEY$
```

```
IF a$ = "s" THEN STOP
LOOP UNTIL range = 0
IF range = 0 THEN PRINT "ERROR rangef failed"
IF range = 0 THEN SYSTEM
END SUB




"C:\Users\Reactor1967\vcns\work\VCFRAME8.BAS"
DECLARE SUB codedown (p#, r#(), range!, c!)
DECLARE SUB inflate (p#, r#(), range!)
DECLARE SUB rangef (p#, r#(), range!)
DECLARE SUB deflate (p#)
REM                     COPYRIGHT C 2002
REM                 Lloyd Dudley Burris
REM ----------------------Initilize R#-------------------------------------
CLS
DIM r#(255)
r#(1) = 1
REM PRINT r#(1)
pt1# = 5
rf# = 100
count = 2
a# = 0
point1:
DO
a# = a# + pt1#
test = (a# / 2) - INT(a# / 2)
r#(count) = a#
IF (r#(count) / 2) - INT(r#(count) / 2) = 0 THEN r#(count) = r#(count) + 1
count = count + 1
IF a$ = "s" THEN STOP
IF count >= 254 THEN EXIT DO
LOOP UNTIL a# = rf#
pt1# = pt1# * 10
rf# = rf# * 10
IF count >= 254 THEN GOTO point2:
IF a# + pt1# > 999999999999999# THEN GOTO point2:
GOTO point1:
point2:
r#(254) = 950000000000001#
r#(255) = 999999999999999#
RANDOMIZE TIMER
p# = 451234567890123#
repeat:
DO
```

```
GOSUB printer:
REM c = 0
tasty = p# + (p# - r#(range)) > (r#(range) + 25000000000001#)
IF tasty = 0 THEN c = 0
IF tasty = 1 THEN c = INT(RND * 2) + 0
IF c = 0 THEN p# = p# + (p# - r#(range) + 1)
IF c = 1 THEN p# = p# + (p# - r#(range) + 2)
CALL rangef(p#, r#(), range)
test = p# > (r#(range) + 25000000000001#)
REM IF test = -1 THEN EXIT DO
ex = (test = -1) AND (range = 253)
IF test = -1 THEN GOSUB printer:
IF test = -1 THEN c = 1
IF test = -1 THEN CALL codedown(p#, r#(), range, c)
IF ex = -1 THEN EXIT DO
IF a$ = "s" THEN STOP
test = 0
LOOP
REM GOSUB printer:
deflate (p#)
INPUT a$
IF a$ = "s" THEN STOP
GOTO repeat:
printer:
CALL rangef(p#, r#(), range)
bt = (p# / 2) - INT(p# / 2)
IF bt >= .5 THEN bt = 1
PRINT p#; range; bt
RETURN

SUB codedown (p#, r#(), range, c)
REM decodeing equation p#numsign - (rodd) + 1 - (p#numsign) for binary 0
REM encoding equation p#numsign - (rodd) + 2 - (p#numsign) for binary 1
redo:
c = 1
p1# = p# - r#(range)
p2# = r#(range + 2) - p1# - 1
REM PRINT p2#; r#(range + 1) - p#
b1 = p2# / 2 - INT(p2# / 2)
b2 = (r#(range + 1) - p#) / 2 - INT((r#(range + 1) - p#) / 2)
IF b1 <> b2 THEN STOP
IF c = 0 THEN p# = p# + 1
IF c = 1 THEN p# = p# + 2
PRINT p#
STOP
END SUB
```

```
SUB codeup (p#, r#(), range, c)
c = p# / 2 - INT(p# / 2)
IF c >= .5 THEN c = 1
IF c = 0 THEN p# = p# + 1
IF c = 1 THEN p# = p# + 2
test# = (r#(range + 1) - 1) - p#
p# = (r#(range - 1) - 1) + test#

REM test# = ABS((r#(range) - 1) - p#)
REM p# = (r#(range + 2) - 1) - test#
REM IF c = 0 THEN test2# = test2# - 1
REM IF c = 1 THEN test2# = test2# - 2
REM PRINT test2#; c
END SUB

SUB deflate (p#)
DO
p# = p# - 50000000000000#
LOOP UNTIL p# < 500000000000000#
END SUB

SUB inflate (p#, r#(), range)
DO
p# = p# + 50000000000000#
REM LOOP UNTIL p# > 550000000000000#
LOOP UNTIL p# >= r#(246) + 25000000000000#
END SUB

SUB rangef (p#, r#(), range)
range = 255
DO
test = r#(range) <= p#
IF test = -1 THEN EXIT DO
range = range - 1
a$ = INKEY$
IF a$ = "s" THEN STOP
LOOP UNTIL range = 0
IF range = 0 THEN PRINT "ERROR rangef failed"
IF range = 0 THEN SYSTEM
END SUB


"C:\Users\Reactor1967\vcns\work\VCFRAME7.BAS"
DECLARE SUB codedown (p#, r#(), range!, c!)
```

```
DECLARE SUB inflate (p#, r#(), range!)
DECLARE SUB rangef (p#, r#(), range!)
DECLARE SUB deflate (p#)
REM                    COPYRIGHT C 2002
REM              Lloyd Dudley Burris
REM ----------------------Initilize R#------------------------------------
CLS
DIM r#(255)
r#(1) = 1
REM PRINT r#(1)
pt1# = 5
rf# = 100
count = 2
a# = 0
point1:
DO
a# = a# + pt1#
test = (a# / 2) - INT(a# / 2)
r#(count) = a#
IF (r#(count) / 2) - INT(r#(count) / 2) = 0 THEN r#(count) = r#(count) + 1
count = count + 1
IF a$ = "s" THEN STOP
IF count >= 254 THEN EXIT DO
LOOP UNTIL a# = rf#
pt1# = pt1# * 10
rf# = rf# * 10
IF count >= 254 THEN GOTO point2:
IF a# + pt1# > 999999999999999# THEN GOTO point2:
GOTO point1:
point2:
r#(254) = 950000000000001#
r#(255) = 999999999999999#
REM -----------------------------End Init r#()-----------------------------
RANDOMIZE TIMER
p# = 451234567890123#
repeat:
count = 0
DO
CALL rangef(p#, r#(), range)
GOSUB printer:
IF p# + (p# - r#(range)) >= r#(range + 1) THEN c = 1 ELSE c = 0
REM c = 0
IF c = 0 THEN p# = p# + (p# - r#(range) + 1)
IF c = 1 THEN p# = p# + (p# - r#(range) + 2)
CALL rangef(p#, r#(), range)
test2 = (range = 254) AND (p# >= r#(range) + 25000000000000#)
```

```
test3 = p# >= r#(range) + 25000000000001#
IF test2 = -1 THEN EXIT DO
REM IF test3 = -1 THEN GOSUB ex:
a$ = INKEY$
IF a$ = "s" THEN STOP
LOOP
CALL rangef(p#, r#(), range)
GOSUB printer:
CALL deflate(p#)
INPUT a$
IF a$ = "s" THEN STOP
GOTO repeat:
printer:
bt = (p# / 2) - INT(p# / 2)
IF bt >= .5 THEN bt = 1
PRINT p#; range; bt
RETURN
ex:
p# = p# - r#(range)
p# = p# - 25000000000001#
p# = p# + r#(range + 1)
CALL rangef(p#, r#(), range)
RETURN:
decode:
PRINT "Coming Soon"

SUB deflate (p#)
DO
p# = p# - 50000000000000#
LOOP UNTIL p# < 500000000000000#
END SUB

SUB inflate (p#, r#(), range)
DO
p# = p# + 50000000000000#
REM LOOP UNTIL p# > 550000000000000#
LOOP UNTIL p# >= r#(246) + 25000000000000#
END SUB

SUB rangef (p#, r#(), range)
range = 255
DO
test = r#(range) <= p#
IF test = -1 THEN EXIT DO
range = range - 1
a$ = INKEY$
```

```
IF a$ = "s" THEN STOP
LOOP UNTIL range = 0
IF range = 0 THEN PRINT "ERROR rangef failed"
IF range = 0 THEN SYSTEM
END SUB



"C:\Users\Reactor1967\vcns\work\VCFRAME6.BAS"
DECLARE SUB down (p#, r2#(), range!)
DECLARE SUB getdata (p#, r#(), range!)
DECLARE SUB up (p#, r2#(), range!, r#())
REM decodeing in the 45 range the last digits will be a 10 or a 11.
REM 11 means flip read the first bit as the data bit then look for the 1
REM to flip. 10 means no data just flip.
REM                     COPYRIGHT C 2002
REM               Lloyd Dudley Burris
REM ---------------------Initilize R#-------------------------------------
CLS
DIM r#(255)
r#(1) = 1
REM PRINT r#(1)
pt1# = 5
rf# = 100
count = 2
a# = 0
point1:
DO
a# = a# + pt1#
test = (a# / 2) - INT(a# / 2)
r#(count) = a#
IF (r#(count) / 2) - INT(r#(count) / 2) = 0 THEN r#(count) = r#(count) + 1
count = count + 1
IF count >= 254 THEN EXIT DO
LOOP UNTIL a# = rf#
pt1# = pt1# * 10
rf# = rf# * 10
IF count >= 254 THEN GOTO point2:
IF a# + pt1# > 999999999999999# THEN GOTO point2:
GOTO point1:
point2:
r#(254) = 950000000000001#
r#(255) = 999999999999999#
REM ----------------------Initilize R2#-------------------------------------
DIM r2#(37)
p# = 100000000000001#
```

```
CLS
count = 1
DO
r2#(count) = p#
p# = p# + 25000000000000#
a$ = INKEY$
IF a$ = "s" THEN STOP
count = count + 1
LOOP UNTIL p# >= 999999999999999#
r2#(count) = 999999999999999#
REM ----------------------Code P# up.---------------------------------------
p# = 452747503729231#
range = 244
RANDOMIZE TIMER
c = 0
repeat:
range = 244
count = 0
count2 = 0
d$ = ""
DO
bt = (p# / 2) - INT(p# / 2)
IF bt >= .5 THEN bt = 1
COLOR 2, 0
PRINT , p#; bt; r#(range); count; range; count2
IF (p# + (p# - r#(range) + 1)) > (r#(range + 1)) THEN c = 1 ELSE c = 0
IF count = 0 THEN c = 1
test = (p# + (p# - r#(range)) > 975000000000000#) AND count2 >= 2
IF test = -1 THEN c = INT(RND * 2) + 0
IF c = 0 THEN p# = p# + (p# - (r#(range)) + 1)
IF c = 1 THEN p# = p# + (p# - (r#(range)) + 2)
IF c = 1 THEN p# = p# - 25000000000000#
IF p# > (r#(range + 1)) THEN range = range + 1
count = count + 1
IF range = 254 THEN count2 = count2 + 1
LOOP UNTIL p# + (p# - r#(range)) >= 999999999999999#
bt = (p# / 2) - INT(p# / 2)
IF bt >= .5 THEN bt = 1
PRINT , p#; bt; r#(range); count; range; count2
PRINT "                                        "
REM a$ = INKEY$
INPUT a$
IF a$ = "s" THEN STOP
IF a$ = "d" THEN GOTO Decodedown:
IF count2 = 1 THEN p# = p# - 525000000000001#
IF count2 >= 2 THEN CALL down(p#, r2#(), range)
```

```
GOTO repeat:
REM -------------------------decode--------------------------------
Decodedown: REM from + to -
REM put your get data here.
IF count2 > 1 THEN CALL getdata(p#, r#(), range)
repeat2:
range = 254
d$ = ""
DO
bt = (p# / 2) - INT(p# / 2)
IF bt >= .5 THEN bt = 1
IF bt = 1 THEN range = range - 1
testy = (range = 244) AND (bt = 1)
IF testy = -1 THEN d$ = d$ + "1"
IF bt = 0 THEN p# = p# - ((p# - r#(range) + 1) / 2)
IF bt = 1 THEN p# = p# - ((p# - r#(range) + 2) / 2)
a$ = INKEY$
IF a$ = "s" THEN STOP
bt = (p# / 2) - INT(p# / 2)
IF bt >= .5 THEN bt = 1
test = (range = 244) AND (bt = 1)
LOOP UNTIL test = -1
bt = (p# / 2) - INT(p# / 2)
IF bt >= .5 THEN bt = 1
IF bt = 1 THEN p# = p# - ((p# - r#(range) + 2) / 2)
bt = (p# / 2) - INT(p# / 2)
IF bt >= .5 THEN bt = 1
IF bt = 1 THEN d$ = "11"
IF bt = 0 THEN d$ = "10"
PRINT p#; bt; r#(range); range; d$
IF p# = 452747503729231# THEN GOTO finish:
IF d$ = "11" THEN CALL up(p#, r2#(), range, r#())
IF d$ = "10" THEN p# = p# + 525000000000001#
IF d$ = "11" THEN PRINT p#; "this came from up"
IF d$ = "10" THEN PRINT p#; "this came from add"
IF d$ = "11" THEN CALL getdata(p#, r#(), range)
INPUT a$
IF a$ = "s" THEN STOP
GOTO repeat2:
finish:
PRINT "Decode Completed Successfully"

SUB down (p#, r2#(), range)
range = 37
store1# = p#
DO
```

```
bt = (p# / 2) - INT(p# / 2)
IF bt >= .5 THEN bt = 1
IF (p# - (r2#(range) - p#)) < (r2#(range - 1)) THEN c = 1 ELSE c = 0
IF c = 0 THEN p# = p# - (r2#(range) - p# + 1)
IF c = 1 THEN p# = p# - (r2#(range) - p# + 2)
IF p# < (r2#(range - 1)) THEN range = range - 1
LOOP UNTIL p# < 475000000000000#
END SUB


SUB getdata (p#, r#(), range)
range = 254
bt = (p# / 2) - INT(p# / 2)
IF bt >= .5 THEN bt = 1
IF bt = 0 THEN p# = p# - ((p# - r#(range) + 1) / 2)
IF bt = 1 THEN p# = p# - ((p# - r#(range) + 2) / 2)
PRINT bt; "This is Your data Lloyd"
c = bt: REM this is your data:
END SUB


SUB up (p#, r2#(), range, r#())
range = 16
DO
bt = (p# / 2) - INT(p# / 2)
IF bt >= .5 THEN bt = 1
IF bt = 1 THEN range = range + 1
IF bt = 0 THEN p# = p# + ((r2#(range) - p# + 1) / 2)
IF bt = 1 THEN p# = p# + ((r2#(range) - p# + 2) / 2)
bt = (p# / 2) - INT(p# / 2)
IF bt >= .5 THEN bt = 1
PRINT p#
test = (range = 37) AND (bt = 1)
LOOP UNTIL test = -1
END SUB




"C:\Users\Reactor1967\vcns\work\VCFRAME5.BAS"
DECLARE SUB down (p#, r2#(), range!)
DECLARE SUB getdata (p#, r#(), range!)
DECLARE SUB up (p#, r2#(), range!)
REM decodeing in the 45 range the last digits will be a 10 or a 11.
REM 11 means flip read the first bit as the data bit then look for the 1
REM to flip. 10 means no data just flip.
REM                     COPYRIGHT C 2002
REM                 Lloyd Dudley Burris
REM ----------------------Initilize R#------------------------------------
```

```
CLS
DIM r#(255)
r#(1) = 1
REM PRINT r#(1)
pt1# = 5
rf# = 100
count = 2
a# = 0
point1:
DO
a# = a# + pt1#
test = (a# / 2) - INT(a# / 2)
r#(count) = a#
IF (r#(count) / 2) - INT(r#(count) / 2) = 0 THEN r#(count) = r#(count) + 1
count = count + 1
IF count >= 254 THEN EXIT DO
LOOP UNTIL a# = rf#
pt1# = pt1# * 10
rf# = rf# * 10
IF count >= 254 THEN GOTO point2:
IF a# + pt1# > 999999999999999# THEN GOTO point2:
GOTO point1:
point2:
r#(254) = 950000000000001#
r#(255) = 999999999999999#
REM ---------------------Initilize R2#-------------------------------------
DIM r2#(37)
p# = 100000000000001#
CLS
count = 1
DO
r2#(count) = p#
p# = p# + 25000000000000#
a$ = INKEY$
IF a$ = "s" THEN STOP
count = count + 1
LOOP UNTIL p# >= 999999999999999#
r2#(count) = 999999999999999#
REM ---------------------Code P# up.-------------------------------------
p# = 452747503729231#
range = 244
RANDOMIZE TIMER
c = 0
REM boy$ = "test.txt"
REM OPEN boy$ FOR OUTPUT AS #1
repeat:
```

```
count = 0
count2 = 0
d$ = ""
DO
bt = (p# / 2) - INT(p# / 2)
IF bt >= .5 THEN bt = 1
COLOR 2, 0
PRINT , p#; bt; r#(range); count; range; count2
IF (p# + (p# - r#(range) + 1)) > (r#(range + 1)) THEN c = 1 ELSE c = 0
IF count = 0 THEN c = 1
test = (p# + (p# - r#(range)) > 975000000000000#) AND count2 >= 2
IF test = -1 THEN c = INT(RND * 2) + 0
IF c = 0 THEN p# = p# + (p# - (r#(range)) + 1)
IF c = 1 THEN p# = p# + (p# - (r#(range)) + 2)
IF p# > (r#(range + 1)) THEN range = range + 1
count = count + 1
IF range = 254 THEN count2 = count2 + 1
LOOP UNTIL p# + (p# - r#(range)) >= 999999999999999#
REM LOOP UNTIL p# >= 987500000000000#
REM made changes here.
bt = (p# / 2) - INT(p# / 2)
IF bt >= .5 THEN bt = 1
PRINT , p#; bt; r#(range); count; range; count2
REM PRINT #1, p#; bt; r#(range); count; range; count2
PRINT "                                    "
REM a$ = INKEY$
INPUT a$
IF a$ = "s" THEN CLOSE
IF a$ = "s" THEN STOP
IF a$ = "d" THEN GOTO Decodedown:
IF count2 = 1 THEN p# = p# - 525000000000001#
store1# = p#
IF count2 >= 2 THEN CALL down(p#, r2#(), range)
store2# = p#
IF count2 >= 2 THEN CALL up(p#, r2#(), range)
IF p# <> store1# THEN STOP
p# = store2#
range = 244
GOTO repeat:
REM -------------------------decode--------------------------------
Decodedown: REM from + to -
REM put your get data here.
IF count2 > 1 THEN CALL getdata(p#, r#(), range)
repeat2:
range = 254
d$ = ""
```

```
DO
bt = (p# / 2) - INT(p# / 2)
IF bt >= .5 THEN bt = 1
IF bt = 1 THEN range = range - 1
testy = (range = 244) AND (bt = 1)
IF testy = -1 THEN d$ = d$ + "1"
IF bt = 0 THEN p# = p# - ((p# - r#(range) + 1) / 2)
IF bt = 1 THEN p# = p# - ((p# - r#(range) + 2) / 2)
a$ = INKEY$
IF a$ = "s" THEN STOP
bt = (p# / 2) - INT(p# / 2)
IF bt >= .5 THEN bt = 1
test = (range = 244) AND (bt = 1)
LOOP UNTIL test = -1
bt = (p# / 2) - INT(p# / 2)
IF bt >= .5 THEN bt = 1
IF bt = 1 THEN p# = p# - ((p# - r#(range) + 2) / 2)
bt = (p# / 2) - INT(p# / 2)
IF bt >= .5 THEN bt = 1
IF bt = 1 THEN d$ = "11"
IF bt = 0 THEN d$ = "10"
PRINT p#; bt; r#(range); range; d$
REM PRINT #1, p#; bt; r#(range); range; d$
IF p# = 452747503729231# THEN GOTO finish:
IF d$ = "11" THEN CALL up(p#, r2#(), range)
IF d$ = "10" THEN p# = p# + 525000000000001#
IF d$ = "11" THEN PRINT p#; "this came from up"
REM IF d$ = "11" THEN PRINT #1, p#; "this came from up"
IF d$ = "10" THEN PRINT p#; "this came from add"
REM IF d$ = "10" THEN PRINT #1, p#; "this came from add"
INPUT a$
IF a$ = "s" THEN CLOSE
IF a$ = "s" THEN STOP
range = 254
IF d$ = "11" THEN CALL getdata(p#, r#(), range)
GOTO repeat2:
finish:
PRINT "Decode Completed Successfully"

SUB down (p#, r2#(), range)
range = 37
DO
bt = (p# / 2) - INT(p# / 2)
IF bt >= .5 THEN bt = 1
IF (p# - (r2#(range) - p#)) < (r2#(range - 1)) THEN c = 1 ELSE c = 0
IF c = 0 THEN p# = p# - (r2#(range) - p# + 1)
```

```
IF c = 1 THEN p# = p# - (r2#(range) - p# + 2)
IF p# < (r2#(range - 1)) THEN range = range - 1
count = count + 1
LOOP UNTIL p# <= 475000000000000#
END SUB

SUB getdata (p#, r#(), range)
range = 254
bt = (p# / 2) - INT(p# / 2)
IF bt >= .5 THEN bt = 1
IF bt = 0 THEN p# = p# - ((p# - r#(range) + 1) / 2)
IF bt = 1 THEN p# = p# - ((p# - r#(range) + 2) / 2)
PRINT bt; "This is Your data Lloyd"
c = bt: REM this is your data:
END SUB

SUB up (p#, r2#(), range)
range = 16
DO
bt = (p# / 2) - INT(p# / 2)
IF bt >= .5 THEN bt = 1
IF bt = 1 THEN range = range + 1
IF bt = 0 THEN p# = p# + ((r2#(range) - p# + 1) / 2)
IF bt = 1 THEN p# = p# + ((r2#(range) - p# + 2) / 2)
LOOP UNTIL range = 37
REM LOOP UNTIL p# > 975000000000000#
END SUB




"C:\Users\Reactor1967\vcns\work\VCFRAME4.BAS"
DECLARE SUB getdata (p#, r#(), range)
DECLARE SUB up (p#, r#(), range)
DECLARE SUB down (p#, r#(), range)
REM decodeing in the 45 range the last digits will be a 10 or a 11.
REM 11 means flip read the first bit as the data bit then look for the 1
REM to flip. 10 means no data just flip.
REM                    COPYRIGHT C 2002
REM              Lloyd Dudley Burris
REM ----------------------Initilize R#------------------------------------
CLS
DIM r#(255)
r#(1) = 1
REM PRINT r#(1)
pt1# = 5
rf# = 100
```

```
count = 2
a# = 0
point1:
DO
a# = a# + pt1#
test = (a# / 2) - INT(a# / 2)
r#(count) = a#
IF (r#(count) / 2) - INT(r#(count) / 2) = 0 THEN r#(count) = r#(count) + 1
count = count + 1
IF count >= 254 THEN EXIT DO
LOOP UNTIL a# = rf#
pt1# = pt1# * 10
rf# = rf# * 10
IF count >= 254 THEN GOTO point2:
IF a# + pt1# > 999999999999999# THEN GOTO point2:
GOTO point1:
point2:
r#(254) = 950000000000001#
r#(255) = 999999999999999#
REM ----------------------Code P# up.---------------------------------------
range = 237
REM p# = 1000
p# = 100000000000001#
RANDOMIZE TIMER
c = 0
repeat:
count = 0
count2 = 0
d$ = ""
DO
bt = (p# / 2) - INT(p# / 2)
IF bt >= .5 THEN bt = 1
COLOR 2, 0
PRINT , p#; bt; r#(range); count; range; count2
IF (p# + (p# - r#(range) + 1)) > (r#(range + 1)) THEN c = 1 ELSE c = 0
IF count = 0 THEN c = 1
test = (p# + (p# - r#(range)) >= 975000000000000#) AND count2 >= 2
IF test = -1 THEN c = INT(RND * 2) + 0
IF c = 0 THEN p# = p# + (p# - (r#(range))) + 1
IF c = 1 THEN p# = p# + (p# - (r#(range))) + 2
IF p# > (r#(range + 1)) THEN range = range + 1
count = count + 1
IF range = 254 THEN count2 = count2 + 1
LOOP UNTIL p# + (p# - (r#(range))) + 2 >= 999999999999999#
bt = (p# / 2) - INT(p# / 2)
IF bt >= .5 THEN bt = 1
```

```
PRINT , p#; bt; r#(range); count; range; count2
PRINT "                                    "
REM a$ = INKEY$
INPUT a$
IF a$ = "s" THEN STOP
IF a$ = "d" THEN GOTO Decodedown:
IF count2 = 1 THEN p# = p# - 525000000000001#
IF count2 >= 2 THEN CALL down(p#, r#(), range)
range = 244
GOTO repeat:
REM --------------------------decode----------------------------------
Decodedown: REM from + to -
REM put your get data here.
IF count2 > 1 THEN CALL getdata(p#, r#(), range)
repeat2:
range = 254
d$ = ""
DO
bt = (p# / 2) - INT(p# / 2)
IF bt >= .5 THEN bt = 1
IF bt = 1 THEN range = range - 1
testy = (range = 244) AND (bt = 1)
IF testy = -1 THEN d$ = d$ + "1"
IF bt = 0 THEN p# = p# - ((p# - r#(range) + 1) / 2)
IF bt = 1 THEN p# = p# - ((p# - r#(range) + 2) / 2)
a$ = INKEY$
IF a$ = "s" THEN STOP
bt = (p# / 2) - INT(p# / 2)
IF bt >= .5 THEN bt = 1
test = (range = 244) AND (bt = 1)
LOOP UNTIL test = -1
bt = (p# / 2) - INT(p# / 2)
IF bt >= .5 THEN bt = 1
REM IF bt = 0 THEN p# = p# - ((p# - r#(range) + 1) / 2): REM added here
IF bt = 1 THEN p# = p# - ((p# - r#(range) + 2) / 2)
bt = (p# / 2) - INT(p# / 2)
IF bt >= .5 THEN bt = 1
IF bt = 1 THEN d$ = "11"
IF bt = 0 THEN d$ = "10"
PRINT p#; "-"; bt; r#(range); range; d$
IF d$ = "11" THEN CALL up(p#, r#(), range)
IF d$ = "10" THEN p# = p# + 525000000000001#
range = 254
IF d$ = "11" THEN PRINT p#; "this came from up"
IF d$ = "10" THEN PRINT p#; "this came from add"
```

```
SUB down (p#, r#(), range)
p# = p# - 25000000000000#
range = 255
DO
bt = (p# / 2) - INT(p# / 2)
IF bt >= .5 THEN bt = 1
a$ = INKEY$
IF a$ = "s" THEN STOP
IF (p# - (r#(range) - p# + 1)) < (r#(range - 1)) THEN c = 1 ELSE c = 0
IF (p# - (r#(range) - p#)) <= 475000000000000# THEN c = 1
IF c = 0 THEN p# = p# - ((r#(range)) - p# + 1)
IF c = 1 THEN p# = p# - ((r#(range)) - p# + 2)
IF p# < (r#(range - 1)) THEN range = range - 1
count = count + 1
LOOP UNTIL p# <= 475000000000000#

END SUB

SUB getdata (p#, r#(), range)
bt = (p# / 2) - INT(p# / 2)
IF bt >= .5 THEN bt = 1
IF bt = 0 THEN p# = p# - ((p# - r#(range) + 1) / 2)
IF bt = 1 THEN p# = p# - ((p# - r#(range) + 2) / 2)
c = bt: REM this is your data:
END SUB

SUB up (p#, r#(), range)
range = 245
DO
bt = (p# / 2) - INT(p# / 2)
IF bt >= .5 THEN bt = 1
IF bt = 1 THEN range = range + 1
IF bt = 0 THEN p# = p# + ((r#(range) - p# + 1) / 2)
IF bt = 1 THEN p# = p# + ((r#(range) - p# + 2) / 2)
IF p# = bingo# THEN STOP
a$ = INKEY$
IF a$ = "s" THEN STOP
LOOP UNTIL p# >= 950000000000000#
REM LOOP UNTIL (p# + 25000000000000#) >= 975000000000000#
p# = p# + 25000000000000#
END SUB
```

"C:\Users\Reactor1967\vcns\work\VCFRAME3.BAS"
REM   COPYRIGHT C 2002

```
REM                    Lloyd Dudley Burris
REM ----------------------Initilize R#-------------------------------------
CLS
DIM r#(255)
r#(1) = 1
REM PRINT r#(1)
pt1# = 5
rf# = 100
count = 2
a# = 0
redoa:
DO
a# = a# + pt1#
test = (a# / 2) - INT(a# / 2)
r#(count) = a#
IF (r#(count) / 2) - INT(r#(count) / 2) = 0 THEN r#(count) = r#(count) + 1
REM PRINT r#(count); count
count = count + 1
REM INPUT a$
REM a$ = INKEY$
IF a$ = "s" THEN STOP
IF count >= 254 THEN EXIT DO
LOOP UNTIL a# = rf#
pt1# = pt1# * 10
rf# = rf# * 10
IF count >= 254 THEN GOTO redob:
IF a# + pt1# > 999999999999999# THEN GOTO redob:
GOTO redoa:
redob:
r#(254) = 950000000000001#
r#(255) = 999999999999999#
REM ----------------------Code P# up.-------------------------------------
range = 237
REM p# = 1000
p# = 100000000000001#
RANDOMIZE TIMER
c = 0
REM file$ = "vcout.txt"
REM OPEN file$ FOR OUTPUT AS #1
repeat:
count = 0
DO
bt = (p# / 2) - INT(p# / 2)
IF bt >= .5 THEN bt = 1
COLOR 2, 0
PRINT , , p#; "+"; bt; r#(range); count
```

```
REM WRITE #1, p#, bt, r#(range)
REM a$ = INKEY$
REM INPUT a$
IF a$ = "s" THEN SYSTEM
IF (p# + (p# - r#(range) + 1)) > (r#(range + 1)) THEN c = 1 ELSE c = 0
IF c = 0 THEN p# = p# + (p# - (r#(range)) + 1)
IF c = 1 THEN p# = p# + (p# - (r#(range)) + 2)
IF a$ = "s" THEN STOP
IF p# > (r#(range + 1)) THEN range = range + 1
count = count + 1
IF range = 254 THEN EXIT DO
LOOP UNTIL p# + (p# - (r#(range)) + 2) >= 999999999999999#
REM CLOSE #1
bt = (p# / 2) - INT(p# / 2)
IF bt >= .5 THEN bt = 1
PRINT , , p#; "+"; bt; r#(range); count
a$ = INKEY$
REM INPUT a$
IF a$ = "s" THEN SYSTEM
IF a$ = "d" THEN GOTO Decodedown:
REM ---------------------Code P# down.----------------------------------
range = range + 1
count = 0
DO
bt = (p# / 2) - INT(p# / 2)
IF bt >= .5 THEN bt = 1
COLOR 2, 0
PRINT , , p#; "-"; bt; r#(range); count
REM a$ = INKEY$
REM INPUT a$
REM IF a$ = "s" THEN STOP
REM c = INT(RND * 2) + 0
IF (p# - (r#(range) - p# + 1)) < (r#(range - 1)) THEN c = 1 ELSE c = 0
IF c = 0 THEN p# = p# - ((r#(range)) - p# + 1)
IF c = 1 THEN p# = p# - ((r#(range)) - p# + 2)
IF p# < (r#(range - 1)) THEN range = range - 1
count = count + 1
IF range <= 238 THEN EXIT DO
LOOP UNTIL p# < 150000000000000#
bt = (p# / 2) - INT(p# / 2)
IF bt >= .5 THEN bt = 1
PRINT , , p#; "-"; bt; r#(range); count
a$ = INKEY$
REM INPUT a$
IF a$ = "s" THEN SYSTEM
IF a$ = "d" THEN GOTO decode2:
```

```
range = range - 1
GOTO repeat:
REM ----------------------------DECODING AREA------------------------------
REM ----------------------------RESTRICTED AREA:PROGRAM AUTHOR
ONLY!!!------
decode1:
REM---------------------------[DECODING + TO -]--------------------------
Decodedown: REM from + to -
DO
bt = (p# / 2) - INT(p# / 2)
IF bt >= .5 THEN bt = 1
IF bt = 1 THEN range = range - 1
COLOR 2, 0
PRINT , , p#; "-"; bt; r#(range)
IF bt = 0 THEN p# = p# - ((p# - r#(range) + 1) / 2)
IF bt = 1 THEN p# = p# - ((p# - r#(range) + 2) / 2)
a$ = INKEY$
REM INPUT a$
IF a$ = "s" THEN STOP
bt = (p# / 2) - INT(p# / 2)
IF bt >= .5 THEN bt = 1
test = (bt = 1) AND range <= 237
IF p# = 100000000000001# THEN GOTO finish:
LOOP UNTIL test = -1
bt = (p# / 2) - INT(p# / 2)
IF bt >= .5 THEN bt = 1
COLOR 2, 0
PRINT , , p#; "-"; bt; r#(range)
REM ---------------------------[DECODING - TO +]--------------------------
range = range + 1
Decodeup: REM from - to +
decode2:
DO
bt = (p# / 2) - INT(p# / 2)
IF bt >= .5 THEN bt = 1
IF bt = 1 THEN range = range + 1
COLOR 2, 0
PRINT , , p#; "+"; bt; r#(range)
IF bt = 0 THEN p# = p# + ((r#(range) - p# + 1) / 2)
IF bt = 1 THEN p# = p# + ((r#(range) - p# + 2) / 2)
a$ = INKEY$
REM INPUT a$
IF a$ = "s" THEN STOP
bt = (p# / 2) - INT(p# / 2)
IF bt >= .5 THEN bt = 1
test = (bt = 1) AND range >= 255
```

```
LOOP UNTIL test = -1
bt = (p# / 2) - INT(p# / 2)
IF bt >= .5 THEN bt = 1
COLOR 2, 0
PRINT , , p#; "+"; bt; r#(range)
range = range - 1
GOTO decode1:
finish:
PRINT , , "DECODE FINISHED!!!!"
SYSTEM




"C:\Users\Reactor1967\vcns\work\VCFRAME2.BAS"
REM                    COPYRIGHT C 2002
REM              Lloyd Dudley Burris
REM ----------------------Initilize R#-------------------------------------
CLS
DIM r#(255)
r#(1) = 1
REM PRINT r#(1)
pt1# = 5
rf# = 100
count = 2
a# = 0
replaced1:
DO
a# = a# + pt1#
test = (a# / 2) - INT(a# / 2)
r#(count) = a#
IF (r#(count) / 2) - INT(r#(count) / 2) = 0 THEN r#(count) = r#(count) + 1
REM PRINT r#(count); count
count = count + 1
REM INPUT a$
REM a$ = INKEY$
IF a$ = "s" THEN STOP
IF count >= 254 THEN EXIT DO
LOOP UNTIL a# = rf#
pt1# = pt1# * 10
rf# = rf# * 10
IF count >= 254 THEN GOTO replaced2:
IF a# + pt1# > 999999999999999# THEN GOTO replaced2:
GOTO replaced1:
replaced2:
r#(254) = 950000000000001#
r#(255) = 999999999999999#
```

```
REM ---------------------Code P# up.--------------------------------------
range = 237
REM p# = 1000
p# = 100000000000001#
RANDOMIZE TIMER
c = 0
REM file$ = "vcout.txt"
REM OPEN file$ FOR OUTPUT AS #1
repeat:
DO
bt = (p# / 2) - INT(p# / 2)
IF bt >= .5 THEN bt = 1
COLOR 2, 0
PRINT , , p#; bt; r#(range)
REM WRITE #1, p#, bt, r#(range)
REM a$ = INKEY$
REM INPUT a$
IF a$ = "s" THEN STOP
IF (p# + (p# - r#(range) + 1)) > (r#(range + 1)) THEN c = 1 ELSE c = 0
IF c = 0 THEN p# = p# + (p# - (r#(range)) + 1)
IF c = 1 THEN p# = p# + (p# - (r#(range)) + 2)
IF a$ = "s" THEN STOP
IF p# > (r#(range + 1)) THEN range = range + 1
IF range = 254 THEN EXIT DO
LOOP UNTIL p# + (p# - (r#(range)) + 2) >= 999999999999999#
REM CLOSE #1
bt = (p# / 2) - INT(p# / 2)
IF bt >= .5 THEN bt = 1
PRINT , , p#; bt; r#(range)
REM a$ = INKEY$
INPUT a$
IF a$ = "s" THEN STOP
IF a$ = "d" THEN GOTO Decodedown:
REM ---------------------Code P# down.----------------------------------
range = range + 1
count = 0
DO
bt = (p# / 2) - INT(p# / 2)
IF bt >= .5 THEN bt = 1
COLOR 2, 0
PRINT p#; bt; r#(range)
REM a$ = INKEY$
REM INPUT a$
IF a$ = "s" THEN STOP
REM c = INT(RND * 2) + 0
IF (p# - (r#(range) - p# + 1)) < (r#(range - 1)) THEN c = 1 ELSE c = 0
```

```
IF c = 0 THEN p# = p# - ((r#(range)) - p# + 1)
IF c = 1 THEN p# = p# - ((r#(range)) - p# + 2)
IF p# < (r#(range - 1)) THEN range = range - 1
count = count + 1
IF range <= 238 THEN EXIT DO
LOOP UNTIL p# < 150000000000000#
bt = (p# / 2) - INT(p# / 2)
IF bt >= .5 THEN bt = 1
PRINT p#; bt; r#(range)
REM a$ = INKEY$
INPUT a$
IF a$ = "s" THEN STOP
IF a$ = "d" THEN GOTO decode2:
range = range - 1
GOTO repeat:
REM ---------------------------DECODING AREA------------------------------
REM ---------------------------RESTRICTED AREA:PROGRAM AUTHOR
ONLY!!!------
decode1:
REM----------------------------[DECODING + TO -]--------------------------
Decodedown: REM from + to -
DO
bt = (p# / 2) - INT(p# / 2)
IF bt >= .5 THEN bt = 1
IF bt = 1 THEN range = range - 1
COLOR 2, 0
PRINT , , p#; bt; r#(range)
IF bt = 0 THEN p# = p# - ((p# - r#(range) + 1) / 2)
IF bt = 1 THEN p# = p# - ((p# - r#(range) + 2) / 2)
a$ = INKEY$
REM INPUT a$
IF a$ = "s" THEN STOP
bt = (p# / 2) - INT(p# / 2)
IF bt >= .5 THEN bt = 1
test = (bt = 1) AND range <= 237
IF p# = 100000000000001# THEN GOTO finish:
LOOP UNTIL test = -1
bt = (p# / 2) - INT(p# / 2)
IF bt >= .5 THEN bt = 1
COLOR 2, 0
PRINT , , p#; bt; r#(range)
REM ---------------------------[DECODING - TO +]--------------------------
range = range + 1
Decodeup: REM from - to +
decode2:
DO
```

```
bt = (p# / 2) - INT(p# / 2)
IF bt >= .5 THEN bt = 1
IF bt = 1 THEN range = range + 1
COLOR 2, 0
PRINT , , p#; bt; r#(range)
IF bt = 0 THEN p# = p# + ((r#(range) - p# + 1) / 2)
IF bt = 1 THEN p# = p# + ((r#(range) - p# + 2) / 2)
a$ = INKEY$
REM INPUT a$
IF a$ = "s" THEN STOP
bt = (p# / 2) - INT(p# / 2)
IF bt >= .5 THEN bt = 1
test = (bt = 1) AND range >= 255
LOOP UNTIL test = -1
bt = (p# / 2) - INT(p# / 2)
IF bt >= .5 THEN bt = 1
COLOR 2, 0
PRINT , , p#; bt; r#(range)
range = range - 1
GOTO decode1:
finish:
PRINT , , "DECODE FINISHED!!!!"
SYSTEM




"C:\Users\Reactor1967\vcns\work\VCFRAME.BAS"
REM GOT TO GET R TO SYNCHRONIZE WITH P
REM THERE IS A GAP BETWEEN 1 AND THE NEXT R#
REM FILL IN THAT GAP OR ELIMANATE THAT RANGE ALL
TOGETHER.
REM ----------------------Initilize R#-------------------------------------
CLS
DIM r#(255)
r#(1) = 1
REM PRINT r#(1)
pt1# = 5
rf# = 100
count = 2
a# = 0
replaced1:
DO
a# = a# + pt1#
test = (a# / 2) - INT(a# / 2)
r#(count) = a#
IF (r#(count) / 2) - INT(r#(count) / 2) = 0 THEN r#(count) = r#(count) + 1
```

```
REM PRINT r#(count); count
count = count + 1
REM INPUT a$
a$ = INKEY$
IF a$ = "s" THEN STOP
IF count >= 254 THEN EXIT DO
LOOP UNTIL a# = rf#
pt1# = pt1# * 10
rf# = rf# * 10
IF count >= 254 THEN GOTO replaced2:
IF a# + pt1# > 999999999999999# THEN GOTO replaced2:
GOTO replaced1:
replaced2:
r#(254) = 950000000000001#
r#(255) = 999999999999999#
REM ----------------------Code P# up.--------------------------------------
range = 38
p# = 1000
RANDOMIZE TIMER
c = 0
repeat:
DO
bt = (p# / 2) - INT(p# / 2)
IF bt >= .5 THEN bt = 1
COLOR 2, 0
PRINT , , p#; bt; r#(range)
a$ = INKEY$
REM INPUT a$
REM IF a$ = "s" THEN STOP
IF (p# + (p# - r#(range) + 1)) > (r#(range + 1)) THEN c = 1 ELSE c = 0
IF c = 0 THEN p# = p# + (p# - (r#(range)) + 1)
IF c = 1 THEN p# = p# + (p# - (r#(range)) + 2)
IF a$ = "s" THEN STOP
IF p# > (r#(range + 1)) THEN range = range + 1
LOOP UNTIL p# + (p# - (r#(range)) + 2) >= 999999999999999#
bt = (p# / 2) - INT(p# / 2)
IF bt >= .5 THEN bt = 1
IF bt = 0 THEN p# = p# + 1: REM make sure we have flip here.
bt = (p# / 2) - INT(p# / 2)
IF bt >= .5 THEN bt = 1
PRINT , , p#; bt; r#(range)
REM INPUT a$
IF a$ = "s" THEN STOP
REM ----------------------Reduce P# down.--------------------------------
range = range + 1
count = 0
```

```
DO
bt = (p# / 2) - INT(p# / 2)
IF bt >= .5 THEN bt = 1
COLOR 2, 0
PRINT p#; bt; r#(range)
a$ = INKEY$
REM INPUT a$
IF a$ = "s" THEN STOP
REM c = INT(RND * 2) + 0
IF (p# - (r#(range) - p# + 1)) < (r#(range - 1)) THEN c = 1 ELSE c = 0
IF c = 0 THEN p# = p# - ((r#(range)) - p# + 1)
IF c = 1 THEN p# = p# - ((r#(range)) - p# + 2)
IF p# < (r#(range - 1)) THEN range = range - 1
count = count + 1
LOOP UNTIL p# < 150000000000000#
bt = (p# / 2) - INT(p# / 2)
IF bt >= .5 THEN bt = 1
IF bt = 0 THEN p# = p# + 1: REM making sure flip is ok here.
bt = (p# / 2) - INT(p# / 2)
IF bt >= .5 THEN bt = 1
PRINT p#; bt; r#(range)
REM INPUT a$
IF a$ = "s" THEN STOP
range = range - 1
GOTO repeat:
Decode:
Decodedown: REM from + to -




SUB down (p#, r#(), range)
store# = p#
p# = p# - 69999999999990#
p# = p# - 2999999999990#
PRINT p# - (r#(range) - p#)
INPUT a$
IF a$ = "s" THEN STOP
p# = store#
END SUB




"C:\Users\Reactor1967\vcns\work\VBPOL1.BAS"
REM changing the polerization of each
REM number to keep it in sink. Will use
REM this for a number to fill in the
```

```
REM p-r gap.
CLS
RANDOMIZE TIMER
redo:
p# = 1
v# = 2
pol = 0
DO
c = INT(RND * 2) + 0
IF c = 0 THEN p# = p# + v# + 1
IF c = 1 THEN p# = p# + v# + 2
REM pol = 0 add 1 for binary 0
REM pol = 0 add 2 for binary 1
t1 = (c = 0) AND (pol = 0)
t2 = (c = 1) AND (pol = 0)
REM pol = 1 add 1 for binary 1
REM pol = 1 add 2 for binary 0
t3 = (c = 0) AND (pol = 1)
t4 = (c = 1) AND (pol = 1)
IF t1 = -1 THEN add# = 1
IF t2 = -1 THEN add# = 2
IF t3 = -1 THEN add# = 2
IF t4 = -1 THEN add# = 1
IF t1 = -1 OR t2 = -1 THEN p$ = "+"
IF t3 = -1 OR t4 = -1 THEN p$ = "-"
v# = v# + add#
test1 = (add# = 1) AND (pol = 0)
test2 = (add# = 1) AND (pol = 1)
IF test1 = -1 THEN pol = 1
IF test2 = -1 THEN pol = 0
PRINT , , "+"; p#; v#; c; pol; p$
a$ = INKEY$
IF a$ = "s" THEN STOP
LOOP UNTIL p# >= 900000000000000#
GOTO redo:




"C:\Users\Reactor1967\vcns\work\VBALANCE.BAS"

p# = 1
add1# = 2
CLS
RANDOMIZE TIMER
redo:
DO
```

```
REM IF (p# / 2) - INT(p# / 2) >= .5 THEN add# = add1#
REM IF (p# / 2) - INT(p# / 2) = 0 THEN add# = add1# + 1
PRINT p#; "="; p# + add1# + 1; p# + add1# + 2; add1#
REM IF p# = 12752 THEN STOP
p# = p# + 1
add1# = add1# + 1
REM INPUT a$
a$ = INKEY$
IF a$ = "s" THEN STOP
LOOP UNTIL add1# = 4
DO
REM IF (p# / 2) - INT(p# / 2) >= .5 THEN add# = add1#
REM IF (p# / 2) - INT(p# / 2) = 0 THEN add# = add1# + 1
PRINT p#; "="; p# + add1# + 1; p# + add1# + 2; add1#
REM IF p# = 12752 THEN STOP
p# = p# + 1
add1# = add1# - 1
REM INPUT a$
a$ = INKEY$
IF a$ = "s" THEN STOP
LOOP UNTIL add1# = 1
GOTO redo:


"C:\Users\Reactor1967\vcns\work\VARCAL.BAS"
CLS
PRINT "...SINK CALCULATOR FOR SINKING AND UNSINKING YOUR
VECTORS..."
a# = 12502
x = 198
PRINT "Subtract you vector number with #1 in col 1."
PRINT "Sink #1 in col 1 with your current varience."
PRINT "Take the product of your subtraction and add it to the"
PRINT "output of the sinked number in #1 col 1."
PRINT "That is your sinked vector."
PRINT "col 1 is your vector number to be sinked"
PRINT "col 2 is your output vector that was sinked"
PRINT "col 3 is your starting varience"
PRINT "col 4 is your ending varience"
PRINT "col 5 is your number of sinks to get your input to your output"
PRINT " "
PRINT "   1    2   3  4  5"
PRINT "-----------------------------"
FOR count = 1 TO 21
b# = a# + x
```

```
store# = a#
sink = 0
DO
store# = store# + 2
sink = sink + 1
LOOP UNTIL b# - store# < 100
PRINT count; a#; store#; x; b# - store#; sink
REM x = x + 1
a# = a# + 1
NEXT count
```

```
"C:\Users\Reactor1967\vcns\work\UNITY.BAS"
v# = 157
r# = 103
CLS
DO
test = (v# + (v# - r#)) >= (r# + 100)
IF test = 0 THEN z = 0
IF test = -1 THEN z = 1
z = INT(RND * 2) + 0
d# = (v# - r#)
IF z = 0 THEN v# = v# + (v# - r#) + 2
IF z = 1 THEN v# = v# + (v# - r#) + 4
COLOR 2, 0
PRINT , , z; v#; r#; v# - r#; d#
REM IF z = 1 THEN r# = r# + 102
IF v# - r# >= 100 THEN r# = r# + 102
REM INPUT a$
a$ = INKEY$
IF a$ = "s" THEN STOP
REM r# = r# + (r# - 2) + 4
LOOP
```

```
"C:\Users\Reactor1967\vcns\work\TWOSTATE.BAS"
r1# = 1
r2# = 2
v# = 2
CLS
DO
z = INT(RND * 2) + 0
test = v# + (v# - r2#) >= (r2# + 100)
```

```
IF z = 0 THEN v# = v# + (v# - r2#) + 2
IF z = 1 THEN v# = v# + (v# - r2#) + 4
COLOR 2, 0
PRINT , , z; v#; v# / 2; r1#; r2#
IF test = -1 THEN r1# = r1# + 104
IF test = -1 THEN r2# = r2# + 104
test = v# + (v# - r2#) >= (r2# + 100)
v# = v# + (v# - r1#) + 2
REM PRINT , , v#;r1#;r2#
IF test = -1 THEN r1# = r1# + 104
IF test = -1 THEN r2# = r2# + 104
z = INT(RND * 2) + 0
test = v# + (v# - r2#) >= (r2# + 100)
IF z = 0 THEN v# = v# + (v# - r1#) + 2
IF z = 1 THEN v# = v# + (v# - r1#) + 4
COLOR 2, 0
PRINT , , z; v#; v# / 2; r1#; r2#
IF test = -1 THEN r1# = r1# + 104
IF test = -1 THEN r2# = r2# + 104
test = v# + (v# - r2#) >= (r2# + 100)
v# = v# + (v# - r1#) + 1
REM PRINT , , v#
IF test = -1 THEN r1# = r1# + 104
IF test = -1 THEN r2# = r2# + 104
a$ = INKEY$
INPUT a$
IF a$ = "s" THEN STOP
LOOP




"C:\Users\Reactor1967\vcns\work\TRYING.BAS"
REM this program teaches how to change the focus. On regular numbers
REM you can alway get a even number by multiplying it by two. Well, that
REM takes the number up in value but what about going down in value. And,
REM what about doing it constructivly. Here learning how to change the
REM focus on the difference between variables, where one variable
REM incleases 2:1 of another variable then switching that focus so that
REM the other variable increases 1:2 and so forth.
a# = 1
b# = 2
CLS
redo:
DO
PRINT a#; b#
a# = a# + 1
```

```
b# = b# + 2
REM INPUT a$
a$ = INKEY$
IF a$ = "s" THEN STOP
test = (a$ = "d") AND (a# / 2 - INT(a# / 2) = 0)
IF test = -1 THEN EXIT DO
LOOP
a$ = ""
DO
PRINT a#; b#
a# = a# + 2
b# = b# + 1
REM INPUT a$
a$ = INKEY$
IF a$ = "s" THEN STOP
test = (a$ = "d") AND (a# / 2 - INT(a# / 2) = 0)
IF test = -1 THEN EXIT DO
LOOP
a$ = ""
GOTO redo:
```


```
"C:\Users\Reactor1967\vcns\work\TRIED.BAS"
v# = 1
r# = 1
flip = 1
CLS
RANDOMIZE TIMER
test$ = "data.txt"
OPEN test$ FOR OUTPUT AS #1
DO
z = INT(RND * 2) + 0
IF z = 0 THEN v# = v# + (v# - r# + 1)
IF z = 1 THEN v# = v# + (v# - r# + 2)
flip = flip * 2
IF flip > 9 THEN carry = 1 ELSE carry = 0
IF carry = 1 THEN flip = flip - 10
IF carry = 1 THEN flip = flip + 1
store# = r#
IF carry = 1 THEN r# = v#
test = (carry = 1) AND (flip > 10)
IF test = -1 THEN STOP
IF (r# / 2) - INT(r# / 2) = 0 THEN r# = r# - 1
COLOR 2, 0
PRINT , , z; v#; r#; flip; r# - store#
```

```
REM WRITE #1, r# - store#
a$ = INKEY$
IF a$ = "s" THEN STOP
LOOP




"C:\Users\Reactor1967\vcns\work\TMP.BAS"
DECLARE SUB reduce (v#)
CLS
RANDOMIZE TIMER
r# = 1
v# = 1
redo:
DO
z = INT(RND * 2) + 0
IF v# + ((v# - r#) * 2) >= 9999# THEN carry = 1 ELSE carry = 0
IF carry = 1 THEN z = 3
IF z = 0 THEN v# = v# + (((v# - r#) * 2) + 1)
IF z = 1 THEN v# = v# + (((v# - r#) * 2) + 2)
IF z = 3 THEN v# = v# + (((v# - r#) * 2) + 3)
COLOR 2, 0
PRINT , , v#; z; (v# / 3) - INT(v# / 3)
REM INPUT a$
a$ = INKEY$
IF a$ = "s" THEN STOP
LOOP UNTIL carry = 1

SUB reduce (v#)
count = count + 1
store# = v#
DO
v# = v# - 3
stop$ = INKEY$
IF stop$ = "s" THEN STOP
LOOP UNTIL v# < 99
av# = av# + (store# - v#)
ave# = av# / count
REM PRINT ave#; v#
END SUB




"C:\Users\Reactor1967\vcns\work\THEORM.BAS"
a# = 1
b# = 0: c# = 0
```

```
CLS
DO
dist# = a#
test = (dist# / 2) - INT(dist# / 2)
IF test = .5 THEN dist# = dist# - 1
dist# = dist# / 2
IF test = .5 THEN c# = dist# + 1 ELSE c# = dist#
b# = dist#
PRINT a#; b#; c#
a# = a# + 1
INPUT a$
IF a$ = "s" THEN STOP
LOOP
```

```
"C:\Users\Reactor1967\vcns\work\TESTFR.BAS"
RANDOMIZE TIMER
DO
p# = INT(RND * 999999999999999#) + 0
LOOP UNTIL p# >= 500000000000000#
PRINT p#
r# = p#
count = 0
DO
count = count + 1
c = INT(RND * 2) + 0
IF c = 0 THEN p# = p# - ((r# - p#) + 1)
IF c = 1 THEN p# = p# - ((r# - p#) + 2)
PRINT p#; count
a$ = INKEY$
IF a$ = "s" THEN STOP
LOOP UNTIL p# - (r# - p#) <= 0
a# = 1
DO
count = count - 1
p# = p# - a#
a# = (a# * 2) + 1
test1 = p# - r# <= 50
test2 = r# - p# <= 50
IF test1 = -1 THEN STOP
IF test2 = -1 THEN STOP
LOOP
```

```
"C:\Users\Reactor1967\vcns\work\TEST4.BAS"
v# = 100
cr# = 0
CLS
RANDOMIZE TIMER
DO
z = INT(RND * 2) + 0
IF z = 1 THEN z = .5
store1# = v#
store2# = cr#
redo:
DO
v# = v# + 1
test = (v# / 2) - INT(v# / 2)
IF test = z THEN EXIT DO
a$ = INKEY$
IF a$ = "s" THEN STOP
LOOP
dist# = v# - store1#
dist# = (dist# * 2) - 1
cr# = v# - dist#
REM GOTO redo:
COLOR 2, 0
PRINT , , z; v#; v# - store1#; cr#; cr# - store2#
INPUT a$
IF a$ = "s" THEN STOP
LOOP
```

```
"C:\Users\Reactor1967\vcns\work\TEST3.BAS"
a# = 1
b# = 2
c# = 3
d# = 4
carry# = 1
RANDOMIZE TIMER
CLS
coding:
x = 0
FOR count = 1 TO 10
PRINT a#; b#; c#; d#; carry#
test1 = (carry# = a#) AND x = 0
test = (carry# = a#) AND x = 1
IF test = -1 THEN z = INT(RND * 2) + 0
test2 = (test = -1) AND z = 0
test3 = (test = -1) AND z = 1
IF test2 = -1 THEN carry# = b#
IF test3 = -1 THEN carry# = c#
IF test1 = -1 THEN carry# = d#
IF test1 = -1 THEN x = 1
a# = a# + 1
b# = b# + 3
c# = c# + 3
d# = d# + 3
NEXT count
test1 = (carry# = a#) AND x = 0
test = (carry# = a#) AND x = 1
IF test = -1 THEN z = INT(RND * 2) + 0
test2 = (test = -1) AND z = 0
test3 = (test = -1) AND z = 1
IF test2 = -1 THEN carry# = b#
IF test3 = -1 THEN carry# = c#
IF test1 = -1 THEN carry# = d#
IF test1 = -1 THEN x = 1
carry# = carry# + 20
a# = a# + 20
INPUT a$
IF a$ = "s" THEN STOP
IF a$ = "d" THEN STOP
GOTO coding:
carry# = carry# - 20
a# = a# - 20
x = 0
FOR count = 1 TO 10
test = (carry# = a#) OR (carry# = b#) OR (carry# = c#)
```

```
NEXT count


"C:\Users\Reactor1967\vcns\work\TEST2.BAS"
liner = 1
CLS
RANDOMIZE TIMER
test$ = "chart"
OPEN test$ FOR INPUT AS #1
z = INT(RND * 2) + 0
INPUT #1, a#, b#, c#, d#, e#, f#
IF z = 0 THEN carry# = d#
IF z = 1 THEN carry# = e#
r# = f#
REM PRINT a#, b#, c#, d#, e#, f#
REM STOP
DO
z = INT(RND * 2) + 0
DO
INPUT #1, a#, b#, c#, d#, e#, f#
test = (b# = r#) AND (carry# = c#)
LOOP UNTIL test = -1
PRINT ; liner, a#; b#; c#; d#; e#; f#
liner = liner + 1
REM INPUT a$
a$ = INKEY$
IF a$ = "s" THEN STOP
IF z = 0 THEN carry# = d#
IF z = 1 THEN carry# = e#
r# = f#
LOOP
CLOSE


"C:\Users\Reactor1967\vcns\work\TEST.BAS"
REM PRINT v4#; v3#; v2#; v1#, v8#; v7#; v6#; v5#, v12#; v11#; v10#; v9#,
v16#; v15#; v14#; v13#
REM PRINT pos$, vec$, c1$, c2$
REM WRITE #1, v16#, v15#, v14#, v13#, v12#, v11#, v10#, v9#, v8#, v7#,
v6#, v5#, v4#, v3#, v2#, v1#
```

```
REM WRITE #1, pos$, vec$, c1$, c2$
RANDOMIZE TIMER
CLS
test$ = "chart1"
OPEN test$ FOR INPUT AS #1
INPUT #1, v16#, v15#, v14#, v13#, v12#, v11#, v10#, v9#, v8#, v7#, v6#, v5#,
v4#, v3#, v2#, v1#
INPUT #1, pos$, vec$, c1$, c2$
DO
z = INT(RND * 2) + 0
IF z = 0 THEN carry$ = c1$
IF z = 1 THEN carry$ = c2$
PRINT pos$, vec$, c1$, c2$, carry$; z
a$ = INKEY$
REM INPUT a$
IF a$ = "s" THEN CLOSE #1
IF a$ = "s" THEN STOP
DO
INPUT #1, v16#, v15#, v14#, v13#, v12#, v11#, v10#, v9#, v8#, v7#, v6#, v5#,
v4#, v3#, v2#, v1#
INPUT #1, pos$, vec$, c1$, c2$
stop$ = INKEY$
IF stop$ = "s" THEN CLOSE #1
IF stop$ = "s" THEN STOP
LOOP UNTIL carry$ = vec$
LOOP




"C:\Users\Reactor1967\vcns\work\TEP.BAS"
d# = 11
a# = 10
b# = 20
c# = 21
p# = 10
RANDOMIZE TIMER
CLS
redo:
DO
z = INT(RND * 2) + 0
test1 = (p# = a#) AND (z = 0)
test2 = (p# = a#) AND (z = 1)
IF test1 = -1 THEN p# = b#
IF test2 = -1 THEN p# = c#
PRINT d#; a#; b#; c#, p#; ABS(a# - b#)
d# = d# + 1: a# = a# + 1
```

```
b# = b# + 2: c# = c# + 2
REM INPUT a$
a$ = INKEY$
IF a$ = "s" THEN STOP
LOOP UNTIL p# = a#
redo2:
p# = b#
PRINT d#; a#; b#; c#, p#; ABS(a# - b#)
DO
z = INT(RND * 2) + 0
test1 = (p# = b#) AND (z = 0)
test2 = (p# = b#) AND (z = 1)
IF test1 = -1 THEN p# = a#
IF test2 = -1 THEN p# = d#
PRINT d#; a#; b#; c#, p#; ABS(a# - b#)
dist# = ABS(p# - b#)
d# = d# + (dist# * 2): a# = a# + (dist# * 2)
b# = b# - dist#: c# = c# - dist#
INPUT a$
REM a$ = INKEY$
IF a$ = "s" THEN STOP
LOOP UNTIL p# = b#
p# = a#
PRINT d#; a#; b#; c#, p#; ABS(a# - b#)
DO
z = INT(RND * 2) + 0
test1 = (p# = a#) AND (z = 0)
test2 = (p# = a#) AND (z = 1)
IF test1 = -1 THEN p# = b#
IF test2 = -1 THEN p# = c#
PRINT d#; a#; b#; c#, p#; ABS(a# - b#)
dist# = ABS(p# - a#)
d# = d# - dist#: a# = a# - dist#
b# = b# + (dist# * 2): c# = c# + (dist# * 2)
INPUT a$
REM a$ = INKEY$
IF a$ = "s" THEN STOP
LOOP UNTIL p# = a#
PRINT d#; a#; b#; c#, p#; ABS(a# - b#)
REM INPUT a$
a$ = INKEY$
IF a$ = "s" THEN STOP
GOTO redo2:
```

```
"C:\Users\Reactor1967\vcns\work\TEMPLATE.BAS"
v# = 110
r# = 75
CLS
RANDOMIZE TIMER
COLOR 2, 0
PRINT , , v#; r#
DO
z = INT(RND * 2) + 0
store2# = r#
r1# = r#
DO
IF z = 0 THEN v1# = v# + (v# - r1# + 1)
IF z = 1 THEN v1# = v# + (v# - r1# + 2)
IF v1# - r1# < 11 THEN EXIT DO
a$ = INKEY$
REM INPUT a$
IF a$ = "s" THEN STOP
IF a$ = "d" THEN EXIT DO
r1# = r1# + 2
LOOP
r# = r1#
dist# = r# - store2#
dist# = (dist# * 2) - 1
cs# = cr#
cr# = r# - dist#
store1# = v#
REM PRINT , , z; v# - store1#; v#; r#; r# - store2#
IF z = 0 THEN v# = v# + (v# - r# + 1)
IF z = 1 THEN v# = v# + (v# - r# + 2)
COLOR 2, 0
count = count + 1
IF count = 3 THEN PRINT , , "--------------------"
IF count = 3 THEN count = 0
PRINT , , z; v# - store1#; v#; r#; r# - store2#; cr#; cr# - cs#
INPUT a$
REM a$ = INKEY$
IF a$ = "s" THEN STOP
IF a$ = "d" THEN EXIT DO
IF a$ = "D" THEN EXIT DO

LOOP


"C:\Users\Reactor1967\vcns\work\TEMP17.BAS"
```

```
v# = 101
r# = 75
CLS
RANDOMIZE TIMER
z = INT(RND * 2) + 0
DO
store1# = v#
IF z = 0 THEN v# = v# + (v# - r# + 1)
IF z = 1 THEN v# = v# + (v# - r# + 2)
COLOR 2, 0
PRINT , , v#; r#; v# - r#
INPUT a$
REM a$ = INKEY$
IF a$ = "s" THEN STOP
z = INT(RND * 2) + 0
d# = 0
store2# = r#
dist# = v# - r#
DO
d# = d# + 1
r1# = v# - d#
IF (r1# / 2) - INT(r1# / 2) = 0 THEN d# = d# + 1
IF (r1# / 2) - INT(r1# / 2) = 0 THEN r1# = v# - d#
IF z = 0 THEN v1# = v# + (v# - r1# + 1)
IF z = 1 THEN v1# = v# + (v# - r1# + 2)
IF (v# - r1#) = (r1# - r#) + 1 THEN EXIT DO
IF (v# - r1#) = (r1# - r#) + 2 THEN EXIT DO
a$ = INKEY$
IF a$ = "s" THEN STOP
LOOP
REM PRINT v1# - r1#
r# = r1#
LOOP


"C:\Users\Reactor1967\vcns\work\TEMP16.BAS"
v# = 102
r# = 95
RANDOMIZE TIMER
CLS
DO
z = INT(RND * 2) + 0
IF z = 0 THEN v# = v# + (v# - r# + 1)
IF z = 1 THEN v# = v# + (v# - r# + 2)
COLOR 2, 0
```

```basic
PRINT , , v#; r#
INPUT a$
REM a$ = INKEY$
IF a$ = "s" THEN STOP
dist# = v# - r#
store# = v# - r#
IF (dist# / 2) - INT(dist# / 2) = .5 THEN x = 1 ELSE x = 0
IF x = 1 THEN dist# = dist# - 1
dist# = dist# / 2
a# = dist#
IF x = 1 THEN b# = (a# + 1)
IF x = 0 THEN b# = a#
IF (v# / 2) - INT(v# / 2) = 0 THEN r# = v# - b# ELSE r# = v# - a#
LOOP
```

```basic
"C:\Users\Reactor1967\vcns\work\TEMP15.BAS"
d# = 2
v# = 100
r# = 95
RANDOMIZE TIMER
CLS
DO
z = INT(RND * 2) + 0
IF z = 0 THEN v# = v# + (v# - r# + 1)
IF z = 1 THEN v# = v# + (v# - r# + 2)
PRINT , , v#; r#; d#
INPUT a$
REM a$ = INKEY$
IF a$ = "s" THEN STOP
dist# = v# - r#
d# = dist#
IF (d# / 2) - INT(d# / 2) = .5 THEN d# = d# - 1
d# = d# / 2
dist# = dist# - d#
r# = v# - dist#
IF (r# / 2) - INT(r# / 2) = 0 THEN r# = r# - 1
LOOP
```

```basic
"C:\Users\Reactor1967\vcns\work\TEMP14.BAS"
DIM matrix1#(50)
```

```
DIM matrix2#(50)
a# = 1: b# = 2
FOR count = 1 TO 50
matrix1#(count) = a#
matrix2#(count) = b#
a# = a# + 2
b# = b# + 2
NEXT count
a# = 1: b# = 1
CLS
RANDOMIZE TIMER
PRINT , , v#; z; r#
redo:
DO
z = INT(RND * 2) + 0
IF z = 0 THEN v# = v# + (v# - r# + 1)
IF z = 1 THEN v# = v# + (v# - r# + 2)
COLOR 2, 0
PRINT , , v#; z; r#; v# - r#
PRINT , , matrix1#(a#); matrix2#(b#)
PRINT , , a#; b#
INPUT a$
REM a$ = INKEY$
IF a$ = "s" THEN STOP
IF a$ = "d" THEN GOTO decode:
test = (v# / 2) - INT(v# / 2)
IF test = 0 THEN r# = v# - matrix1#(a#)
IF test = .5 THEN r# = v# - matrix2#(b#)
IF test = 0 THEN a# = a# + 1
IF test = .5 THEN b# = b# + 1
IF a# = 50 THEN a# = 1
IF b# = 50 THEN b# = 1
LOOP UNTIL v# + (v# - r#) >= 99999
decode:
REM DO
dist# = (v# - r#)
IF (dist# / 2) - INT(dist# / 2) = .5 THEN dist# = dist# - 1
dist# = dist# / 2
dist# = dist# + 1
v# = v# - dist#
test = (v# / 2) - INT(v# / 2)
IF test = 0 THEN dist# = a#
IF test = .5 THEN dist# = b#
```

```
"C:\Users\Reactor1967\vcns\work\TEMP13.BAS"
DIM matrix1#(50)
DIM matrix2#(50)
a# = 1: b# = 2
FOR count = 1 TO 50
matrix1#(count) = a#
matrix2#(count) = b#
a# = a# + 2
b# = b# + 2
NEXT count
a# = 1: b# = 1
CLS
RANDOMIZE TIMER
PRINT , , v#; z; r#
redo:
DO
z = INT(RND * 2) + 0
IF z = 0 THEN v# = v# + (v# - r# + 1)
IF z = 1 THEN v# = v# + (v# - r# + 2)
COLOR 2, 0
PRINT , , v#; z; r#; v# - r#
REM INPUT a$
a$ = INKEY$
IF a$ = "s" THEN STOP
test = (v# / 2) - INT(v# / 2)
IF test = 0 THEN r# = v# - matrix2#(a#)
IF test = .5 THEN r# = v# - matrix1#(b#)
IF test = 0 THEN a# = a# + 1
IF test = .5 THEN b# = b# + 1
IF a# = 50 THEN a# = 1
IF b# = 50 THEN b# = 1
LOOP UNTIL v# + (v# - r#) >= 99999
DO
z = INT(RND * 2) + 0
IF z = 0 THEN v# = v# - (r# - v# + 1)
IF z = 1 THEN v# = v# - (r# - v# + 2)
COLOR 2, 0
PRINT , , v#; z; r#; r# - v#
REM INPUT a$
a$ = INKEY$
IF a$ = "s" THEN STOP
test = (v# / 2) - INT(v# / 2)
IF test = 0 THEN r# = v# + matrix2#(a#)
IF test = .5 THEN r# = v# + matrix1#(b#)
IF test = 0 THEN a# = a# + 1
IF test = .5 THEN b# = b# + 1
```

```
IF a# = 50 THEN a# = 1
IF b# = 50 THEN b# = 1
LOOP UNTIL v# - (r# - v#) <= 9999
GOTO redo:
```



```
"C:\Users\Reactor1967\vcns\work\TEMP12.BAS"
v# = 20
d# = 1
r# = v# - d#
RANDOMIZE TIMER
CLS
DO
z = INT(RND * 2) + 0
IF z = 0 THEN v# = v# + (v# - r# + 1)
IF z = 1 THEN v# = v# + (v# - r# + 2)
COLOR 2, 0
PRINT , , v#; r#; z; v# - r#; d#
d# = d# + 1
r# = v# - d#
IF (r# / 2) - INT(r# / 2) = 0 THEN r# = r# + 1
REM INPUT a$
a$ = INKEY$
IF a$ = "s" THEN STOP
LOOP
```



```
"C:\Users\Reactor1967\vcns\work\TEMP11.BAS"
REM Its getting blatenly and a long time coming apparent that to do
REM this I need to know in advance what im going to code a 1 or a zero.
REM then generate an appropiate r# at a relative distance from v# that
REM will somehow tell me how much to subtract to get my previous R#
REM I can just use my equations to get my previous P#. My other
REM alternative is to also do an equation to r# relative to P#
REM that gives me my previous R# as well as my next R# if coding.
v# = 1
r# = 1
CLS
RANDOMIZE TIMER
DO
z = INT(RND * 2) + 0
vs# = v#
IF z = 0 THEN v# = v# + (v# - r# + 1)
```

```
IF z = 1 THEN v# = v# + (v# - r# + 2)
COLOR 2, 0
PRINT , , v#; r#; v# - r#; v# - vs#
REM INPUT a$
a$ = INKEY$
IF a$ = "s" THEN STOP
REM IF a$ = "d" THEN GOTO decode:
dist# = v# - vs#
IF (dist# / 2) - INT(dist# / 2) = .5 THEN dist# = dist# - 1
IF x = 0 THEN r# = r# + dist#
IF x = 1 THEN r# = r# + dist# + 2
IF v# - r# >= 100 THEN x = 1
IF v# - r# <= 4 THEN x = 0
LOOP


"C:\Users\Reactor1967\vcns\work\TEMP10.BAS"
COLOR 2, 0
v# = 200
d1# = 1
d2# = 2
r# = v# - d1#
CLS
RANDOMIZE TIMER
PRINT , , v#; z; r#; d#; r# - rs#
DO
z = INT(RND * 2) + 0
vs# = v#
IF z = 0 THEN v# = r# + ((v# - r#) * 2) + 1
IF z = 1 THEN v# = r# + ((v# - r#) * 2) + 2
COLOR 2, 0
PRINT v#; z; r#; v# - vs#; r# - rs#; d1#; d2#
REM INPUT a$
a$ = INKEY$
IF a$ = "s" THEN STOP
IF a$ = "d" THEN GOTO Decode:
d1# = d1# + 2
d2# = d2# + 2
rs# = r#
IF z = 0 THEN r# = v# - d1#
IF z = 1 THEN r# = v# - d2#
LOOP
Decode:
```

```
"C:\Users\Reactor1967\vcns\work\TEMP9.BAS"
DIM matrix1#(50)
DIM matrix2#(50)
a# = 1: b# = 2
FOR count = 1 TO 50
matrix1#(count) = a#
matrix2#(count) = b#
a# = a# + 2
b# = b# + 2
NEXT count
a# = 1: b# = 1
CLS
RANDOMIZE TIMER
PRINT , , v#; z; r#
redo:
DO
z = INT(RND * 2) + 0
IF z = 0 THEN v# = v# + (v# - r# + 1)
IF z = 1 THEN v# = v# + (v# - r# + 2)
COLOR 2, 0
PRINT , , v#; z; r#; v# - r#
REM INPUT a$
a$ = INKEY$
IF a$ = "s" THEN STOP
test = (v# / 2) - INT(v# / 2)
IF test = 0 THEN r# = v# - matrix2#(a#)
IF test = .5 THEN r# = v# - matrix1#(b#)
IF test = 0 THEN a# = a# + 1
IF test = .5 THEN b# = b# + 1
IF a# = 50 THEN a# = 1
IF b# = 50 THEN b# = 1
LOOP UNTIL v# + (v# - r#) >= 99999
DO
z = INT(RND * 2) + 0
IF z = 0 THEN v# = v# - (r# - v# + 1)
IF z = 1 THEN v# = v# - (r# - v# + 2)
COLOR 2, 0
PRINT , , v#; z; r#; v# - r#
REM INPUT a$
a$ = INKEY$
IF a$ = "s" THEN STOP
test = (v# / 2) - INT(v# / 2)
IF test = 0 THEN r# = v# - matrix2#(a#)
IF test = .5 THEN r# = v# - matrix1#(b#)
IF test = 0 THEN a# = a# + 1
```

```
IF test = .5 THEN b# = b# + 1
IF a# = 50 THEN a# = 1
IF b# = 50 THEN b# = 1
LOOP UNTIL v# - (r# - v#) <= 0
GOTO redo:




"C:\Users\Reactor1967\vcns\work\TEMP8.BAS"
v# = 4
r# = 3
d# = v# - r#
d1# = ((v# + (v# - r# + 1)) - r#)
d2# = ((v# + (v# - r# + 2)) - r#)
CLS
RANDOMIZE TIMER
x = 0
DO
z = INT(RND * 2) + 0
store2# = v#
IF z = 0 THEN v# = v# + (v# - r# + 1)
IF z = 1 THEN v# = v# + (v# - r# + 2)
COLOR 2, 0
PRINT , "z"; z; "v"; v#; "r"; r#; "d"; d#; "d1"; d1#; "d2"; d2#; "vr"; v# - r#;
"rr"; r# - store#; "vv"; v# - store2#
REM PRINT , , z; v#; r#
INPUT a$
REM a$ = INKEY$
IF a$ = "s" THEN STOP
IF a$ = "d" THEN GOTO Decode:
IF x = 0 THEN d# = d# + 1
IF x = 1 THEN d# = d# - 1
REM IF d# = 50 THEN x = 1
REM IF d# = 1 THEN x = 0
d1# = (d# * 2) + 1
d2# = (d# * 2) + 2
store# = r#
IF z = 0 THEN r# = v# - d1#
IF z = 1 THEN r# = v# - d2#
LOOP
Decode:
DO
dist# = v# - r#
test = (dist# / 2) - INT(dist# / 2)
IF test = .5 THEN dist# = dist# - 1
```

```
dist# = dist# / 2
dist# = dist# + 1
store1# = v#
v# = v# - dist#
store2# = v#
dist# = store1# - store2#
test = (store1# / 2) - INT(store1# / 2)
IF test = 0 THEN dist# = dist# - 1
IF test = .5 THEN dist# = dist# - 2
test = (store2# / 2) - INT(store2# / 2)
IF test = 0 THEN dist# = dist# - 1
IF test = .5 THEN dist# = dist# - 2
r# = r# - dist#
PRINT v#; r#
INPUT a$
IF a$ = "s" THEN STOP
LOOP




"C:\Users\Reactor1967\vcns\work\TEMP7.BAS"
v# = 1
r# = 1
d# = v# - r#
d1# = ((v# + (v# - r# + 1)) - r#)
d2# = ((v# + (v# - r# + 2)) - r#)
CLS
RANDOMIZE TIMER
DO
z = INT(RND * 2) + 0
d1# = ((v# + (v# - r# + 1)) - r#)
d2# = ((v# + (v# - r# + 2)) - r#)
IF z = 0 THEN v# = v# + (v# - r# + 1)
IF z = 1 THEN v# = v# + (v# - r# + 2)
COLOR 2, 0
REM LOCATE 12
PRINT , "      "; "z"; z; "v"; v#; "r"; r#; "d"; d#; "d1"; d1#; "d2"; d2#; "vr"; v# -
r#
test1 = ((v# - (d# + 1)) / 2) - INT((v# - (d# + 1)) / 2)
test2 = ((v# - (d# + 2)) / 2) - INT((v# - (d# + 2)) / 2)
IF test1 = .5 THEN d# = d# + 1
IF test2 = .5 THEN d# = d# + 2
r# = v# - d#
INPUT a$
REM a$ = INKEY$
IF a$ = "s" THEN STOP
```

```
"C:\Users\Reactor1967\vcns\work\TEMP6.BAS"
r# = 3
v# = 5
d# = 2
CLS
RANDOMIZE TIMER
redo:
DO
z = INT(RND * 2) + 0
test1 = (v# / 2) - INT(v# / 2)
test2 = (test1 = 0) AND (z = 0)
test3 = (test1 = 0) AND (z = 1)
test4 = (test1 = .5) AND (z = 0)
test5 = (test1 = .5) AND (z = 1)
IF test2 = -1 THEN v# = v# + (v# - r# + 2)
IF test3 = -1 THEN v# = v# + (v# - r# + 1)
IF test4 = -1 THEN v# = v# + (v# - r# + 1)
IF test5 = -1 THEN v# = v# + (v# - r# + 2)
COLOR 2, 0
PRINT , , z; v#; r#; d#; v# - r#
REM INPUT a$
a$ = INKEY$
IF a$ = "s" THEN STOP
REM IF ((v# / 2) - INT(v# / 2) = 0) THEN d# = d# + 2
REM IF ((v# / 2) - INT(v# / 2) = .5) THEN d# = d# + 3
IF x = 0 THEN d# = d# + 1
IF x = 1 THEN d# = d# - 1
IF d# = 100 THEN x = 1
IF d# = 1 THEN x = 0
r# = v# - d#
LOOP UNTIL v# + (v# - r# + 1) >= 99999
PRINT "----------------------------------------"
r# = v# + d#
DO
z = INT(RND * 2) + 0
test1 = (v# / 2) - INT(v# / 2)
test2 = (test1 = 0) AND (z = 0)
test3 = (test1 = 0) AND (z = 1)
test4 = (test1 = .5) AND (z = 0)
test5 = (test1 = .5) AND (z = 1)
IF test2 = -1 THEN v# = v# - (r# - v# + 2)
IF test3 = -1 THEN v# = v# - (r# - v# + 1)
```

```
IF test4 = -1 THEN v# = v# - (r# - v# + 1)
IF test5 = -1 THEN v# = v# - (r# - v# + 2)
COLOR 2, 0
PRINT , , z; v#; r#; d#; r# - v#
REM INPUT a$
a$ = INKEY$
IF a$ = "s" THEN STOP
REM IF ((v# / 2) - INT(v# / 2) = 0) THEN d# = d# + 2
REM IF ((v# / 2) - INT(v# / 2) = .5) THEN d# = d# + 3
IF x = 0 THEN d# = d# + 1
IF x = 1 THEN d# = d# - 1
IF d# = 100 THEN x = 1
IF d# = 1 THEN x = 0
r# = v# + d#
LOOP UNTIL v# - (r# - v# + 1) <= 0
r# = v# - d#
GOTO redo:




"C:\Users\Reactor1967\vcns\work\TEMP5.BAS"
v# = 2
d1# = 1
d2# = 2
RANDOMIZE TIMER
CLS
DO
z = INT(RND * 2) + 0
IF z = 0 THEN v# = v# + (v# - r# + 1)
IF z = 1 THEN v# = v# + (v# - r# + 2)
COLOR 2, 0
PRINT , , z; v#; r#; d#, d1#; d2#
INPUT a$
REM a$ = INKEY$
IF a$ = "s" THEN STOP
IF a$ = "d" THEN EXIT DO
IF z = 0 THEN d# = d1#
IF z = 1 THEN d# = d2#
r# = v# - d#
d1# = d1# + 2
d2# = d2# + 2
LOOP
d1# = d1# - 4
d2# = d2# - 4
PRINT v#; d#, d1#; d2#
```

```
decode:
REM DO
test = (v# / 2) - INT(v# / 2)
IF test = 0 THEN d# = d# + 2
IF test = 1 THEN d# = d# + 1
v# = v# - d#
PRINT v#; d#
PRINT d1#; d2#




"C:\Users\Reactor1967\vcns\work\TEMP4.BAS"
v# = 1
r# = 1
d# = 1
RANDOMIZE TIMER
CLS
x = 0
DO
z = INT(RND * 2) + 0
IF z = 0 THEN v# = v# + (v# - r# + 1)
IF z = 1 THEN v# = v# + (v# - r# + 2)
COLOR 2, 0
PRINT , , v#; r#; d#
INPUT a$
REM a$ = INKEY$
IF a$ = "s" THEN STOP
IF x = 0 THEN d# = d# + 1
IF x = 1 THEN d# = d# - 1
REM IF d# = 100 THEN x = 1
REM IF d# = 1 THEN x = 0
IF ((v# / 2) - INT(v# / 2)) = ((d# / 2) - INT(d# / 2)) THEN d# = d# + 1
r# = v# - d#
LOOP




"C:\Users\Reactor1967\vcns\work\TEMP3.BAS"
v# = 2
r# = 1
RANDOMIZE TIMER
CLS
DO
z = INT(RND * 2) + 0
IF z = 0 THEN v# = v# + (v# - r# + 1)
IF z = 1 THEN v# = v# + (v# - r# + 2)
```

```
COLOR 2, 0
PRINT , , v#; r#; v# - r#
r# = v# - 1
IF z = 1 THEN r# = v# - 2
IF z = 0 THEN r# = v# - 1
REM INPUT a$
a$ = INKEY$
IF a$ = "s" THEN STOP
LOOP
```

```
"C:\Users\Reactor1967\vcns\work\TEMP2.BAS"
v# = 50
b# = 1
CLS
RANDOMIZE TIMER
DO
COLOR 2, 0
PRINT , , v#; b#; v# - b#; dist#; count
a$ = INKEY$
REM INPUT a$
IF a$ = "s" THEN STOP
z = INT(RND * 2) + 0
IF z = 0 THEN v# = v# + b# + 1
IF z = 1 THEN v# = v# + b# + 2
store# = b#
count = 1
DO
b# = b# + 25
t1 = (b# / 2) - INT(b# / 2)
t2 = (v# / 2) - INT(v# / 2)
test1 = (t1 = 0) AND (t2 = .5)
test2 = (t1 = .5) AND (t2 = 0)
IF test1 = -1 THEN EXIT DO
IF test2 = -1 THEN EXIT DO
stop$ = INKEY$
IF stop$ = "s" THEN STOP
count = count + 1
LOOP
dist# = b# - store#
LOOP
```

```
"C:\Users\Reactor1967\vcns\work\TEMP.BAS"
```

```
v# = 1
r# = 1
dist# = 0
CLS
DO
z = INT(RND * 2) + 0
dist# = v# - r#
IF z = 0 THEN v# = v# + (v# - r# + 2)
IF z = 1 THEN v# = v# + (v# - r# + 4)
COLOR 2, 0
r# = r# + dist#
PRINT , z; v#; r#; dist#; v# / 4; (v# - 1) / 4
a$ = INKEY$
INPUT a$
IF a$ = "s" THEN STOP
LOOP
```

"C:\Users\Reactor1967\vcns\work\SYMETTRY.BAS"
```
REM distance between v# & r# has to be mathmatically in relationship to
REM the N value of v# to have symetry or however the hell you spell it.
a# = 0
b# = 2
c# = 4
CLS
DO
PRINT a#; b#; c#, a# / 4
INPUT a$
IF a$ = "s" THEN STOP
a# = a# + 2
b# = b# + 4
c# = c# + 4
LOOP
```

"C:\Users\Reactor1967\vcns\work\SYMETERY.BAS"
```
r# = 1
a# = 1
b# = a# + ((a# - r#) * 2) + 1
c# = a# + ((a# - r#) * 2) + 2
d# = a# + ((a# - r#) * 2) + 3
CLS
DO
PRINT a#; b#; c#; d#; r#
```

```
PRINT a#; b# / 3; c# / 3; d# / 3
PRINT b# - a#
REM z = INT(RND * 3) + 0
REM IF z = 0 THEN a# = b#
REM IF z = 1 THEN a# = c#
REM IF z = 2 THEN a# = d#
a# = a# + 1
b# = a# + ((a# - r#) * 2) + 1
c# = a# + ((a# - r#) * 2) + 2
d# = a# + ((a# - r#) * 2) + 3
INPUT a$
IF a$ = "s" THEN STOP
IF a$ = "d" THEN INPUT "Please Enter your New r value."; r#
LOOP




"C:\Users\Reactor1967\vcns\work\SUCCESS3.BAS"
a# = 1
b# = 2
c# = 3
t# = 1
CLS
st# = b#
DO
test = (a# / 2) - INT(a# / 2)
IF test = .5 THEN test = 1
PRINT ABS(a# - b#); test; a#; b#; c#
REM INPUT a$
a$ = INKEY$
IF a$ = "s" THEN STOP
IF a$ = "d" THEN GOTO decode:
IF (ABS(a# - b#) * 2) >= 100 THEN st# = c# ELSE st# = b#
distance# = st# - a#
a# = a# + distance#
b# = b# + (distance# * 2)
c# = c# + (distance# * 2)
test = (a# / 2) - INT(a# / 2)
IF test = .5 THEN a# = a# + 100
LOOP
decode:
PRINT "---------------"
PRINT a#; b#; c#
DO
test = (a# / 2) - INT(a# / 2)
IF test = .5 THEN a# = a# - 100
```

```
IF test = 0 THEN distance# = b# - a#
IF test = .5 THEN distance# = c# - a#
b# = b# - distance#
c# = c# - distance#
distance# = distance# / 2
a# = a# - distance#
test = (a# / 2) - INT(a# / 2)
IF test = .5 THEN test = 1
PRINT ABS(a# - b#); test; a#; b#; c#
IF a# = 1 THEN PRINT "Decode Complete."
IF a# = 1 THEN SYSTEM
REM INPUT a$
a$ = INKEY$
IF a$ = "s" THEN STOP
LOOP




"C:\Users\Reactor1967\vcns\work\SUCCESS1.BAS"
DECLARE SUB chart2 (a#, e#, f#, g#, n!, chart!)
DECLARE SUB chart1 (a#, b#, c#, d#, n!, chart!)
over:
a# = 1
b# = 2: REM .6 for decode
c# = 3: REM  0 for decode
d# = 4: REM .3 for decode
e# = 1000: REM  .3 for decode
f# = 1001: REM  .6 for decode
g# = 1002: REM   0 for decode
flip = 0
CLS
RANDOMIZE TIMER
redo:
PRINT a#; b#; c#; d#
PRINT "Entering chart 1"
DO
chart = 1
z = INT(RND * 2) + 0
IF z = 0 THEN st# = b#
IF z = 1 THEN st# = c#
IF z = 2 THEN st# = d#
IF flip = 1 THEN st# = d#
flip = 0
distance# = st# - a#
a# = a# + distance#
```

```
b# = b# + (distance# * 3)
c# = c# + (distance# * 3)
d# = d# + (distance# * 3)
PRINT a#; b#; c#; d#
REM INPUT a$
a$ = INKEY$
IF a$ = "s" THEN STOP
IF a$ = "d" THEN GOTO decode:
IF a# >= 300000000000000# THEN GOTO decode:
LOOP UNTIL b# >= e#
flip = 1
PRINT "----------------------"
PRINT "Entering chart 2"
PRINT a#; e#; f#; g#
DO
chart = 2
z = INT(RND * 2) + 0
IF z = 0 THEN st# = e#
IF z = 1 THEN st# = f#
IF z = 2 THEN st# = g#
IF flip = 1 THEN st# = g#
flip = 0
distance# = st# - a#
a# = a# + distance#
e# = e# + (distance# * 3)
f# = f# + (distance# * 3)
g# = g# + (distance# * 3)
PRINT a#; e#; f#; g#
REM INPUT a$
a$ = INKEY$
IF a$ = "s" THEN STOP
IF a$ = "d" THEN GOTO decode:
IF a# >= 300000000000000# THEN GOTO decode:
LOOP UNTIL e# >= b#
flip = 1
PRINT "----------------------"
GOTO redo:
REM b# = 2: REM .6 for decode
REM c# = 3: REM  0 for decode
REM d# = 4: REM .3 for decode
REM e# = 1000: REM  .3 for decode
REM f# = 1001: REM  .6 for decode
REM g# = 1002: REM   0 for decode
decode:
INPUT a$
IF a$ = "s" THEN STOP
```

```
spaceandtime:
z = 99
test# = (a# / 3) - INT(a# / 3)
test# = INT(test# * 10)
test1 = (chart = 1) AND test# = 6: REM binary 0
test2 = (chart = 1) AND test# = 0: REM binary 1
test3 = (chart = 1) AND test# = 3: REM flip to chart 2
test4 = (chart = 2) AND test# = 3: REM binary 0
test5 = (chart = 2) AND test# = 6: REM binary 1
test6 = (chart = 2) AND test# = 0: REM flip to chart 1
IF test1 = -1 THEN n = 0
IF test2 = -1 THEN n = 1
IF test3 = -1 THEN chart = 1
IF test3 = -1 THEN n = 99
IF test4 = -1 THEN n = 0
IF test5 = -1 THEN n = 1
IF test6 = -1 THEN chart = 2
IF test6 = -1 THEN n = 99
PRINT a#; "data = "; n; "chart ="; chart
IF a# = 1 THEN PRINT "DECODE COMPLETE"
REM IF a# = 1 THEN GOTO over:
IF a# = 1 THEN SYSTEM
REM INPUT a$
a$ = INKEY$
IF a$ = "s" THEN STOP
REM chart2 (a#, e#, f#, g#, n, chart)
REM chart1 (a#, b#, c#, d#, n, chart)
IF chart = 1 THEN z = 0
IF chart = 1 THEN GOSUB chart1:
IF z = 0 THEN GOTO spaceandtime:
IF chart = 2 THEN z = 1
IF chart = 2 THEN GOSUB chart2:
IF z = 1 THEN GOTO spaceandtime:
chart1:
IF chart = 1 THEN CALL chart1(a#, b#, c#, d#, n, chart)
RETURN
chart2:
IF chart = 2 THEN CALL chart2(a#, e#, f#, g#, n, chart)
RETURN

SUB chart1 (a#, b#, c#, d#, n, chart)
IF n = 0 THEN distance# = (b# - a#) / 3
IF n = 1 THEN distance# = (c# - a#) / 3
IF n = 99 THEN distance# = (d# - a#) / 3
IF n = 99 THEN chart = 2
a# = a# - distance#
```

```basic
b# = b# - (distance# * 3)
c# = c# - (distance# * 3)
d# = d# - (distance# * 3)
EXIT SUB
END SUB


SUB chart2 (a#, e#, f#, g#, n, chart)
IF n = 0 THEN distance# = (e# - a#) / 3
IF n = 1 THEN distance# = (f# - a#) / 3
IF n = 99 THEN distance# = (g# - a#) / 3
IF n = 99 THEN chart = 1
a# = a# - distance#
e# = e# - (distance# * 3)
f# = f# - (distance# * 3)
g# = g# - (distance# * 3)
EXIT SUB
END SUB




"C:\Users\Reactor1967\vcns\work\SUCCESS0.BAS"
REM                     COPYRIGHT C 2002
REM               Lloyd Dudley Burris
REM ----------------------Initilize R#-------------------------------------
CLS
DIM r#(255)
r#(1) = 1
REM PRINT r#(1)
pt1# = 5
rf# = 100
count = 2
a# = 0
replaced1:
DO
a# = a# + pt1#
test = (a# / 2) - INT(a# / 2)
r#(count) = a#
IF (r#(count) / 2) - INT(r#(count) / 2) = 0 THEN r#(count) = r#(count) + 1
REM PRINT r#(count); count
count = count + 1
REM INPUT a$
REM a$ = INKEY$
IF a$ = "s" THEN STOP
IF count >= 254 THEN EXIT DO
LOOP UNTIL a# = rf#
pt1# = pt1# * 10
```

```
rf# = rf# * 10
IF count >= 254 THEN GOTO replaced2:
IF a# + pt1# > 999999999999999# THEN GOTO replaced2:
GOTO replaced1:
replaced2:
r#(254) = 950000000000001#
r#(255) = 999999999999999#
REM ---------------------Code P# up.--------------------------------------
range = 237
REM p# = 1000
p# = 100000000000001#
RANDOMIZE TIMER
c = 0
REM file$ = "vcout.txt"
REM OPEN file$ FOR OUTPUT AS #1
repeat:
count = 0
DO
bt = (p# / 2) - INT(p# / 2)
IF bt >= .5 THEN bt = 1
COLOR 2, 0
PRINT , , p#; "+"; bt; r#(range); count
REM WRITE #1, p#, bt, r#(range)
REM a$ = INKEY$
REM INPUT a$
IF a$ = "s" THEN SYSTEM
IF (p# + (p# - r#(range) + 1)) > (r#(range + 1)) THEN c = 1 ELSE c = 0
IF c = 0 THEN p# = p# + (p# - (r#(range)) + 1)
IF c = 1 THEN p# = p# + (p# - (r#(range)) + 2)
IF a$ = "s" THEN STOP
IF p# > (r#(range + 1)) THEN range = range + 1
count = count + 1
IF range = 254 THEN EXIT DO
LOOP UNTIL p# + (p# - (r#(range)) + 2) >= 999999999999999#
REM CLOSE #1
bt = (p# / 2) - INT(p# / 2)
IF bt >= .5 THEN bt = 1
PRINT , , p#; "+"; bt; r#(range); count
a$ = INKEY$
REM INPUT a$
IF a$ = "s" THEN SYSTEM
IF a$ = "d" THEN GOTO Decodedown:
REM ---------------------Code P# down.---------------------------------
range = range + 1
count = 0
DO
```

```
bt = (p# / 2) - INT(p# / 2)
IF bt >= .5 THEN bt = 1
COLOR 2, 0
PRINT p#; "-"; bt; r#(range); count
REM a$ = INKEY$
REM INPUT a$
REM IF a$ = "s" THEN STOP
REM c = INT(RND * 2) + 0
IF (p# - (r#(range) - p# + 1)) < (r#(range - 1)) THEN c = 1 ELSE c = 0
IF c = 0 THEN p# = p# - ((r#(range)) - p# + 1)
IF c = 1 THEN p# = p# - ((r#(range)) - p# + 2)
IF p# < (r#(range - 1)) THEN range = range - 1
count = count + 1
IF range <= 238 THEN EXIT DO
LOOP UNTIL p# < 150000000000000#
bt = (p# / 2) - INT(p# / 2)
IF bt >= .5 THEN bt = 1
PRINT p#; "-"; bt; r#(range); count
a$ = INKEY$
REM INPUT a$
IF a$ = "s" THEN SYSTEM
IF a$ = "d" THEN GOTO decode2:
range = range - 1
GOTO repeat:
REM ---------------------------DECODING AREA-------------------------------
REM ---------------------------RESTRICTED AREA:PROGRAM AUTHOR
ONLY!!!------
decode1:
REM---------------------------[DECODING + TO -]--------------------------
Decodedown: REM from + to -
DO
bt = (p# / 2) - INT(p# / 2)
IF bt >= .5 THEN bt = 1
IF bt = 1 THEN range = range - 1
COLOR 2, 0
PRINT , , p#; "-"; bt; r#(range)
IF bt = 0 THEN p# = p# - ((p# - r#(range) + 1) / 2)
IF bt = 1 THEN p# = p# - ((p# - r#(range) + 2) / 2)
a$ = INKEY$
REM INPUT a$
IF a$ = "s" THEN STOP
bt = (p# / 2) - INT(p# / 2)
IF bt >= .5 THEN bt = 1
test = (bt = 1) AND range <= 237
IF p# = 100000000000001# THEN GOTO finish:
LOOP UNTIL test = -1
```

```
bt = (p# / 2) - INT(p# / 2)
IF bt >= .5 THEN bt = 1
COLOR 2, 0
PRINT , , p#; "-"; bt; r#(range)
REM ----------------------------[DECODING - TO +]--------------------------
range = range + 1
Decodeup: REM from - to +
decode2:
DO
bt = (p# / 2) - INT(p# / 2)
IF bt >= .5 THEN bt = 1
IF bt = 1 THEN range = range + 1
COLOR 2, 0
PRINT , , p#; "+"; bt; r#(range)
IF bt = 0 THEN p# = p# + ((r#(range) - p# + 1) / 2)
IF bt = 1 THEN p# = p# + ((r#(range) - p# + 2) / 2)
a$ = INKEY$
REM INPUT a$
IF a$ = "s" THEN STOP
bt = (p# / 2) - INT(p# / 2)
IF bt >= .5 THEN bt = 1
test = (bt = 1) AND range >= 255
LOOP UNTIL test = -1
bt = (p# / 2) - INT(p# / 2)
IF bt >= .5 THEN bt = 1
COLOR 2, 0
PRINT , , p#; "+"; bt; r#(range)
range = range - 1
GOTO decode1:
finish:
PRINT , , "DECODE FINISHED!!!!"
SYSTEM




"C:\Users\Reactor1967\vcns\work\SPECIAL.BAS"
REM Lesson learned here. R and V have to be measured in some way to
REM             control them. Using speed, time, acceleration
REM             and deacceleration seems the most proper since
REM             a non-linear numerical system uses time to store
REM             numerical values in a numerical base.
REM Key here is to Learn to control the acceleration and de-acceleration of
REM both v# and r#. By doing that you can control r# and v# and code and
REM decode binary data if using base two. If not then base 3 or what ever.
v# = 1
r# = 1
```

```
dist# = v# - r#
count = 0
CLS
RANDOMIZE TIMER
again:
DO
count = count + 1
m2# = r#
DO
IF (v# - r#) >= dist# THEN r# = r# + 2
REM a$ = INKEY$
REM IF a$ = "s" THEN STOP
LOOP UNTIL v# - r# < dist#
IF (v# - r#) > 500 THEN x = 0
IF (v# - r#) < 10 THEN x = 1
IF x = 1 THEN r# = r# - 4
dist# = v# - r#
z = INT(RND * 2) + 0
m1# = v#
IF z = 0 THEN v# = v# + (v# - r#) + 1
IF z = 1 THEN v# = v# + (v# - r#) + 2
COLOR 2, 0
PRINT "Speed +r"; r# - m2#; "+r"; r#; "N"; z; "+v"; v#; "Speed +v"; v# - m1#;
"Distance v,r"; v# - r#; "+Time"; count
a$ = INKEY$
IF a$ = "s" THEN INPUT b$
IF b$ = "s" THEN STOP
IF b$ = "o" THEN EXIT DO
IF v# + (v# - r#) > 999999 THEN EXIT DO
a$ = ""
b$ = ""
LOOP
a$ = ""
b$ = ""
switch# = v# - r#
r# = v# + switch#
DO
count = count + 1
m2# = r#
DO
IF (r# - v#) >= dist# THEN r# = r# - 2
REM a$ = INKEY$
REM IF a$ = "s" THEN STOP
LOOP UNTIL r# - v# < dist#
IF (r# - v#) > 500 THEN x = 0
IF (r# - v#) < 10 THEN x = 1
```

```
IF x = 1 THEN r# = r# + 4
dist# = r# - v#
z = INT(RND * 2) + 0
m1# = v#
IF z = 0 THEN v# = v# - (r# - v#) - 1
IF z = 1 THEN v# = v# - (r# - v#) - 2
COLOR 2, 0
PRINT "Speed -r"; m2# - r#; "-r"; r#; "N"; z; "-v"; v#; "Speed -v"; m1# - v#;
"Distance v,r"; r# - v#; "+Time"; count
a$ = INKEY$
IF a$ = "s" THEN INPUT b$
IF b$ = "s" THEN STOP
IF b$ = "o" THEN EXIT DO
IF count = 1 THEN EXIT DO: REM Even go using Negative V is possible this
program avoids it to keep things simple.
IF v# - (r# - v#) < 0 THEN EXIT DO
a$ = ""
b$ = ""
LOOP
a$ = ""
b$ = ""
switch# = r# - v#
r# = v# - switch#
GOTO again:



"C:\Users\Reactor1967\vcns\work\SNKITALL.BAS"
CLS
a# = 1
b# = 2
c# = 3
r# = 0
PRINT a#; b#; c#; r#
r# = r# + 1
redo:
DO
FOR count1 = 1 TO 2
a# = a# + 1
b# = b# + 2
c# = c# + 2
PRINT a#; b#; c#; r#
NEXT count1
r# = r# + 1
a$ = INKEY$
REM INPUT a$
```

```
IF a$ = "s" THEN STOP
LOOP UNTIL a# >= 1000
REM sink it all make it numerically correct.
REM not keeping the coordinates and variables
REM right when sinking.
a# = a# - 500
b# = b# - 1000
c# = c# - 1000
r# = r# - 250
INPUT a$
IF a$ = "s" THEN STOP
GOTO redo:



"C:\Users\Reactor1967\vcns\work\SINK8.BAS"
a# = 1
b# = 2
c# = 3
t# = 1
CLS
st# = b#
DO
test = (a# / 2) - INT(a# / 2)
IF test = .5 THEN test = 1
PRINT ABS(a# - b#); test; a#; b#; c#
REM INPUT a$
a$ = INKEY$
IF a$ = "s" THEN STOP
IF a$ = "d" THEN GOTO decode:
IF (ABS(a# - b#) * 2) >= 100 THEN st# = c# ELSE st# = b#
distance# = st# - a#
a# = a# + distance#
b# = b# + (distance# * 2)
c# = c# + (distance# * 2)
test = (a# / 2) - INT(a# / 2)
IF test = .5 THEN a# = a# + 100
LOOP
decode:
PRINT "--------------"
PRINT a#; b#; c#
DO
test = (a# / 2) - INT(a# / 2)
IF test = .5 THEN a# = a# - 100
IF test = 0 THEN distance# = b# - a#
IF test = .5 THEN distance# = c# - a#
```

```
b# = b# - distance#
c# = c# - distance#
distance# = distance# / 2
a# = a# - distance#
test = (a# / 2) - INT(a# / 2)
IF test = .5 THEN test = 1
PRINT ABS(a# - b#); test; a#; b#; c#
IF a# = 1 THEN PRINT "Decode Complete."
IF a# = 1 THEN SYSTEM
REM INPUT a$
a$ = INKEY$
IF a$ = "s" THEN STOP
LOOP
```

```
"C:\Users\Reactor1967\vcns\work\SINK7.BAS"
a# = 1
b# = 2
c# = 3
d# = 4
CLS
RANDOMIZE TIMER
DO
z = INT(RND * 2) + 0
IF z = 0 THEN st# = b#
IF z = 1 THEN st# = c#
IF flip = 1 THEN st# = d#
distance# = st# - a#
a# = a# + distance#
b# = b# + (distance# * 3)
c# = c# + (distance# * 3)
d# = d# + (distance# * 3)
PRINT a#; b#; c#; d#; flip
IF b# - a# >= 4 THEN flip = 1 ELSE flip = 0
IF b# - a# >= 4 THEN a# = a# + 6
IF b# - a# >= 4 THEN a# = a# + 6
INPUT a$
REM a$ = INKEY$
IF a$ = "s" THEN STOP
IF a$ = "d" THEN GOTO decode:
LOOP UNTIL b# >= 999999999999999#
decode:
test1# = INT(((a# / 3) - INT(a# / 3)) * 10) = 6: REM b#
test2# = INT(((a# / 3) - INT(a# / 3)) * 10) = 0: REM c#
```

```
test3# = INT(((a# / 3) - INT(a# / 3)) * 10) = 3: REM d#
PRINT "b#"; test1#
PRINT "c#"; test2#
PRINT "d#"; test3#


"C:\Users\Reactor1967\vcns\work\SINK6.BAS"
DECLARE SUB chart2 (a#, e#, f#, g#, n!, chart!)
DECLARE SUB chart1 (a#, b#, c#, d#, n!, chart!)
over:
a# = 1
b# = 2: REM .6 for decode
c# = 3: REM  0 for decode
d# = 4: REM .3 for decode
e# = 1000: REM  .3 for decode
f# = 1001: REM  .6 for decode
g# = 1002: REM   0 for decode
flip = 0
CLS
RANDOMIZE TIMER
redo:
PRINT a#; b#; c#; d#
PRINT "Entering chart 1"
DO
chart = 1
z = INT(RND * 2) + 0
IF z = 0 THEN st# = b#
IF z = 1 THEN st# = c#
IF z = 2 THEN st# = d#
IF flip = 1 THEN st# = d#
flip = 0
distance# = st# - a#
a# = a# + distance#
b# = b# + (distance# * 3)
c# = c# + (distance# * 3)
d# = d# + (distance# * 3)
PRINT a#; b#; c#; d#
REM INPUT a$
a$ = INKEY$
IF a$ = "s" THEN STOP
IF a$ = "d" THEN GOTO decode:
IF (a# * 3) >= 999999999999999# THEN GOTO decode:
LOOP UNTIL b# >= e#
flip = 1
PRINT "----------------------"
```

```
PRINT "Entering chart 2"
PRINT a#; e#; f#; g#
DO
chart = 2
z = INT(RND * 2) + 0
IF z = 0 THEN st# = e#
IF z = 1 THEN st# = f#
IF z = 2 THEN st# = g#
IF flip = 1 THEN st# = g#
flip = 0
distance# = st# - a#
a# = a# + distance#
e# = e# + (distance# * 3)
f# = f# + (distance# * 3)
g# = g# + (distance# * 3)
PRINT a#; e#; f#; g#
REM INPUT a$
a$ = INKEY$
IF a$ = "s" THEN STOP
IF a$ = "d" THEN GOTO decode:
IF (a# * 3) >= 999999999999999# THEN GOTO decode:
LOOP UNTIL e# >= b#
flip = 1
PRINT "----------------------"
GOTO redo:
REM b# = 2: REM .6 for decode
REM c# = 3: REM  0 for decode
REM d# = 4: REM .3 for decode
REM e# = 1000: REM  .3 for decode
REM f# = 1001: REM  .6 for decode
REM g# = 1002: REM   0 for decode
decode:
z = 99
test# = (a# / 3) - INT(a# / 3)
test# = INT(test# * 10)
test1 = (chart = 1) AND test# = 6: REM binary 0
test2 = (chart = 1) AND test# = 0: REM binary 1
test3 = (chart = 1) AND test# = 3: REM flip to chart 2
test4 = (chart = 2) AND test# = 3: REM binary 0
test5 = (chart = 2) AND test# = 6: REM binary 1
test6 = (chart = 2) AND test# = 0: REM flip to chart 1
IF test1 = -1 THEN n = 0
IF test2 = -1 THEN n = 1
IF test3 = -1 THEN chart = 1
IF test3 = -1 THEN n = 99
IF test4 = -1 THEN n = 0
```

```
IF test5 = -1 THEN n = 1
IF test6 = -1 THEN chart = 2
IF test6 = -1 THEN n = 99
PRINT a#; "data = "; n; "chart ="; chart
IF a# = 1 THEN PRINT "DECODE COMPLETE"
IF a# = 1 THEN GOTO over:
IF a# = 1 THEN SYSTEM
REM INPUT a$
a$ = INKEY$
IF a$ = "s" THEN STOP
REM chart2 (a#, e#, f#, g#, n, chart)
REM chart1 (a#, b#, c#, d#, n, chart)
IF chart = 1 THEN z = 0
IF chart = 1 THEN GOSUB chart1:
IF z = 0 THEN GOTO decode:
IF chart = 2 THEN z = 1
IF chart = 2 THEN GOSUB chart2:
IF z = 1 THEN GOTO decode:
chart1:
IF chart = 1 THEN CALL chart1(a#, b#, c#, d#, n, chart)
RETURN
chart2:
IF chart = 2 THEN CALL chart2(a#, e#, f#, g#, n, chart)
RETURN

SUB chart1 (a#, b#, c#, d#, n, chart)
IF n = 0 THEN distance# = (b# - a#) / 3
IF n = 1 THEN distance# = (c# - a#) / 3
IF n = 99 THEN distance# = (d# - a#) / 3
IF n = 99 THEN chart = 2
a# = a# - distance#
b# = b# - (distance# * 3)
c# = c# - (distance# * 3)
d# = d# - (distance# * 3)
EXIT SUB
END SUB

SUB chart2 (a#, e#, f#, g#, n, chart)
IF n = 0 THEN distance# = (e# - a#) / 3
IF n = 1 THEN distance# = (f# - a#) / 3
IF n = 99 THEN distance# = (g# - a#) / 3
IF n = 99 THEN chart = 1
a# = a# - distance#
e# = e# - (distance# * 3)
f# = f# - (distance# * 3)
g# = g# - (distance# * 3)
```

```
EXIT SUB
END SUB




"C:\Users\Reactor1967\vcns\work\SINK5.BAS"
a# = 1
b# = 2
c# = 3
d# = 50
e# = 51
CLS
RANDOMIZE TIMER
redo:
DO
s = INT(RND * 2) + 0
IF s = 0 THEN st# = b#
IF s = 1 THEN st# = c#
distance# = st# - a#
a# = a# + distance#
b# = b# + (distance# * 2)
c# = c# + (distance# * 2)
PRINT a#; b#; c#, ABS(a# - b#); ABS(b# - d#)
REM INPUT a$
a$ = INKEY$
IF a$ = "s" THEN STOP
r# = a# - ABS(a# - b#) - 1
LOOP UNTIL a# + (a# - r#) >= d#
b# = b# + 2
c# = c# + 2
IF ABS(b# - d#) <= 0 THEN STOP
PRINT "----------------------"
DO
s = INT(RND * 2) + 0
IF s = 0 THEN st# = d#
IF s = 1 THEN st# = e#
distance# = st# - a#
a# = a# + distance#
d# = d# + (distance# * 2)
e# = e# + (distance# * 2)
PRINT a#; d#; e#, ABS(a# - d#); ABS(b# - d#)
REM INPUT a$
a$ = INKEY$
IF a$ = "s" THEN STOP
r# = a# - ABS(a# - d#) - 1
LOOP UNTIL a# + (a# - r#) >= b#
```

```
d# = d# + 2
e# = e# + 2
IF ABS(b# - d#) <= 0 THEN STOP
PRINT "-----------------------"
GOTO redo:
```

```
"C:\Users\Reactor1967\vcns\work\SINK4.BAS"
r# = 1
v# = 1
b# = v# + (v# - r# + 1)
c# = c# + (v# - r# + 2)
rate# = ((v# - b#) * 3) - 10
CLS
RANDOMIZE TIMER
count2# = 0
redo:
FOR count = 1 TO 4
d = INT(RND * 2) + 0
IF d = 0 THEN st# = b#
IF d = 1 THEN st# = c#
v# = st#
b# = v# + (v# - r# + 1)
c# = c# + (v# - r# + 2)
rate# = ((b# - v#) * 2) - 10
LOCATE 15, 20
PRINT v# - (b# - v#); v#; b#; c#; b# - v#; rate#
NEXT count
storage# = r#
r# = r# + rate#
speed# = speed# + (r# - storage#)
count2# = count2# + 1
LOCATE 16, 20
PRINT speed# / count2#
REM INPUT a$
a$ = INKEY$
IF a$ = "s" THEN STOP
GOTO redo:
```

```
"C:\Users\Reactor1967\vcns\work\SINK3.BAS"
CLS
v# = 1
r# = 1
```

```
RANDOMIZE TIMER
redo:
DO
d = INT(RND * 2) + 0
IF d = 0 THEN v# = v# + (v# - r# + 1)
IF d = 1 THEN v# = v# + (v# - r# + 2)
PRINT v#; v# - r#, count
IF (v# - r#) > 50 THEN GOSUB sink:
a$ = INKEY$
IF a$ = "s" THEN STOP
a$ = INKEY$
IF a$ = "s" THEN STOP
count = count + 1
LOOP
sink:
count = 1
a# = 100
b# = 50
DO
REM PRINT a#; b#
a# = a# - 1
b# = b# - 1
a$ = INKEY$
IF a$ = "s" THEN STOP
IF a# = 50 THEN EXIT DO
LOOP UNTIL a# = (v# - r#)
v# = b# + r#
RETURN
GOTO redo:




"C:\Users\Reactor1967\vcns\work\SINK2.BAS"
a# = 1
b# = 2
c# = 3
CLS
speed# = 0
count# = 0
redo:
rate# = (a# - (b# - a#))
```

```
DO
d = INT(RND * 2) + 0
LOCATE 15, 25
PRINT a# - (b# - a#); a#; b#; c#; ABS(a# - b#)
IF d = 0 THEN st# = b#
IF d = 1 THEN st# = c#
DO
a# = a# + 1
b# = b# + 2
c# = c# + 2
a$ = INKEY$
IF a$ = "s" THEN STOP
LOOP UNTIL a# = st#
a$ = INKEY$
IF a$ = "s" THEN STOP
LOOP UNTIL ABS(a# - b#) >= 50
LOCATE 15, 25
PRINT a# - (b# - a#); a#; b#; c#; ABS(a# - b#)
DO
a# = a# + 2
a$ = INKEY$
IF a$ = "s" THEN STOP
LOOP UNTIL ABS(a# - b#) <= 10
REM PRINT a# - (b# - a#); a#; b#; c#; ABS(a# - b#)
LOCATE 16, 25
PRINT a# - (b# - a#) - rate#
speed# = speed# + (a# - (b# - a#) - rate#)
count# = count# + 1
LOCATE 17, 25
PRINT speed# / count#
REM INPUT a$
a$ = INKEY$
IF a$ = "s" THEN STOP
IF count# >= 999999999999999# THEN count = 1
IF count# = 1 THEN speed = 0
GOTO redo:


"C:\Users\Reactor1967\vcns\work\SINK.BAS"
a# = 1
b# = 2
c# = 3
CLS
RANDOMIZE TIMER
PRINT a#; b#; c#
```

cl

```
count = 0
redo:
DO
c = INT(RND * 2) + 0
IF c = 0 THEN st# = b#
IF c = 1 THEN st# = c#
distance# = ABS(a# - st#)
a# = st#
b# = b# + (distance# * 2)
c# = c# + (distance# * 2)
PRINT a#; b#; c#; ABS(a# - b#)
INPUT a$
IF a$ = "d" THEN EXIT DO
IF a$ = "s" THEN STOP
count = count + 1
LOOP
down# = 1
DO
down# = down# * 2
a# = a# + down#
LOOP UNTIL ((down# * 2) + a#) >= b#
GOTO redo:




"C:\Users\Reactor1967\vcns\work\SETUP.BAS"
p# = 1
p1# = 2
p2# = 3
CLS
RANDOMIZE TIMER
DO
PRINT p#; p1#; p2#
p# = p# + 1
p1# = p1# + 2
p2# = p2# + 2
INPUT a$
IF a$ = "s" THEN STOP
LOOP




"C:\Users\Reactor1967\vcns\work\SELFC.BAS"
a# = 20
CLS
DIM matrix#(100)
```

```
FOR count = 1 TO 100
matrix#(count) = count
NEXT count
RANDOMIZE TIMER
CLS
p = 6
fp = 0
DO
PRINT , , z; a#; r#; x#
INPUT a$
REM a$ = INKEY$
IF a$ = "s" THEN STOP
IF a$ = "d" THEN GOTO decode:
test1# = (a# / 2) - INT(a# / 2)
store# = p
IF p >= 85 THEN fp = 1
IF p <= 20 THEN fp = 0
DO
IF fp = 0 THEN p = p + 1
IF fp = 1 THEN p = p - 1
x# = matrix#(p)
test2# = (x# / 2) - INT(x# / 2)
IF test2# <> test1# THEN EXIT DO
LOOP

move# = p - store#
dist# = x#
r# = a# - x#
z = INT(RND * 2) + 0
dist2# = a# - r#
IF move# = 1 THEN a# = a# + (a# - r# + 1)
IF move# = 2 THEN a# = a# + (a# - r# + 2)

LOOP
decode:

DO
IF p + 1 >= 85 THEN fp = 1
IF p - 1 <= 20 THEN fp = 0
IF fp = 0 THEN test = (a# / 2) - INT(a# / 2) ELSE test = 6
IF fp = 1 THEN test2 = (a# / 2) - INT(a# / 2) ELSE test2 = 6
a# = a# - x#
IF test = 0 THEN a# = a# - 1
IF test = .5 THEN a# = a# - 2
IF test = 0 THEN x# = x# - 1
IF test = .5 THEN x# = x# - 2
```

```
IF test2 = 0 THEN a# = a# + 1
IF test2 = .5 THEN a# = a# + 2
IF test2 = 0 THEN x# = x# + 1
IF test2 = .5 THEN x# = x# + 2
PRINT , , test; a#; x#
IF a# = 20 THEN PRINT , , "Decode Complete"
IF a# = 20 THEN SYSTEM
REM a$ = INKEY$
INPUT a$
IF a$ = "s" THEN STOP
LOOP




"C:\Users\Reactor1967\vcns\work\SEEMSG.BAS"
v# = 10
r# = 1
d1# = 1
d2# = 2
RANDOMIZE TIMER
CLS
DO
z = INT(RND * 2) + 0
IF z = 0 THEN r2# = v# - d1#
IF z = 1 THEN r2# = v# - d2#
IF z = 0 THEN v# = v# + (v# - r# + 1)
IF z = 1 THEN v# = v# + (v# - r# + 2)
COLOR 2, 0
PRINT , , v#; r#; d1#; d2#; v# - r#
r# = r2#
d1# = d1# + 2
d2# = d2# + 2
a$ = INKEY$
IF a$ = "s" THEN STOP
LOOP




"C:\Users\Reactor1967\vcns\work\RINCF.BAS"
p# = 125789549584951#
r# = p# - 25000000000000#
z# = 125789549584951#
RANDOMIZE TIMER
CLS
redo:
start = 1
```

```
DO
PRINT p#; r#; r# - store#
c = INT(RND * 2) + 0
IF p# + (p# - r#) >= 900000000000000# THEN c = 1
store# = r#
IF c = 0 THEN p# = p# + (p# - r# + 1)
IF c = 1 THEN p# = p# + (p# - r# + 2)
t# = p# - z#
r# = (z# - 25000000000000#) + t#
IF (r# / 2) - INT(r# / 2) = 0 THEN r# = r# - 1
a$ = INKEY$
REM INPUT a$
IF a$ = "s" THEN STOP
LOOP UNTIL p# > 925000000000000#
PRINT p#; r#; p# - r#
PRINT "---------------"
INPUT a$
IF a$ = "s" THEN STOP
r# = r# + 50000000000000#
DO
PRINT p#; r#; r# - p#
c = INT(RND * 2) + 0
IF p# - (r# - p#) < 15000000000000# THEN c = 1
IF c = 0 THEN p# = p# + (p# - r# + 1)
IF c = 1 THEN p# = p# + (p# - r# + 2)
r# = r# - 25000000000000#
REM a$ = INKEY$
INPUT a$
IF a$ = "s" THEN STOP
LOOP UNTIL p# < 150000000000000#
GOTO redo:


"C:\Users\Reactor1967\vcns\work\RINC.BAS"
DIM cach#(100)
r# = 100000000000001#
p# = 125000000000002#
count = 1
CLS
DO
p# = p# + (p# - r# + 1)
PRINT p#
cach#(count) = p#
IF p# >= r# + 50000000000000# THEN r# = r# + 50000000000000#
count = count + 1
```

```
p# = p# + (p# - r# + 2)
PRINT p#
cach#(count) = p#
IF p# >= r# + 50000000000000# THEN r# = r# + 50000000000000#
count = count + 1
LOOP UNTIL p# + (p# - r#) >= 992000000000000#
CLS
r# = 150000000000001#
FOR count = 1 TO 84
p# = cach#(count)
PRINT p#; p# + (p# - r# + 1); p# + (p# - r# + 2)
IF p# + (p# - r#) >= r# + 500000000000000# THEN r# = r# +
500000000000000#
INPUT a$
IF a$ = "s" THEN STOP
NEXT count
```

```
"C:\Users\Reactor1967\vcns\work\REXPRMNT.BAS"
v# = 100
r# = 99
CLS
DO
PRINT v# + (v# - r# + 1); v# + (v# - r# + 2)
PRINT r#; v#
INPUT a$
IF a$ = "s" THEN STOP
REM r# = r# - 2: REM change uncomment these rem statements 1 at a time to
experiment with r#
r# = r# + 2: REM change uncomment these rem statements 1 at a time to
experiment with r#
LOOP
```

```
"C:\Users\Reactor1967\vcns\work\REXPMNT.BAS"
v# = 100
r# = 99
CLS
DO
PRINT v# + (v# - r# + 1); v# + (v# - r# + 2)
PRINT r#; v#
INPUT a$
IF a$ = "s" THEN STOP
r# = r# - 2: REM change uncomment these rem statements 1 at a time to
experiment with r#
REM r# = r# + 2:REM change uncomment these rem statements 1 at a time to
experiment with r#
LOOP
```

```
"C:\Users\Reactor1967\vcns\work\REWIND4.BAS"
a# = 12501
r# = 10001
CLS
RANDOMIZE TIMER
redo:
DO
PRINT , , "data+"; a#; r#; a# - r#
c = INT(RND * 2) + 0
IF c = 0 THEN a# = a# + (a# - r# + 1)
IF c = 1 THEN a# = a# + (a# - r# + 2)
r# = r# + 2500
IF (r# - a#) > 2600 THEN PRINT "          THIS IS A replaced3ING RED
FLAG!!!"
IF (a# - r#) > 2600 THEN r# = r# + 102
REM INPUT a$
a$ = INKEY$
IF a$ = "s" THEN STOP
LOOP UNTIL a# + (a# - r#) > 99999
r# = (a# + (a# - r#))
DO
PRINT , , "data-"; a#; r#; r# - a#
c = INT(RND * 2) + 0
IF c = 0 THEN a# = a# - (r# - a# + 1)
IF c = 1 THEN a# = a# - (r# - a# + 2)
r# = r# - 2500
IF (r# - a#) > 2600 THEN PRINT "          THIS IS A replaced3ING RED
```

```
FLAG!!!"
IF (r# - a#) > 2600 THEN r# = r# - 102
REM INPUT a$
a$ = INKEY$
IF a$ = "s" THEN STOP
LOOP UNTIL a# - (r# - a#) < 20000
r# = a# - (r# - a#)
GOTO redo:




"C:\Users\Reactor1967\vcns\work\REWIND3.BAS"
REM in most of my programs I try to code and
REM stay in sink at the same time. Here I code
REM then resink the numbers then code again.
a# = 1
b# = 98
c# = 99
CLS
RANDOMIZE TIMER
count# = 0
redo:
tms = 0
DO
REM PRINT "     "
REM CLS
REM LOCATE 20, 15
PRINT d; "a"; a#; "b"; b#; "c"; c#; "Varance"; ABS(a# - b#); "R ="; a# - (b# -
a# - 1)
REM LOCATE 21, 16
PRINT "bits"; count#; "Rate"; a# - storeage#; "bytes"; INT(count# / 4);
"resink"; resink
INPUT a$
REM a$ = INKEY$
IF a$ = "s" THEN STOP
IF a$ = "d" THEN GOTO decode:
IF tms = 0 THEN tms = 1 ELSE tms = 0
add# = INT(b# - a# - 1) / 2
add# = INT(add# / 4)
IF tms = 0 THEN add# = add# + 2
IF tms = 1 THEN add# = add# - 2
IF b# - a# > 100 THEN GOSUB it:
storeage# = st#
d = INT(RND * 2) + 0
IF d = 0 THEN st# = b#
IF d = 1 THEN st# = c#
```

```
count# = count# + 1
DO
IF a# = st# THEN EXIT DO
a# = a# + 1
b# = b# + 2
c# = c# + 2
a$ = INKEY$
IF a$ = "s" THEN STOP
LOOP
LOOP
it:
resink = 0
DO
a# = a# + 4
REM b# = b# + 2
REM c# = c# + 2
resink = resink + 1
a$ = INKEY$
IF a$ = "s" THEN STOP
LOOP UNTIL resink = add#
RETURN
GOTO redo
decode:
resink = 0
DO
a# = a# - 4
REM b# = b# - 2
REM c# = c# - 2
resink = resink + 1
a$ = INKEY$
IF a$ = "s" THEN STOP
LOOP UNTIL ABS(b# - a#) > 293
PRINT d; "a"; a#; "b"; b#; "c"; c#; "V"; ABS(a# - b#); resink
INPUT a$
IF a$ = "s" THEN STOP
GOTO redo:


"C:\Users\Reactor1967\vcns\work\REWIND2.BAS"
REM in most of my programs I try to code and
REM stay in sink at the same time. Here I code
REM then resink the numbers then code again.
a# = 1
b# = 98
c# = 99
```

```
CLS
RANDOMIZE TIMER
count# = 0
redo:
DO
REM PRINT "    "
REM CLS
LOCATE 20, 15
PRINT d; "a"; a#; "b"; b#; "c"; c#; "Varance"; ABS(a# - b#); "R ="; a# - (b# -
a# - 1)
LOCATE 21, 16
PRINT "bits"; count#; "Rate"; a# - storeage#; "bytes"; INT(count# / 4);
"resink"; resink
REM INPUT a$
a$ = INKEY$
IF a$ = "s" THEN STOP
IF a$ = "d" THEN GOTO decode:
IF ABS(a# - b#) > 100 THEN GOTO it:
storeage# = st#
d = INT(RND * 2) + 0
IF d = 0 THEN st# = b#
IF d = 1 THEN st# = c#
count# = count# + 1
DO
IF a# = st# THEN EXIT DO
a# = a# + 1
b# = b# + 2
c# = c# + 2
LOOP
LOOP
it:
resink = 0
DO
a# = a# + 4
REM b# = b# + 2
REM c# = c# + 2
resink = resink + 1
a$ = INKEY$
IF a$ = "s" THEN STOP
LOOP UNTIL ABS(a# - b#) < 100
GOTO redo
decode:
resink = 0
DO
a# = a# - 4
REM b# = b# - 2
```

```
REM c# = c# - 2
resink = resink + 1
a$ = INKEY$
IF a$ = "s" THEN STOP
LOOP UNTIL ABS(b# - a#) > 293
PRINT d; "a"; a#; "b"; b#; "c"; c#; "V"; ABS(a# - b#); resink
INPUT a$
IF a$ = "s" THEN STOP
GOTO redo:



"C:\Users\Reactor1967\vcns\work\REWIND1.BAS"
REM in most of my programs I try to code and
REM stay in sink at the same time. Here I code
REM then resink the numbers then code again.
a# = 1
b# = 98
c# = 99
CLS
RANDOMIZE TIMER
count# = 0
redo:
DO
REM PRINT "      "
REM CLS
LOCATE 20, 15
PRINT d; "a"; a#; "b"; b#; "c"; c#; "Varance"; ABS(a# - b#); "R ="; a# - (b# -
a# - 1)
LOCATE 21, 16
PRINT "bits"; count#; "Rate"; a# - storeage#; "bytes"; INT(count# / 4);
"resink"; resink
REM INPUT a$
a$ = INKEY$
IF a$ = "s" THEN STOP
IF a$ = "d" THEN GOTO decode:
IF ABS(a# - b#) > 100 THEN GOTO it:
storeage# = st#
d = INT(RND * 2) + 0
IF d = 0 THEN st# = b#
IF d = 1 THEN st# = c#
count# = count# + 1
DO
IF a# = st# THEN EXIT DO
a# = a# + 1
b# = b# + 2
```

```
c# = c# + 2
LOOP
LOOP
it:
resink = 0
DO
a# = a# + 4
REM b# = b# + 2
REM c# = c# + 2
resink = resink + 1
a$ = INKEY$
IF a$ = "s" THEN STOP
LOOP UNTIL ABS(a# - b#) < 100
GOTO redo
decode:
resink = 0
DO
a# = a# - 4
REM b# = b# - 2
REM c# = c# - 2
resink = resink + 1
a$ = INKEY$
IF a$ = "s" THEN STOP
LOOP UNTIL ABS(b# - a#) > 293
PRINT d; "a"; a#; "b"; b#; "c"; c#; "V"; ABS(a# - b#); resink
INPUT a$
IF a$ = "s" THEN STOP
GOTO redo:


"C:\Users\Reactor1967\vcns\work\REDUCEC.BAS"
r1# = 999999999999999#
r# = 500000000000000#
v = 1
p# = 1
RANDOMIZE TIMER
store# = 1
CLS
GOTO start:
ohmygod:
p# = p# + 1
start:
count = 0
p# = store#
DO
```

```
IF count = 1 THEN c = INT(RND * 2) + 0 ELSE c = 0
PRINT , , p#; c; count
IF c = 0 THEN p# = p# + (p# - v + 1)
IF c = 1 THEN p# = p# + (p# - v + 2)
count = count + 1
LOOP UNTIL p# >= 500000000000000#
p# = p# + 1
PRINT , , p#; c; count
PRINT "                                    "
IF count < 2 THEN STOP
a$ = INKEY$
REM INPUT a$
IF a$ = "s" THEN STOP
r1# = 999999999999999#
repeat:
DO
store# = p#
p# = p# - (r1# - p# + 1)
IF p# < 0 THEN GOTO ohmygod:
a$ = INKEY$
IF a$ = "s" THEN STOP
LOOP UNTIL p# < INT(r1# / 2)
p# = p# + 1
r1# = INT(r1# / 2)
GOTO repeat:


"C:\Users\Reactor1967\vcns\work\REDUCEA.BAS"
RANDOMIZE TIMER
CLS
DO
c# = INT(RND * 999999999999999#) + 0
a$ = INKEY$
IF a$ = "s" THEN STOP
LOOP UNTIL c# > 500000000000000#
PRINT c#
z = 1
repeat:
count = 0
a# = 1
DO
c# = c# - a#
a# = (a# * 2) + 1
a$ = INKEY$
IF a$ = "s" THEN STOP
```

```
IF count = (count1 - 1) THEN EXIT DO
count = count + 1
LOOP UNTIL c# - a# <= 0
IF z = 1 THEN count1 = count
z = 0
PRINT c#; count
INPUT a$
IF a$ = "s" THEN STOP
IF count1 = 0 THEN STOP
GOTO repeat:


"C:\Users\Reactor1967\vcns\work\RDVRCHRT.BAS"
REM Lesson Learned "One R hands off to another R"
REM Just like passing the stick on a team of runners
REM Derive mathmaticly how to handoff between R's"
REM Be able to tell what R went to and its choices
REM to goto. Make an R chart just like a vector chart.
REM Hell, just write everything to the hard disk as a
REM chart and read it from there. See if that works.
DIM a#(500)
DIM b#(500)
DIM c#(500)
DIM r1#(500)
DIM r2#(500)
test$ = "test.txt"
OPEN test$ FOR INPUT AS #1
CLS
FOR count = 1 TO 500
INPUT #1, count1, rs#, v#, a$, v1#, v2#, r#, dist#
PRINT count1; rs#; v#; a$; v1#; v2#; r#; dist#
a#(count) = v#
r1#(count) = rs#
b#(count) = v1#
c#(count) = v2#
r2#(count) = r#
REM INPUT a$
a$ = INKEY$
IF a$ = "s" THEN STOP
NEXT count
CLOSE #1
PRINT "Array initialized"
INPUT a$
IF a$ = "s" THEN STOP
RANDOMIZE TIMER
```

```
carry# = a#(1)
r# = r1#(1)
g = 1
DO
z = INT(RND * 2) + 0
IF z = 0 THEN carry# = b#(g)
IF z = 1 THEN carry# = c#(g)
r# = r2#(g)
COLOR 2, 0
PRINT , , carry#; r#; carry# - r#
REM INPUT a$
a$ = INKEY$
IF a$ = "s" THEN STOP
FOR count = 1 TO 500
test = (carry# = a#(count)) AND (r# = r1#(count))
IF test = -1 THEN EXIT FOR
NEXT count
g = count
IF g >= 500 THEN STOP
LOOP




"C:\Users\Reactor1967\vcns\work\RDINF.BAS"
p1# = 999999999999999#
p2# = 500000000000001#
RANDOMIZE TIMER
CLS
repeat:
DO
p# = INT(RND * 999999999999999#)
a$ = INKEY$
IF a$ = "s" THEN STOP
LOOP UNTIL p# >= 975000000000000#
p# = p# - 25000000000000#
PRINT , , p#
store1# = p#
DO
p# = p# - 50000000000000#
LOOP UNTIL p# < 500000000000000#
PRINT , , p#
IF p# > 475000000000000# THEN STOP
DO
p# = p# + 50000000000000#
LOOP UNTIL p# > 950000000000000#
IF p# > 999999999999999# THEN p# = p# - 50000000000000#
```

```
PRINT , , p#
IF p# <> store1# THEN STOP
PRINT "                    "
a$ = INKEY$
REM INPUT a$
IF a$ = "s" THEN STOP
GOTO repeat:
```

```
"C:\Users\Reactor1967\vcns\work\RDECODE.BAS"
v# = 1
r# = 1
c# = 1
CLS
DO
z = INT(RND * 2) + 0
PRINT v# - r#, v#; r#
v# = v# + (v# - r# + 1)
PRINT v# - r#, v#; r#
v# = v# + 1
PRINT v# - r#, v#; r#
REM c# = c# + 1
v# = v# + 2
r# = r# + 2
IF (r# / 2) - INT(r# / 2) = 0 THEN r# = r# - 1
INPUT a$
IF a$ = "s" THEN STOP
LOOP
```

```
"C:\Users\Reactor1967\vcns\work\RCONTROL.BAS"
REM this program puts every even number to another even number within
REM a specific range and every odd number to another odd number within
REM a specif range. This is used as a chart to get the distance from d - r
REM use it to generate another distance between d - r. The charts are a
REM reversable process so that it is possible to decode the same way.
a = 100
b = 50
c = 99
d = 49
CLS
redo:
DO
PRINT a; b; c; d
```

```basic
a = a - 2
b = b - 2
c = c - 2
d = d - 2
REM INPUT a$
a$ = INKEY$
IF a$ = "s" THEN STOP
LOOP UNTIL b = 2
INPUT a$
REM a$ = INKEY$
IF a$ = "s" THEN STOP
DO
PRINT b; a; c; d
a = a + 2
b = b + 2
c = c + 2
d = d + 2
a$ = INKEY$
IF a$ = "s" THEN STOP
LOOP UNTIL b = 50
PRINT b; a
INPUT a$
REM a$ = INKEY$
IF a$ = "s" THEN STOP
GOTO redo:
```

```basic
"C:\Users\Reactor1967\vcns\work\RCODE2.BAS"
REM Keeping the Symetery
REM r# = (r# * base) + n1 b -1 or n2 b-1 + b-1 or n3 b-1 + b-1 + b-1
r# = 0
a# = 1
b# = a# + r# + 1
c# = a# + r# + 2
d# = a# + r# + 3
CLS
RANDOMIZE TIMER
DO
PRINT a#; b#; c#; d#, r#
PRINT (b# / 3) - INT(b# / 3)
INPUT a$
IF a$ = "s" THEN STOP
z = INT(RND * 3) + 0
IF z = 0 THEN a# = b#
```

```
IF z = 1 THEN a# = c#
IF z = 2 THEN a# = d#
IF z = 3 THEN a# = e#
r# = r# * 3
IF z = 0 THEN r# = r# + 2
IF z = 1 THEN r# = r# + 4
IF z = 2 THEN r# = r# + 6
b# = a# + r# + 1
c# = a# + r# + 2
d# = a# + r# + 3
LOOP
```

```
"C:\Users\Reactor1967\vcns\work\RCODE1.BAS"
r# = 1
v1# = 1
v2# = (r# * 2) + 1
v3# = (r# * 2) + 2
v4# = (r# * 2) + 3
RANDOMIZE TIMER
CLS
DO
PRINT v1#; v2#; v3#; v4#; r#
z = INT(RND * 3) + 0
IF z = 0 THEN v1# = v2#
IF z = 1 THEN v1# = v3#
IF z = 2 THEN v1# = v4#
REM IF r# >= 1000 THEN v1# = v4#
REM IF r# >= 1000 THEN EXIT DO
r# = r# + 2
v2# = v1# + (r# * 2) + 1
v3# = v1# + (r# * 2) + 2
v4# = v1# + (r# * 2) + 3
REM a$ = INKEY$
INPUT a$
IF a$ = "s" THEN STOP
LOOP
PRINT v1#; v2#; v3#; v4#; r#
```

```
"C:\Users\Reactor1967\vcns\work\RCHART2.BAS"
v1# = 1
CLS
DO
```

```
r1# = ((INT(v1# / 6)) * 6) + 1
IF r1# > v1# THEN r1# = r1# - 6
v2# = v1# + (v1# - r1# + 1)
v3# = v1# + (v1# - r1# + 2)
r2# = ((INT(v2# / 6)) * 6) + 1
IF r2# > v2# THEN r2# = r2# - 6
r3# = ((INT(v3# / 6)) * 6) + 1
IF r3# > v3# THEN r3# = r3# - 6
PRINT v1#; r1#; "="; v2#; r2#; v3#; r3#
v1# = v1# + 1
INPUT a$
IF a$ = "s" THEN STOP
LOOP




"C:\Users\Reactor1967\vcns\work\RCHART.BAS"
CLS
DIM back(9999): REM get previous r#
DIM p(9999): REM current p#
DIM num1(9999): REM forward p#
DIM num2(9999): REM forward p#
DIM forward(9999): REM forward r#
COLOR 0, 1
CLS
RANDOMIZE TIMER
r# = 1
FOR count# = 1 TO 9999
store1# = r#
p1# = count# + (count# - r# + 1)
p2# = count# + (count# - r# + 2)
IF count# - r# > 2 THEN r# = r# + 4
REM LOCATE 11, 31
REM PRINT "POWERING UP VCNS"
REM LOCATE 12, 25
p(count#) = count#: num1(count#) = p1#: num2(count#) = p2#:
forward(count#) = r#
back(p1#) = forward(count#): back(p2#) = forward(count#)
PRINT back(count#); p(count#); num1(count#); num2(count#);
forward(count#)
INPUT a$
REM a$ = INKEY$
IF a$ = "s" THEN STOP
NEXT count#
CLS
LOCATE 11, 30
```

```
PRINT "VCNS IS NOW STARTED"
LOCATE 12, 12
PRINT "HIT ANY KEY TO SAVE UNLIMITED UNLIMITED BYTES OF
DATA!!!"
INPUT a$
p# = 1
r# = 1
DO
c = INT(RND * 2) + 0
GOSUB getit:
PRINT p#, r#
INPUT a$
IF a$ = "s" THEN STOP
LOOP
getit:
FOR count = 1 TO 9999
test = back(count) = r# AND p(count) = p#
IF test = -1 THEN EXIT FOR
NEXT count
IF count = 10000 THEN STOP
IF c = 0 THEN p# = num1(count)
IF c = 1 THEN p# = num2(count)
r# = forward(count)
RETURN




"C:\Users\Reactor1967\vcns\work\RACCV.BAS"
r# = 1
a# = 1
b# = a# + (a# - r# + 1)
c# = a# + (a# - r# + 2)
CLS
RANDOMIZE TIMER
count# = 1
speed# = 0
DO
d = INT(RND * 2) + 0
storage# = a#
IF d = 0 THEN a# = b#
IF d = 1 THEN a# = c#
b# = a# + (a# - r# + 1)
c# = a# + (a# - r# + 2)
speed# = speed# + (a# - storage#)
PRINT a#; speed# / count#
count# = count# + 1
```

```
REM INPUT a$
a$ = INKEY$
IF a$ = "s" THEN STOP
LOOP UNTIL a# + (a# - r#) >= 999999999999999#




"C:\Users\Reactor1967\vcns\work\QUIRKY.BAS"
v# = 1
RANDOMIZE TIMER
CLS
DO
z = INT(RND * 2) + 0
store# = v#
v# = v# + 4
IF (v# / 2) - INT(v# / 2) = z THEN v# = v# + 0
IF (v# / 2) - INT(v# / 2) <> z THEN v# = v# + 1
COLOR 2, 0
dist# = v# - store#
dist# = (dist# * 2) - 1
r# = v# - dist#
PRINT , , z; v#; r#
INPUT a$
REM a$ = INKEY$
IF a$ = "s" THEN STOP
IF a$ = "d" THEN EXIT DO
LOOP
decode:
DO
PRINT v#; r#
INPUT a$
IF a$ = "s" THEN STOP
dist# = v# - r#
IF (dist# / 2) - INT(dist# / 2) = .5 THEN dist# = dist# - 1
dist# = dist# / 2
dist# = dist# + 1
store# = v#
v# = v# - dist#
r# = r# - dist#
LOOP




"C:\Users\Reactor1967\vcns\work\PUTIT2.BAS"
DECLARE SUB reduce (p#)
```

```
COMMON SHARED r1#, r2#, m#, c
p# = 1
r1# = 1
r2# = 999999999999999#
m# = 500000000000001#
CLS
RANDOMIZE TIMER
repeat:
count = 0
store# = p#
DO
PRINT p#; p# - store#; count
REM c = INT(RND * 2) + 0
c = 0
IF c = 0 THEN p# = p# + (p# - r1# + 1)
IF c = 1 THEN p# = p# + (p# - r1# + 2)
count = count + 1
LOOP UNTIL p# > m#
p# = p# + 1
PRINT p#; p# - store#; count
INPUT a$
IF a$ = "s" THEN STOP
IF a$ = "d" THEN GOTO decode:
CALL reduce(p#)
GOTO repeat:
decode:
store# = p#
PRINT p#; store# - p#
DO
c = (p# / 2) - INT(p# / 2)
IF c = 0 THEN p# = p# - ((p# - r1# + 1) / 2)
IF c >= .5 THEN EXIT DO
PRINT p#; store# - p#
INPUT a$
IF a$ = "s" THEN STOP
LOOP
p# = p# + m#
GOTO decode:

SUB Magic (p#, store#)
END SUB

SUB reduce (p#)
DO
p# = p# - (r2# - p# + 1)
LOOP UNTIL p# < m#
```

```
p# = p# + 1
END SUB




"C:\Users\Reactor1967\vcns\work\PUTIT.BAS"
COMMON SHARED r1#, r2#, m#, c
p# = 1
r1# = 1
r2# = 999999999999999#
m# = 500000000000000#
CLS
repeat:
count = 0
p# = p# + (p# - r1# + 2)
store# = p#
count = count + 1
IF p# > m# THEN GOTO skip:
IF p# > m# THEN PRINT p#
DO
PRINT p#
p# = p# + (p# - r1# + 1)
count = count + 1
LOOP UNTIL p# >= m#
p# = p# + 1
skip:
PRINT p#
DO
p# = p# - (r2# - p# + 1)
LOOP UNTIL p# < m#
IF count >= 3 THEN p# = p# + 1
INPUT a$
IF a$ = "s" THEN STOP
PRINT p#
GOTO repeat:

SUB Magic (p#, store#)
END SUB




"C:\Users\Reactor1967\vcns\work\P-R.BAS"
p1# = 3
p2# = 4
p3# = 5
r1# = 2
```

```
r2# = 3
r3# = 4
CLS
redo:
DO
PRINT , , r1#; "="; r2#; r3#
r1# = r1# + 1
r2# = r2# + 2
r3# = r3# + 2
REM INPUT a$
a$ = INKEY$
IF a$ = "s" THEN STOP
LOOP UNTIL r1# = 101
DO
PRINT , , r1#; "="; r2#; r3#
r1# = r1# - 2
r2# = r2# - 4
r3# = r3# - 4
REM INPUT a$
a$ = INKEY$
IF a$ = "s" THEN STOP
LOOP UNTIL r1# < 5
PRINT , , r1#; "="; r2#; r3#
GOTO redo:


"C:\Users\Reactor1967\vcns\work\POINTS.BAS"
p# = 1
p1# = 2
p2# = 3
p3# = 2
p4# = 3
CLS
RANDOMIZE TIMER
count = 0
DO
PRINT p#; p1#; p2#; count
c = INT(RND * 2) + 0
a1# = p1#
a2# = p2#
IF c = 0 THEN p# = p# + (p1# - p#)
IF c = 1 THEN p# = p# + (p2# - p#)
IF c = 0 THEN p1# = p1# + a1#
IF c = 0 THEN p2# = p2# + a1#
IF c = 1 THEN p1# = p1# + a2# + 1
```

```
IF c = 1 THEN p2# = p2# + a2# + 1
REM INPUT a$
a$ = INKEY$
IF a$ = "s" THEN STOP
count = count + 1
LOOP UNTIL p# + (p2# - p#) > 999999999999999#
count = 0
PRINT "------------------------"
DO
PRINT p#; p3#; p4#; count
c = INT(RND * 2) + 0
a1# = p3#
a2# = p4#
st# = p#
IF c = 0 THEN p# = p# - (p# - p3#)
IF c = 1 THEN p# = p# - (p# - p4#)
IF c = 0 THEN p3# = p3# + a1#
IF c = 0 THEN p4# = p4# + a1#
IF c = 1 THEN p3# = p3# + a2# + 1
IF c = 1 THEN p4# = p4# + a2# + 1
INPUT a$
REM a$ = INKEY$
IF a$ = "s" THEN STOP
count = count + 1
LOOP UNTIL p# + (p2# - p#) > 999999999999999#


"C:\Users\Reactor1967\vcns\work\PLAN3.BAS"
CLS
RANDOMIZE TIMER
ma# = 0
mb# = 0
mc# = 0
over:
a# = 1
b# = 1
c# = 3
redo:
DO
d = INT(RND * 2) + 0
IF d = 0 THEN st# = b#
IF d = 1 THEN st# = c#
distance# = st# - a#
a# = a# + distance#
```

```
b# = b# + (distance# * 2)
c# = c# + (distance# * 2)
PRINT a#; b#; c#
REM INPUT a$
a$ = INKEY$
IF a$ = "s" THEN STOP
LOOP UNTIL b# + ((b# - a#) * 2) >= 999999999999999#
d = INT(RND * 2) + 0
IF d = 0 THEN st# = b#
IF d = 1 THEN st# = c#
distance# = st# - a#
a# = a# + distance#
PRINT a#
distance# = 999999999999999# - a#
a# = 999999999999999# - distance#
b# = 999999999999999# - distance#
c# = 999999999999997# - distance#
REM INPUT a$
a$ = INKEY$
IF a$ = "s" THEN STOP
DO
d = INT(RND * 2) + 0
IF d = 0 THEN st# = b#
IF d = 1 THEN st# = c#
distance# = a# - st#
a# = a# - distance#
b# = b# - (distance# * 2)
c# = c# - (distance# * 2)
PRINT a#; b#; c#
REM INPUT a$
a$ = INKEY$
IF a$ = "s" THEN STOP
LOOP UNTIL b# - ((a# - b#) * 2) < 0
d = INT(RND * 2) + 0
IF d = 0 THEN st# = b#
IF d = 1 THEN st# = c#
distance# = a# - st#
a# = a# - distance#
PRINT a#
distance# = a# - 1
a# = 1 + distance#
b# = 1 + distance#
c# = 3 + distance#
GOTO redo:
```

```
"C:\Users\Reactor1967\vcns\work\PLAN2.BAS"
a# = 1
b# = 1
c# = 3
CLS
REM GOTO chart:
DO
PRINT a#; b#; c#
d = INT(RND * 2) + 0
IF d = 0 THEN st# = b#
IF d = 1 THEN st# = c#
dt# = st# - a#
a# = a# + dt#
b# = b# + (dt# * 2)
c# = c# + (dt# * 2)
REM INPUT a$
IF a$ = "s" THEN STOP
LOOP UNTIL a# >= 1000
SYSTEM




chart:
DO
PRINT a#; b#; c#
a# = a# + 2
b# = b# + 4
c# = c# + 4
INPUT a$
IF a$ = "s" THEN STOP
LOOP



"C:\Users\Reactor1967\vcns\work\PLAN.BAS"
REM this chart is for building a base two numerical chart to
REM generate these distances from a to b as vectors in and of
REM themselfs. Doing this as a base two chart will enable a base
REM two chart to be used with a base three chart. The base three
```

```
REM chart will be for generating vectors. The base two chart will
REM be for  generating coordinates. The third value in the base
REM three chart will be used for telling then to code up or down
REM in the base two chart. The base 3 chart will in essence be
REM actually coding binary. The base two charts vectors will be
REM used as actual  distances from a to b in the base three
REM chart. The base three chart will use numerical ranges to
REM tell when to flip from - to + and vise versa with the base
REM three charts third value as said will be used to tell when
REM to flip the base two chart.
CLS
a# = 1
b# = 2
c# = 3
d# = 4
DO
PRINT "a#"; a#; "<"; b# - a#; ">"; b#; c#; d#
a# = a# + 1
b# = b# + 3
c# = c# + 3
d# = d# + 3
INPUT a$
IF a$ = "s" THEN STOP
LOOP




"C:\Users\Reactor1967\vcns\work\PCHART.BAS"
DECLARE SUB ot (r1#(), p#(), r2#(), crd#())
DECLARE SUB Parsing (r1#(), p#(), r2#(), crd#())
REM this program creats a chart that shows what r that p came
REM from and what r p goes to. It can be used for coding or
REM decoding p. Its the latest and greatest in my work on vcns.
DECLARE SUB rcalc (a#, rt#)
c# = 1
r# = 1
rt# = 1
a# = 1
DIM r1#(1000)
DIM p#(1000)
DIM r2#(1000)
DIM crd#(1000)
RANDOMIZE TIMER
CLS
PRINT "Initilizing VCNS......"
x = 1
```

```
FOR count = 1 TO 500
a# = c#
CALL rcalc(a#, rt#)
r# = rt#
r1#(x) = r#
a# = a# + (a# - r# + 1)
p#(x) = a#
CALL rcalc(a#, rt#)
r# = rt#
r2#(x) = r#
x = x + 1
a# = c#
CALL rcalc(a#, rt#)
r# = rt#
r1#(x) = r#
a# = a# + (a# - r# + 2)
p#(x) = a#
CALL rcalc(a#, rt#)
r# = rt#
r2#(x) = r#
c# = c# + 1
x = x + 1
NEXT count
PRINT "Parsing VCNS...."
CALL Parsing(r1#(), p#(), r2#(), crd#())
REM CALL ot(r1#(), p#(), r2#(), crd#())
PRINT "VCNS IS POWERED UP!!!!!!!"
INPUT a$
IF a$ = "s" THEN STOP
v = 1
DO
a# = p#(v)
r# = r2#(v)
z = INT(RND * 2) + 0
IF z = 0 THEN a# = a# + (a# - r# + 1)
IF z = 1 THEN a# = a# + (a# - r# + 2)
count = v
DO
count = count + 1
IF count = 1001 THEN GOTO decode:
test = (p#(count) = a#) AND (r1#(count) = r#) AND (crd#(count) = v)
p$ = INKEY$
IF p$ = "s" THEN STOP
LOOP UNTIL test = -1
v = count
PRINT r1#(v); p#(v); r2#(v); crd#(v)
```

```
INPUT a$
IF a$ = "s" THEN STOP
LOOP
decode:
PRINT "Decoding...."
redo2:
PRINT r1#(v); p#(v); r2#(v); crd#(v)
v = crd#(v)
INPUT a$
IF a$ = "s" THEN STOP
IF p#(v) = 0 THEN SYSTEM
GOTO redo2:


SUB ot (r1#(), p#(), r2#(), crd#())
FOR count = 1 TO 1000
PRINT r1#(count); p#(count); r2#(count), crd#(count)
INPUT a$
IF a$ = "s" THEN STOP
NEXT count
END SUB

SUB Parsing (r1#(), p#(), r2#(), crd#())
position1 = 1
redo:
a# = p#(position1)
r# = r2#(position1)
a# = a# + (a# - r# + 1)
count = position1
DO
count = count + 1
test = (p#(count) = a#) AND r1#(count) = r#
IF test = -1 THEN EXIT DO
IF count = 1000 THEN EXIT SUB
LOOP
crd#(count) = position1
a# = p#(position1)
r# = r2#(position1)
a# = a# + (a# - r# + 2)
count = position1
DO
count = count + 1
test = (p#(count) = a#) AND r1#(count) = r#
IF test = -1 THEN EXIT DO
IF count = 1000 THEN EXIT SUB
LOOP
```

```
crd#(count) = position1
position1 = position1 + 1
IF position1 = 1000 THEN EXIT SUB
a$ = INKEY$
IF a$ = "s" THEN STOP
GOTO redo:
END SUB

SUB rcalc (a#, rt#)
t# = 0
rt# = 1
DO
FOR count = 1 TO 10
IF t# = a# THEN EXIT FOR
t# = t# + 1
NEXT count
IF t# = a# THEN EXIT DO
rt# = rt# + 10
LOOP
END SUB




"C:\Users\Reactor1967\vcns\work\OMEGA4.BAS"
a# = 10001
r1# = 10001
r2# = 10000
RANDOMIZE TIMER
CLS
DO
PRINT , , z; a#; r1#; r2#
z = INT(RND * 2) + 0
dist# = a# - r#
IF z = 0 THEN a# = a# + (a# - r# + 1)
IF z = 1 THEN a# = a# + (a# - r# + 2)
r1# = r1# + dist#
r2# = r2# + dist#
test1 = (r1# / 2) - INT(r1# / 2)
test2 = (r2# / 2) - INT(r2# / 2)
IF test1 >= .5 THEN r# = r1#
IF test2 >= .5 THEN r# = r2#
REM INPUT a$
a$ = INKEY$
IF a$ = "s" THEN STOP
LOOP
```

```
"C:\Users\Reactor1967\vcns\work\OMEGA3.BAS"
d# = 1
v# = 100
r# = v# - d#
RANDOMIZE TIMER
CLS
redo:
DO
PRINT v#; v# + (v# - r# + 1); v# + (v# - r# + 2); d#
z = INT(RND * 2) + 0
IF z = 0 THEN v# = v# + (v# - r# + 1)
IF z = 1 THEN v# = v# + (v# - r# + 2)
d# = d# + 1
r# = v# - d#
REM INPUT a$
a$ = INKEY$
IF a$ = "s" THEN STOP
LOOP UNTIL d# >= 250
```

```
"C:\Users\Reactor1967\vcns\work\OMEGA2.BAS"
r# = 1
v1# = 1
v2# = (r# * 2) + 1
v3# = (r# * 2) + 2
v4# = (r# * 2) + 3
RANDOMIZE TIMER
CLS
redo:
DO
PRINT v1#; v2#; v3#; v4#; r#
z = INT(RND * 2) + 0
IF z = 0 THEN v1# = v2#
IF z = 1 THEN v1# = v3#
IF r# >= 1000 THEN v1# = v#
r# = r# + 2
v2# = v1# + (r# * 2) + 1
v3# = v1# + (r# * 2) + 2
v4# = v1# + (r# * 2) + 3
a$ = INKEY$
REM INPUT a$
IF a$ = "s" THEN STOP
```

```
REM IF r# >= 1000 THEN EXIT DO
LOOP
REM DO
PRINT v1#; v2#; v3#; v4#; r#
find# = (r# * 2) + 3
test1 = v2# - find# = v1#
test2 = v3# - find# = v1#
test3 = v4# - find# = v1#
IF test1 = -1 THEN n = 1
IF test2 = -1 THEN n = 2
IF test3 = -1 THEN n = 3
```

```
"C:\Users\Reactor1967\vcns\work\OMEGA1.BAS"
DECLARE SUB findd (r#, a#)
CLS
RANDOMIZE TIMER
r# = 1
a# = 1
x = 0
b# = a# + ((a# - r#) * 3) + 1
c# = a# + ((a# - r#) * 3) + 2
d# = a# + ((a# - r#) * 3) + 3
e# = a# + ((a# - r#) * 3) + 4
DO
PRINT a#; b#; c#; d#; e#; a# - r#; r#
z = INT(RND * 3) + 0
IF a# + ((a# - r#) * 3) + 4 >= 50 THEN z = 3
test = (a# + ((a# - r#) * 3) + 4 >= 1000) AND ((a# + ((a# - r#) * 3) + 4) <=
2000)
IF z = 0 THEN a# = b#
IF z = 1 THEN a# = c#
IF z = 2 THEN a# = d#
IF z = 3 THEN a# = e#
IF test = -1 THEN CALL findd(r#, a#)
b# = a# + ((a# - r#) * 3) + 1
c# = a# + ((a# - r#) * 3) + 2
d# = a# + ((a# - r#) * 3) + 3
e# = a# + ((a# - r#) * 3) + 4
INPUT a$
REM a$ = INKEY$
IF a$ = "s" THEN STOP
```

```
LOOP

SUB findd (r#, a#)
a1# = 2000
b1# = 2000
dist1# = a# - r#
FOR count = 1 TO 1000
IF dist1# = a1# THEN EXIT FOR
a1# = a1# - 1
b1# = b1# - 2
NEXT count
dist2# = b1#
REM r# = a# - dist2#
a# = a# + dist2#
r# = a# - dist2#
END SUB




"C:\Users\Reactor1967\vcns\work\ODDCHRT.BAS"
a# = 1
b# = 3
c# = 5
CLS
DO
PRINT a#; b#; c#
a# = a# + 2
b# = b# + 4
c# = c# + 4
INPUT a$
IF a$ = "s" THEN STOP
LOOP




"C:\Users\Reactor1967\vcns\work\NOTAG.BAS"
p1# = 12501
p2# = 2
CLS
RANDOMIZE TIMER
redo:
DO
c = INT(RND * 2) + 0
store# = p1#
IF c = 0 THEN p1# = p1# - (p2# + 1)
IF c = 1 THEN p1# = p1# - (p2# + 2)
```

```
add# = store# - p1#
p2# = p2# + add#
PRINT , , p1#; p2#
a$ = INKEY$
IF a$ = "s" THEN STOP
LOOP UNTIL p2# >= p1#
PRINT , , "----------------"
DO
c = INT(RND * 2) + 0
store# = p2#
IF c = 0 THEN p2# = p2# - (p1# + 1)
IF c = 1 THEN p2# = p2# - (p1# + 2)
add# = store# - p2#
p1# = p1# + add#
PRINT , , p2#; p1#
a$ = INKEY$
IF a$ = "s" THEN STOP
LOOP UNTIL p1# >= p2#
PRINT , , "----------------"
GOTO redo:


"C:\Users\Reactor1967\vcns\work\NEWVCNS.BAS"
CLS
RANDOMIZE TIMER
p# = 1
r# = 1
count = 1
store = count
DO
PRINT p#; r#; "count ="; count; store
c = INT(RND * 2) + 0
IF c = 1 THEN p# = p# + (p# - r# + 2)
IF c = 0 THEN p# = p# + (p# - r# + 1)
IF r# + count < p# THEN store = count
IF r# + count < p# THEN r# = r# + count
count = count + 1
REM a$ = INKEY$
INPUT a$
IF a$ = "s" THEN STOP
LOOP UNTIL p# >= 500000000000000#
PRINT p#; r#; "count ="; count
```

```
"C:\Users\Reactor1967\vcns\work\NEWEQ2.BAS"
REM this program is for studing the development of new equations.
REM even though this chart was intended for different purposes than
REM usual it still has the same equation as base three even though
REM it was being used as base two
REM base 4
REM v# = v# + ((v# - r#) * 2) + N
REM Except here b1 and b2 are even and odd divided by two are odd.
REM             b3 and b4 are even and odd divided by two are even
REM this gives us a duel purpose of coding two bits of binary data per
REM calculation of v#
REM thing to figure out is how and when to increment R with more
REM than base two equations using the same method over and over
REM and coding a binary 1 and 0 in the second bit for system
REM data and coding binary data in the first bit.
v2# = 3
v1# = 2
v3# = 4
v4# = 5
v# = 1
CLS
DO
PRINT ; v#; " = "; v1#; v1# / 2; "|"; v2#; v2# / 2; "|"; v3#; v3# / 2; "|"; v4#; v4#
/ 2
a$ = INKEY$
INPUT a$
IF a$ = "s" THEN STOP
v# = v# + 1
v1# = v1# + 4
v2# = v2# + 4
v3# = v3# + 4
v4# = v4# + 4
LOOP




"C:\Users\Reactor1967\vcns\work\NEWEQ.BAS"
RANDOMIZE TIMER
p# = 1
b = 2
CLS
r# = 1
redo:
DO
c = INT(RND * 3) + 1
store# = p#
```

```
p# = p# + ((p# - r#) * b) + c
IF p# >= 999999999999999# THEN EXIT DO
a$ = INKEY$
IF a$ = "s" THEN STOP
PRINT , , p#; c
LOOP UNTIL p# + (p# - r#) >= 999999999999999#
IF p# >= 999999999999999# THEN p# = store#
PRINT , "------------------------------------------------"
DO
p# = p# - 50000000000000#
a$ = INKEY$
IF a$ = "s" THEN STOP
LOOP UNTIL p# < 200000000000000#
GOTO redo:




"C:\Users\Reactor1967\vcns\work\NEW1.BAS"
CLS
v# = 99
r# = 1
dist# = 50
count1# = 1
count2# = 1
test1$ = "bank"
OPEN test1$ FOR OUTPUT AS #1
WRITE #1, count1#, r#, v#
CLOSE #1
DO
test1$ = "bank"
OPEN test1$ FOR INPUT AS #1
DO
INPUT #1, a#, b#, c#
stop$ = INKEY$
IF stop$ = "s" THEN STOP
LOOP UNTIL a# = count1#
CLOSE #1
count1# = count1# + 1
v# = c#
REM IF (r# / 2) - INT(r# / 2) = 0 THEN r# = r# - 1
r# = 1
redo:
DO
v1# = v# + (v# - r# + 1)
```

```
v2# = v# + (v# - r# + 2)
test = (v1# - r#) <= dist#
r# = r# + 2
IF test = -1 THEN EXIT DO
stop$ = INKEY$
IF stop$ = "s" THEN STOP
LOOP
IF r# > v# THEN dist# = dist# + 1
IF r# > v# THEN r# = 1
IF r# > v# THEN GOTO redo:
test1$ = "bank"
OPEN test1$ FOR INPUT AS #1
DO
INPUT #1, a#, b#, c#
test = ((v1# = c#) AND (r# = b#)) OR ((v2# = c#) AND (r# = b#))
IF test = -1 THEN EXIT DO
stop$ = INKEY$
IF stop$ = "s" THEN STOP
LOOP UNTIL a# = count2#
CLOSE #1
IF test = -1 THEN GOTO redo:
COLOR 2, 0
PRINT , , v#; r#; v1#; v2#; v# - r#
a$ = INKEY$
REM INPUT a$
IF a$ = "s" THEN STOP
count2# = count2# + 1
test1$ = "bank"
OPEN test1$ FOR APPEND AS #1
WRITE #1, count2#, r#, v1#
count2# = count2# + 1
WRITE #1, count2#, r#, v1#
CLOSE #1
test1$ = "vcns"
OPEN test1$ FOR OUTPUT AS #1
WRITE #1, count1#, count2#
CLOSE #1
LOOP UNTIL v# > 999


"C:\Users\Reactor1967\vcns\work\MODELFR.BAS"
REM Model Program for floating R. This is just a model of what
REM a floating R would look like in conjunction with a real vector
REM coding data chart. This is still on the drawing board. You have to
REM know when and how much to increment and decrement R, be able to flip
```

```
REM and have a second r2 ect....
REM Solution If R is to increment by two code a 1 into the next p.
REM If R is to increment by 1 then code a 0 into the next p.
REM And vice Versa.
p# = 1
r# = 1
c = 1
CLS
RANDOMIZE TIMER
count# = 1
DO
c = INT(RND * 2) + 0
IF c = 0 THEN p# = p# + (p# - r# + 1)
IF c = 1 THEN p# = p# + (p# - r# + 2)
IF c = 0 THEN r# = r# + 1
IF c = 1 THEN r# = r# + 2
PRINT , ; "R"; r#, "P"; p#, "C"; c; "Bits"; count#
PRINT
"
_____

_____"
a$ = INKEY$
REM INPUT a$
IF a$ = "s" THEN STOP
count# = count# + 1
LOOP UNTIL p# >= 999999999999999#




"C:\Users\Reactor1967\vcns\work\LOWHIGH.BAS"
a# = 100
CLS
REM this is used to spit the distance up for
REM v#-r# so can plot the distance to next R
REM if knowing how this distance is divided can
REM get my chart to decode. Use the distance to
REM v#-r2# to calculate the distance from v# - r1#.
REM this will help the distance to uniformly go
REM up or down which will help the chart to go
REM uniformly up or down.
DO
test = (a# / 2) - INT(a# / 2)
d# = a#
IF test = .5 THEN d# = d# - 1
```

```
d# = d# / 2
IF test = .5 THEN e# = d# + 1 ELSE e# = d#
COLOR 2, 0
PRINT , , a#; "low"; d#; "low or High"; e#; d# + e#
IF d# + e# <> a# THEN STOP
REM a$ = INKEY$
INPUT a$
IF a$ = "s" THEN STOP
a# = a# + 1
LOOP
```

```
"C:\Users\Reactor1967\vcns\work\LOOKAT.BAS"
r# = 1
v# = 1
CLS
RANDOMIZE TIMER
redo:
DO
z = INT(RND * 2) + 0
store# = v#
IF z = 0 THEN v# = v# + ABS(v# - r#) + 1
IF z = 1 THEN v# = v# + ABS(v# - r#) + 2
COLOR 2, 0
PRINT , , r#; v#; ABS(v# - r#); (v# - store#); z
r# = r# + 2
REM a$ = INKEY$
INPUT a$
REM IF a$ = "s" THEN EXIT DO
IF a$ = "s" THEN STOP
LOOP
DO
r# = r# - 2
dist# = ABS(v# - r#)
IF (dist# / 2) - INT(dist# / 2) = .5 THEN dist# = dist# - 1
dist# = dist# / 2
v# = v# - dist#
PRINT r#; v#
INPUT a$
IF a$ = "s" THEN STOP
LOOP
```

```
"C:\Users\Reactor1967\vcns\work\LG.BAS"
```

```
v# = 1
r# = 1
dt# = 1000
RANDOMIZE TIMER
CLS
redo:
DO
test = (v# + (v# - r#) >= (r# + 100))
IF test = 0 THEN z = 0
IF test = -1 THEN z = 1
REM z = INT(RND * 2) + 0
IF z = 0 THEN v# = v# + (v# - r# + 1)
IF z = 1 THEN v# = v# + (v# - r# + 2)
COLOR 2, 0
PRINT , , z; v#; r#; v# - r#
a$ = INKEY$
REM INPUT a$
IF a$ = "s" THEN STOP
IF test = -1 THEN r# = r# + 102
LOOP UNTIL (v# - r# <= 25) AND (((v# + (v# - r# + 1)) - r#) >= 25)
COLOR 2, 0
PRINT , , z; v#; r#; v# - r#
z = INT(RND * 2) + 0
IF z = 0 THEN v# = v# + (v# - r# + 1)
IF z = 1 THEN v# = v# + (v# - r# + 2)
COLOR 2, 0
PRINT , , z; v#; r#; v# - r#; "---------"; dt#
INPUT z$
IF z$ = "s" THEN STOP
GOTO redo:


"C:\Users\Reactor1967\vcns\work\LBSPEC4.BAS"
v# = 100
CLS
RANDOMIZE TIMER
DO
z = INT(RND * 2) + 0
rs# = r#
test = (v# / 2) - INT(v# / 2)
IF test = 0 THEN r# = v# - 25
IF test = .5 THEN r# = v# - 26
IF z = 0 THEN v# = v# + (v# - r# + 1)
IF z = 1 THEN v# = v# + (v# - r# + 2)
COLOR 2, 0
```

```
PRINT , , v#; r#; v# - r#; r# - rs#
REM a$ = INKEY$
INPUT a$
IF a$ = "s" THEN STOP
LOOP




"C:\Users\Reactor1967\vcns\work\LBSPEC3.BAS"
v# = 1
r# = 1
CLS
RANDOMIZE TIMER
DO
z = INT(RND * 2) + 0
test1 = (z = 0) AND (r# / 2) - INT(r# / 2) = 0
test2 = (z = 1) AND (r# / 2) - INT(r# / 2) = 0
test3 = (z = 0) AND (r# / 2) - INT(r# / 2) = .5
test4 = (z = 1) AND (r# / 2) - INT(r# / 2) = .5
IF test1 = -1 THEN v# = v# + (v# - r# + 2)
IF test2 = -1 THEN v# = v# + (v# - r# + 1)
IF test3 = -1 THEN v# = v# + (v# - r# + 1)
IF test4 = -1 THEN v# = v# + (v# - r# + 2)
COLOR 2, 0
PRINT , , z; v#; r#
IF test1 = -1 THEN r# = r# + 2
IF test2 = -1 THEN r# = r# + 1
IF test3 = -1 THEN r# = r# + 1
IF test4 = -1 THEN r# = r# + 2
REM IF (r# / 2) - INT(r# / 2) = 0 THEN r# = r# - 1
a$ = INKEY$
REM INPUT a$
IF a$ = "s" THEN STOP
LOOP




"C:\Users\Reactor1967\vcns\work\LBSPEC2.BAS"
v# = 5
r# = 1
RANDOMIZE TIMER
CLS
d2# = 0
DO
z = INT(RND * 2) + 0
vs# = v#
```

```
IF z = 0 THEN v# = v# + (v# - r# + 1)
IF z = 1 THEN v# = v# + (v# - r# + 2)
COLOR 2, 0
PRINT , , v#; r#; vs# - r#
INPUT a$
IF a$ = "s" THEN STOP
d# = v# - r#
IF z = 0 THEN d# = d# - 1
IF z = 1 THEN d# = d# - 2
d# = (d# / 2)
r# = (v# - d#)
LOOP
```


```
"C:\Users\Reactor1967\vcns\work\LBSPEC1.BAS"
v# = 5
r# = 1
CLS
RANDOMIZE TIMER
DO
z = INT(RND * 2) + 0
vs# = v#
IF z = 0 THEN v# = v# + (v# - r# + 1)
IF z = 1 THEN v# = v# + (v# - r# + 2)
COLOR 2, 0
PRINT , , z; ; v# - vs#; v#; vs# - r#; r#; r# - rs#; v# - r#
REM PRINT , , z; v#; r#
INPUT a$
REM a$ = INKEY$
IF a$ = "s" THEN STOP
d# = v# - r#
IF z = 0 THEN d# = d# - 1
IF z = 1 THEN d# = d# - 2
d# = d# / 2
rs# = r#
r# = v# - d#
LOOP
```


```
"C:\Users\Reactor1967\vcns\work\LB67402.BAS"
r# = 1
v# = 1
CLS
```

```
RANDOMIZE TIMER
d# = 0
DO
z = INT(RND * 2) + 0
IF z = 0 THEN v# = v# + (v# - r# + 1)
IF z = 1 THEN v# = v# + (v# - r# + 2)
PRINT v#; r#; d#; v# - r#; r# - rs#
INPUT a$
IF a$ = "s" THEN STOP
d# = d# + 1
rs# = r#
r# = v# - d#
LOOP




"C:\Users\Reactor1967\vcns\work\LB57402.BAS"
v# = 110
r# = 109
CLS
RANDOMIZE TIMER
COLOR 2, 0
PRINT , , v#; r#
DO
z = INT(RND * 2) + 0
test = (v# / 2) - INT(v# / 2)
rs# = r#
IF test = .5 THEN r# = v# - 4
IF test = 0 THEN r# = v# - 5
IF z = 0 THEN v# = v# + (v# - r# + 1)
IF z = 1 THEN v# = v# + (v# - r# + 2)
COLOR 2, 0
PRINT , , v#; r#; "vr"; v# - r#; "rs"; r# - rs#
INPUT a$
REM a$ = INKEY$
IF a$ = "s" THEN STOP
LOOP




"C:\Users\Reactor1967\vcns\work\LB47502.BAS"
v# = 100
CLS
RANDOMIZE TIMER
```

```
DO
z = INT(RND * 2) + 0
d# = 0
DO
d# = d# + 1
test = v# - d#
IF z = 0 THEN dist# = (d# * 2) + 1
IF z = 1 THEN dist# = (d# * 2) + 2
v1# = (v# - d#) + dist#
test2 = ((v1# - test) / 2) - INT((v1# - test) / 2) = .5
IF test2 = -1 THEN EXIT DO
a$ = INKEY$
IF a$ = "s" THEN STOP
LOOP


"C:\Users\Reactor1967\vcns\work\LB47402.BAS"
v# = 110
r# = 109
CLS
RANDOMIZE TIMER
COLOR 2, 0
PRINT , , v#; r#
dist# = v# - r#
DO
z = INT(RND * 2) + 0
test = (v# / 2) - INT(v# / 2)
d = z: IF d = 1 THEN d = .5
store1# = dist#
IF test <> d THEN dist# = dist# + 1 ELSE dist# = dist# + 2
IF z = 0 THEN v# = v# + (v# - r# + 1)
IF z = 1 THEN v# = v# + (v# - r# + 2)
COLOR 2, 0
PRINT , , v#; r#; "vr"; v# - r#; "rs"; r# - rs#
rs# = r#
r# = v# - dist#
INPUT a$
REM a$ = INKEY$
IF a$ = "s" THEN STOP
LOOP


"C:\Users\Reactor1967\vcns\work\LB37503.BAS"
v# = 100
```

```
r# = 75
cr# = 25
CLS
RANDOMIZE TIMER
DO
z = INT(RND * 2) + 0
r1# = r#
rs# = r#
DO
r1# = r1# + 2
IF z = 0 THEN v1# = v# + (v# - r1# + 1)
IF z = 1 THEN v1# = v# + (v# - r1# + 2)
dist# = r1# - r#
dist# = (dist# * 2) - 1
cr1# = r1# - dist#
IF (cr1# / 2) - INT(cr1# / 2) = 0 THEN cr1# = cr1# + 1
test1 = ((v1# - v#) - (cr1# - cr#) = 0) OR ((v1# - v#) - (cr1# - cr#) = 1) OR
((v1# - v#) - (cr1# - cr#) = 2)
IF test1 = -1 THEN EXIT DO
PRINT v1# - v#; cr1# - cr#
INPUT stop$
REM stop$ = INKEY$
IF stop$ = "s" THEN STOP
LOOP
r# = r1#
vs# = v#
IF z = 0 THEN v# = v# + (v# - r# + 1)
IF z = 1 THEN v# = v# + (v# - r# + 2)
PRINT v#; r#
INPUT a$
IF a$ = "s" THEN STOP
LOOP


"C:\Users\Reactor1967\vcns\work\LB37502.BAS"
v# = 100
r# = 99
rs# = 97
CLS
RANDOMIZE TIMER
DO
z = INT(RND * 2) + 0
rs# = r#
r# = v# - 3
```

```
REM r# = v# - (r# - rs#)
REM IF (v# / 2) - INT(v# / 2) = 0 THEN r# = v# - 3
REM IF (v# / 2) - INT(v# / 2) = .5 THEN r# = v# - 2
REM IF (r# / 2) - INT(r# / 2) = 0 THEN r# = r# - 1
vs# = v#
IF z = 0 THEN v# = v# + (v# - r# + 1)
IF z = 1 THEN v# = v# + (v# - r# + 2)
COLOR 2, 0
PRINT , , z; v#; r#; v# - ((v# - vs#) + (v# - r#)); v# - r#
INPUT a$
IF a$ = "s" THEN STOP
LOOP


"C:\Users\Reactor1967\vcns\work\LB37402.BAS"
v# = 110
r# = 75
CLS
RANDOMIZE TIMER
COLOR 2, 0
PRINT , , v#; r#
DO
z = INT(RND * 2) + 0
store2# = r#
r1# = r#
rs# = r#
DO
IF z = 0 THEN v1# = v# + (v# - r1# + 1)
IF z = 1 THEN v1# = v# + (v# - r1# + 2)
IF v1# - r1# < 11 THEN EXIT DO
a$ = INKEY$
REM INPUT a$
IF a$ = "s" THEN STOP
IF a$ = "d" THEN EXIT DO
r1# = r1# + 2
LOOP
r# = r1#
dist# = r# - store2#
dist# = (dist# * 2) - 1
cs# = cr#
cr# = r# - dist#
store1# = v#
vs# = v#
IF z = 0 THEN v# = v# + (v# - r# + 1)
IF z = 1 THEN v# = v# + (v# - r# + 2)
```

```
COLOR 2, 0
PRINT , , "d"; z; "vv"; v# - vs#; v#; r#; "vsr"; vs# - r#; "rr"; r# - rs#; "vr"; v# -
r#
INPUT a$
REM a$ = INKEY$
IF a$ = "s" THEN STOP
IF a$ = "d" THEN EXIT DO
IF a$ = "D" THEN EXIT DO

LOOP




"C:\Users\Reactor1967\vcns\work\LB27602.BAS"
v# = 1
r# = 1
CLS
RANDOMIZE TIMER
d# = 0
DO
z = INT(RND * 2) + 0
IF z = 0 THEN v# = v# + (v# - r# + 1)
IF z = 1 THEN v# = v# + (v# - r# + 2)
COLOR 2, 0
PRINT , , v#; r#; v# - r#; d#; r# - rs#
d# = d# + 1
rs# = r#
r# = v# - d#
REM IF (r# / 2) - INT(r# / 2) = 0 THEN d# = d# + 1
REM IF (r# / 2) - INT(r# / 2) = 0 THEN r# = v# - d#
REM a$ = INKEY$
INPUT a$
IF a$ = "s" THEN STOP
LOOP




"C:\Users\Reactor1967\vcns\work\LB27502.BAS"
v# = 100
r# = 95
CLS
RANDOMIZE TIMER
DO
z = INT(RND * 2) + 0
r1# = r#
vs# = v#
```

```
DO
r1# = r1# + 1
IF z = 0 THEN v1# = v# + (v# - r1# + 1)
IF z = 1 THEN v1# = v# + (v# - r1# + 2)
test1 = v1# - ((v1# - v#) + (v1# - r1#)) = r#
test2 = v1# - ((v1# - v#) + (v1# - r1#)) + 1 = r#
IF test1 = -1 THEN EXIT DO
IF test2 = -1 THEN EXIT DO
stop$ = INKEY$
IF stop$ = "s" THEN STOP
LOOP
rs# = r#
r# = r1#
IF (r# / 2) - INT(r# / 2) = 0 THEN r# = r# - 1
IF z = 0 THEN v# = v# + (v# - r# + 1)
IF z = 1 THEN v# = v# + (v# - r# + 2)
COLOR 2, 0
dist# = r# - rs#
dist# = (dist# * 2) - 1
crs# = cr#
cr# = r# - dist#
IF (cr# / 2) - INT(cr# / 2) = 0 THEN cr# = cr# + 1
PRINT , , z; v#; r#; v# - vs#; v# - r#; v# - rs#; cr#; cr# - crs#
INPUT a$
IF a$ = "s" THEN STOP
LOOP




"C:\Users\Reactor1967\vcns\work\LB27402.BAS"
REM Use example from template. Control v# - r#
REM before you code using z. Figure out what you
REM what a 0 or 1 then make decisions from there
REM how far to code the dist up or down between
REM v# - r# before you code:
CLS
v# = 2
r# = 1
RANDOMIZE TIMER
DO
z = INT(RND * 2) + 0
vs# = v#
dist# = v# - r#
IF z = 0 THEN v# = v# + (v# - r# + 1)
IF z = 1 THEN v# = v# + (v# - r# + 2)
```

```
COLOR 2, 0
PRINT , , z; v#; r#; "vsr"; vs# - r#; "vr"; v# - r#; "rr"; r# - rs#
INPUT a$
IF a$ = "s" THEN STOP
rs# = r#
r# = r# + dist#
IF (r# / 2) - INT(r# / 2) = 0 THEN r# = r# - 1
LOOP
decode:
```


```
"C:\Users\Reactor1967\vcns\work\LB17602.BAS"
CLS
v# = 1
r# = 1
d# = INT(v# / 100)
RANDOMIZE TIMER
DO
z = INT(RND * 2) + 0
IF z = 0 THEN v# = v# + (v# - r# + 1)
IF z = 1 THEN v# = v# + (v# - r# + 2)
COLOR 2, 0
PRINT , , v#; r#; r# - rs#; v# - r#
rs# = r#
IF (v# - r#) > 100 THEN r# = (v# - 25)
IF (r# / 2) - INT(r# / 2) = 0 THEN r# = (v# - 26)
a$ = INKEY$
REM INPUT a$
IF a$ = "s" THEN STOP
LOOP
```


```
"C:\Users\Reactor1967\vcns\work\LB17502.BAS"
v# = 100
r# = 75
CLS
RANDOMIZE TIMER
DO
dist# = v# - r#
test = (dist# / 2) - INT(dist# / 2)
IF test = .5 THEN dist# = dist# - 2
IF test = 0 THEN dist# = dist# - 1
IF (dist# / 2) - INT(dist# / 2) = .5 THEN x = 1 ELSE x = 0
IF x = 1 THEN dist# = dist# - 1
```

```
dist# = dist# / 2
a# = dist#
IF x = 1 THEN b# = a# + 1
IF x = 0 THEN b# = a#
test = (v# / 2) - INT(v# / 2) = (a# / 2) - INT(a# / 2)
rs# = r#
IF test = -1 THEN r# = v# - b#
IF test = 0 THEN r# = v# - a#
z = INT(RND * 2) + 0
vs# = v#
IF z = 0 THEN v# = v# + (v# - r# + 1)
IF z = 1 THEN v# = v# + (v# - r# + 2)
COLOR 2, 0
PRINT , , v#; r#; vs# - r#; v# - r#; r# - rs#
IF (r# / 2) - INT(r# / 2) = 0 THEN STOP
INPUT a$
REM a$ = INKEY$
IF a$ = "s" THEN STOP
LOOP
```

```
"C:\Users\Reactor1967\vcns\work\LB7402.BAS"
REM Use example from template. Control v# - r#
REM before you code using z. Figure out what you
REM what a 0 or 1 then make decisions from there
REM how far to code the dist up or down between
REM v# - r# before you code:
CLS
v# = 2
r# = 1
RANDOMIZE TIMER
DO
z = INT(RND * 2) + 0
vs# = v#
dist# = v# - r#
IF z = 0 THEN v# = v# + (v# - r# + 1)
IF z = 1 THEN v# = v# + (v# - r# + 2)
COLOR 2, 0
PRINT , , z; v#; r#; v# - r#; r# - rs#
REM a$ = INKEY$
INPUT a$
IF a$ = "s" THEN STOP
IF a$ = "d" THEN GOTO decode:
test1 = (v# / 2) - INT(v# / 2)
```

```basic
test2 = (dist# / 2) - INT(dist# / 2)
IF test1 = test2 THEN dist# = dist# + 1
rs# = r#
r# = v# - dist#
LOOP
decode:




"C:\Users\Reactor1967\vcns\work\LB2.BAS"
DECLARE SUB rangef1 (p#, r1#(), range!)
DECLARE SUB up (p#, r#)
REM in range2 and range3 put if p# < range then range = #
REM                    COPYRIGHT C 2002
REM                 Lloyd Dudley Burris
REM ----------------------Initilize R2#------------------------------------
CLS
DIM r1#(255)
DIM r2#(7999)
RANDOMIZE TIMER
r1#(1) = 1
REM PRINT r#1(1)
pt1# = 5
rf# = 100
count = 2
a# = 0
point1:
DO
a# = a# + pt1#
test = (a# / 2) - INT(a# / 2)
r1#(count) = a#
IF (r1#(count) / 2) - INT(r1#(count) / 2) = 0 THEN r1#(count) = r1#(count) + 1
count = count + 1
IF a$ = "s" THEN STOP
IF count >= 254 THEN EXIT DO
LOOP UNTIL a# = rf#
pt1# = pt1# * 10
rf# = rf# * 10
IF count >= 254 THEN GOTO point2:
IF a# + pt1# > 999999999999999# THEN GOTO point2:
GOTO point1:
point2:
r1#(254) = 950000000000001#
r1#(255) = 999999999999999#
REM --------------------------------------------------------------------
RANDOMIZE TIMER
```

```
repeat:
DO
p# = INT(RND * 999999999999999#) + 1
REM test = (p# > 425000000000001#) AND (p# < 450000000000000#)
CALL rangef1(p#, r1#(), range)
test = p# >= r1#(range) + 25000000000000#
a$ = INKEY$
 IF a$ = "s" THEN STOP
LOOP UNTIL test = -1
IF p# >= 950000000000000# THEN GOTO repeat:
CALL rangef1(p#, r1#(), range)
store# = range
PRINT "------------------------------"
REM p# = 125000004999999#
PRINT p#
store1# = p#
t# = p#
t# = t# - 25000000000000#
t# = 999999999999999# - t#
t# = FIX(t# / 10000000#)
t# = t# * 10000000#
t# = 999999999999999# - t#
store3# = t#
p# = p# + (p# - t# + 2)
CALL rangef1(p#, r1#(), range)
PRINT p#; p# - store1#; t#
PRINT "------------------------------"
a$ = INKEY$
IF a$ = "s" THEN STOP
IF p# > (r1#(range) + 25000000000000#) THEN STOP
REM IF range - 1 = store# THEN GOTO repeat:

SUB deflate (p#)
DO
p# = p# - 50000000000000#
LOOP UNTIL p# < 200000000000000#
END SUB

SUB down (p#, r1#)
END SUB

SUB inflate (p#, r1#(), range)
DO
p# = p# + 50000000000000#
LOOP UNTIL p# >= r1#(254) + 25000000000000#
END SUB
```

```
SUB rangef1 (p#, r1#(), range)
range = 255
DO
test = r1#(range) <= p#
IF test = -1 THEN EXIT DO
range = range - 1
LOOP UNTIL range = 0
IF range = 0 THEN PRINT "ERROR rangef failed"
IF range = 0 THEN SYSTEM
END SUB


SUB up (p#, r#)
END SUB




"C:\Users\Reactor1967\vcns\work\LB1.BAS"
REM goal of project to date is to create a vector chart
REM that uses both vectors and ranges.
DECLARE SUB rangef (p#, r#(), range!)
DECLARE SUB changeu (p#)
DECLARE SUB changed (p#)
REM                    COPYRIGHT C 2002
REM                 Lloyd Dudley Burris
REM ---------------------Initilize R#------------------------------------
CLS
DIM r#(255)
r#(1) = 1
REM PRINT r#(1)
pt1# = 5
rf# = 100
count = 2
a# = 0
replaced1:
DO
a# = a# + pt1#
test = (a# / 2) - INT(a# / 2)
r#(count) = a#
IF (r#(count) / 2) - INT(r#(count) / 2) = 0 THEN r#(count) = r#(count) + 1
PRINT r#(count); count
count = count + 1
INPUT a$
REM a$ = INKEY$
IF a$ = "s" THEN STOP
IF count >= 254 THEN EXIT DO
```

```
LOOP UNTIL a# = rf#
pt1# = pt1# * 10
rf# = rf# * 10
IF count >= 254 THEN GOTO replaced2:
IF a# + pt1# > 999999999999999# THEN GOTO replaced2:
GOTO replaced1:
replaced2:
r#(254) = 950000000000001#
r#(255) = 999999999999999#
REM ----------------------Code P# up.---------------------------------------
p# = 456789876543212#
RANDOMIZE TIMER
c = 0
repeat:
count = 0
silly = 0
DO
bt = (p# / 2) - INT(p# / 2)
IF bt >= .5 THEN bt = 1
COLOR 2, 0
CALL rangef(p#, r#(), range)
PRINT , p#; "+"; bt; r#(range); range
IF p# >= 750000000000000# THEN c = INT(RND * 2) + 0 ELSE c = 0
IF silly = 0 THEN c = 1
silly = 1
p# = p# + 22500000000000#
IF c = 0 THEN p# = p# + (p# - (r#(range)) + 1)
IF c = 1 THEN p# = p# + (p# - (r#(range)) + 2)
REM CALL changeu(p#)
IF a$ = "s" THEN STOP
test = (p# + p# - r#(range) + 22500000000000#) >= 800000000000000#
count = count + 1
LOOP UNTIL test = -1
bt = (p# / 2) - INT(p# / 2)
IF bt >= .5 THEN bt = 1
CALL rangef(p#, r#(), range)
PRINT , p#; "+"; bt; r#(range); range
INPUT a$
IF a$ = "s" THEN STOP
IF a$ = "d" THEN GOTO decode:
p# = p# - 400000000000000#
CALL rangef(p#, r#(), range)
GOTO repeat:
decode:
DO
bt = (p# / 2) - INT(p# / 2)
```

```
IF bt >= .5 THEN bt = 1
CALL rangef(p#, r#(), range)
PRINT , p#; "+"; bt; r#(range); range
CALL changed(p#)
CALL rangef(p#, r#(), range)
IF c = 0 THEN p# = p# - ((p# - (r#(range)) + 1) / 2)
IF c = 1 THEN p# = p# - ((p# - (r#(range)) + 2) / 2)
PRINT p#
STOP
LOOP

finish:
PRINT "YOUR A GOD LLOYD BURRIS"

SUB changed (p#)
store# = p#
store# = store# / 100000000000000#
d# = INT(store#)
store# = store# - d#
store# = store# * 10
store# = INT(store#)
test3 = store# >= 0 AND p# <= 4
IF test3 = -1 THEN GOTO test3:
test4 = store# >= 5 AND p# <= 9
GOTO test4:
test3:
p# = p# - 50000000000000#
EXIT SUB
test4:
p# = p# + 50000000000000#
PRINT p#
END SUB

SUB changeu (p#)
p# = 109843275844454#
STOP
store# = p#
store# = store# / 100000000000000#
d# = INT(store#)
store# = store# - d#
store# = store# * 10
store# = INT(store#)


END SUB
```

```
SUB rangef (p#, r#(), range)
range = 255
DO
test = r#(range) <= p#
IF test = -1 THEN EXIT DO
range = range - 1
a$ = INKEY$
IF a$ = "s" THEN STOP
LOOP UNTIL range = 0
IF range = 0 THEN PRINT "ERROR rangef failed"
IF range = 0 THEN SYSTEM
END SUB
```

```
"C:\Users\Reactor1967\vcns\work\LATER.BAS"
v# = 5
r# = 1
CLS
RANDOMIZE TIMER
d# = 1
DO
z = INT(RND * 2) + 0
IF z = 0 THEN v# = v# + (v# - r# + 1)
IF z = 1 THEN v# = v# + (v# - r# + 2)
COLOR 2, 0
PRINT , , d#; v#; r#; v# - r#
INPUT a$
IF a$ = "s" THEN STOP
r# = v# - d#
IF r# / 2 - INT(r# / 2) = 0 THEN d# = d# + 1
IF r# / 2 - INT(r# / 2) = 0 THEN r# = v# - d#
d# = d# + 1
LOOP
```

```
"C:\Users\Reactor1967\vcns\work\KILLME.BAS"
r# = 1
a# = 1
b# = 2
c# = 3
d# = 4
RANDOMIZE TIMER
CLS
PRINT a#; b#; c#; d#
```

```
DO
c = INT(RND * 3) + 0
IF c = 0 THEN a# = b#
IF c = 1 THEN a# = c#
IF c = 2 THEN a# = d#
b# = a# + ((a# - r#) * 2) + 1
c# = a# + ((a# - r#) * 2) + 2
d# = a# + ((a# - r#) * 2) + 3
a# = a# + ((a# - r#) * 2)
r# = a# - (ABS(a# - b#))
PRINT c; a#; b#; c#; d#; r#
REM INPUT a$
a$ = INKEY$
IF a$ = "s" THEN STOP
LOOP




"C:\Users\Reactor1967\vcns\work\KILL.BAS"
v# = 100
r# = 75
CLS
RANDOMIZE TIMER
z = INT(RND * 2) + 0
DO
IF z = 0 THEN v# = v# + (v# - r# + 1)
IF z = 1 THEN v# = v# + (v# - r# + 2)
COLOR 2, 0
PRINT , , v#; r#
INPUT a$
IF a$ = "s" THEN STOP
z = INT(RND * 2) + 0
d# = 1
DO
r1# = v# - d#
IF z = 0 THEN v1# = v# + (v# - r1# + 1)
IF z = 1 THEN v1# = v# + (v# - r1# + 2)
IF (v1# - v#) >= (r1# - r#) THEN EXIT DO
a$ = INKEY$
IF a$ = "s" THEN STOP
d# = d# + 1
IF ((v# - d#) / 2) - INT((v# - d#) / 2) = 0 THEN d# = d# + 1
LOOP
r# = r1#
LOOP
```

```
"C:\Users\Reactor1967\vcns\work\KEEPME2.BAS"
REM Here like binary numbers make the rate of acceleration of R go up
REM exception to the rule is  1 to a 0 makes the rate go down.
REM 0 to 1 makes the rate go up. Or a higher in value N to the same or
REM higher in value N rate goes up. to a higher to a lower in value N
REM makes the rate go down.
REM subtract dist# from R# to get previous R.
REM use previous r# like this (v# - r#) if odd subtract 1
REM ((v# - r#)/2) This gives the previous dist# up or down.
REM if going up add one to tell previous dist#. It will be even so
REM you may or will have to add one more if the dist# is odd.
REM dist# has to be even. (Hint - The dist# will always be + or - 2 or stay
REM the same.) If this is odd + or 1 one.
REM + 1
REM subtract from v to get previous v.
r# = 1
v# = 2
CLS
RANDOMIZE TIMER
DO
z = INT(RND * 2) + 0
store# = v#
IF z = 0 THEN v# = v# + (v# - r# + 1)
IF z = 1 THEN v# = v# + (v# - r# + 2)
COLOR 2, 0
PRINT , , v#; r#; dist#
dist# = v# - store#
IF (dist# / 2) - INT(dist# / 2) = .5 THEN dist# = dist# - 1
r# = r# + dist#
INPUT a$
REM a$ = INKEY$
IF a$ = "s" THEN STOP
LOOP




"C:\Users\Reactor1967\vcns\work\KEEPME!.BAS"
REM USE THE NUMBERS THAT REPEAT.
REM THEY ARE HOW MUCH YOU HAD AND
REM AND SUBTRACT TO GET TO THE NEXT
REM REFERENCE POINT. THEY WILL ALWAYS
REM BE EVEN BECAUSE R IS ODD.
REM THERE IS A PATTERN HERE.
```

```
REM DO NOT LOOSE THIS PROGRAM!!!.
REM subtract v#,ref# to tell if got data bit or code bit.
REM use code bit to tell if go down 2 or 4 on ref.
CLS
LOCATE 15, 25
COLOR 2, 0
REM PRINT "Non-linear numerical system demonstration."
REM INPUT a$
REM IF a$ = "s" THEN STOP
v# = 1
r# = 2
RANDOMIZE TIMER
CLS
DO
z = INT(RND * 2) + 0
h1# = v#
ref# = v# - r#
IF z = 0 THEN v# = v# + r# + 1
IF z = 1 THEN v# = v# + r# + 2
h2# = v#
test = (h1# / 2) - INT(h1# / 2) = (h2# / 2) - INT(h2# / 2)
IF test = -1 THEN r# = r# + 2 ELSE r# = r# + 1
PRINT , , v#; ref#; (v# - r#) - ref#; r#
INPUT a$
REM a$ = INKEY$
IF a$ = "s" THEN STOP
LOOP




"C:\Users\Reactor1967\vcns\work\JUMP3.BAS"
a# = 1
b# = 2
c# = 3
j# = 1
CLS
DO
PRINT a#; b#; c#; ABS(a# - b#)
z = INT(RND * 2) + 0
IF z = 0 THEN st# = b#
IF z = 1 THEN st# = c#
over:
s1# = a#
s2# = b#
s3# = c#
count = 0
```

```
x = 0
DO
a# = a# + 1
b# = b# + 2
c# = c# + 2
IF count > 125 THEN x = 1
IF x = 1 THEN EXIT DO
a$ = INKEY$
IF a$ = "s" THEN STOP
LOOP UNTIL a# = st#
IF x = 1 THEN st# = s1# + 100
IF x = 1 THEN b# = s2#
IF x = 1 THEN c# = s3#
IF x = 1 THEN GOTO over:
distance# = a# - j#
IF distance# >= 100 THEN flip = 1
IF flip = 1 THEN j# = j# + 200
IF flip = 1 THEN a# = a# + 100
flip = 0
INPUT a$
IF a$ = "s" THEN STOP
IF a$ = "d" THEN GOTO decode:
LOOP
decode:
DO
REM PRINT a#; b#; c#
s1# = a#
s2# = b#
s3# = c#
over2:
count = 0
st# = a#
x = 0
DO
REM PRINT a#; b#; c#
a# = a# - 1
b# = b# - 2
c# = c# - 2
a$ = INKEY$
IF a$ = "s" THEN STOP
count = count + 1
IF count > 125 THEN EXIT DO
IF st# = b# THEN EXIT DO
IF st# = c# THEN EXIT DO
LOOP
IF count > 125 THEN x = 1
```

```
IF x = 1 THEN a# = s1# - 100
IF x = 1 THEN b# = s2#
IF x = 1 THEN c# = s3#
IF x = 1 THEN GOTO over2:
PRINT a#; b#; c#
IF a# = 1 THEN PRINT "Decode Successful!"
IF a# < 0 THEN PRINT "System Crash!!!"
IF a# = 1 THEN SYSTEM
IF a# = 0 THEN SYSTEM
INPUT a$
IF a$ = "s" THEN STOP
LOOP
```

```
"C:\Users\Reactor1967\vcns\work\JUMP2.BAS"
a# = 1
b# = 2
c# = 3
j# = 100
CLS
RANDOMIZE TIMER
DO
dt = INT(RND * 2) + 0
IF dt = 0 THEN s# = b#
IF dt = 1 THEN s# = c#
d# = s# - a#
a# = a# + d#
b# = b# + (d# * 2)
c# = c# + (d# * 2)
IF a# >= j# THEN flip = 1
IF flip = 1 THEN a# = a# + 100
IF flip = 1 THEN j# = j# + 200
PRINT a#; b#; c#, j#, flip, ABS(a# - j#)
flip = 0
REM INPUT a$
a$ = INKEY$
IF a$ = "s" THEN STOP
LOOP
```

```
"C:\Users\Reactor1967\vcns\work\JUMP.BAS"
a# = 1
b# = 2
c# = 3
```

```
j# = 100
CLS
DO
IF flip = 0 THEN PRINT j#, a#; b#; c#
IF flip = 1 THEN PRINT j#, a#; b#; c#; "Jump"
flip = 0
a# = a# + 1
b# = b# + 2
c# = c# + 2
IF a# >= j# THEN flip = 1
IF flip = 1 THEN a# = a# + 100
IF flip = 1 THEN j# = j# + 200
INPUT a$
REM a$ = INKEY$
IF a$ = "s" THEN STOP
LOOP




"C:\Users\Reactor1967\vcns\work\JMPPNT.BAS"
RANDOMIZE TIMER
a# = 1
b# = 2
c# = 3
d# = 4
carry# = 1
CLS
track# = 3
DO
z = INT(RND * 3) + 0
test1 = (z = 0) AND (carry# = a#)
test2 = (z = 1) AND (carry# = a#)
test3 = (z = 2) AND (carry# = a#)
test4 = (carry# = a#)
IF test1 = -1 THEN carry# = b#
IF test2 = -1 THEN carry# = c#
IF test3 = -1 THEN carry# = d#
IF test4 = -1 THEN track# = carry# - a#
PRINT a#; b#; c#; d#; carry#; "next Jump="; track#
PRINT carry# - a#
REM INPUT a$
a$ = INKEY$
IF a$ = "s" THEN STOP
a# = a# + 1
REM test5 = (carry# = a#) AND (track# >= 81)
REM IF test5 = -1 THEN r# = r# + 162
```

```
b# = a# + ((a# - r#) * 2) + 1
c# = a# + ((a# - r#) * 2) + 2
d# = a# + ((a# - r#) * 2) + 3
LOOP




"C:\Users\Reactor1967\vcns\work\ITWORKS.BAS"
DECLARE SUB rangef1 (P#, r1#(), range!)
DECLARE SUB deflate (P#, s#)
REM in range2 and range3 put if p# < range then range = #
REM                    COPYRIGHT C 2002
REM                 Lloyd Dudley Burris
REM ---------------------Initilize R2#------------------------------------
CLS
DIM r1#(255)
DIM r2#(7999)
RANDOMIZE TIMER
r1#(1) = 1
REM PRINT r#1(1)
pt1# = 5
rf# = 100
count = 2
a# = 0
point1:
DO
a# = a# + pt1#
test = (a# / 2) - INT(a# / 2)
r1#(count) = a#
IF (r1#(count) / 2) - INT(r1#(count) / 2) = 0 THEN r1#(count) = r1#(count) + 1
count = count + 1
IF a$ = "s" THEN STOP
IF count >= 254 THEN EXIT DO
LOOP UNTIL a# = rf#
pt1# = pt1# * 10
rf# = rf# * 10
IF count >= 254 THEN GOTO point2:
IF a# + pt1# > 999999999999999# THEN GOTO point2:
GOTO point1:
point2:
r1#(254) = 950000000000001#
r1#(255) = 999999999999999#
REM ------------------------------------------------------------------------
RANDOMIZE TIMER
P# = 135984323123456#
repeat:
```

```
P# = P# / 10000000
PRINT P#
P# = INT(P#)
PRINT P#


SUB deflate (P#, s#)
DO
P# = P# - 50000000000000#
LOOP UNTIL P# < 150000000000000#
DO
s# = s# - 50000000000000#
LOOP UNTIL s# < 150000000000000#
END SUB

SUB rangef1 (P#, r1#(), range)
range = 255
DO
test = r1#(range) <= P#
IF test = -1 THEN EXIT DO
range = range - 1
LOOP UNTIL range = 0
IF range = 0 THEN PRINT "ERROR rangef failed"
IF range = 0 THEN SYSTEM
END SUB



"C:\Users\Reactor1967\vcns\work\INTR2.BAS"
DIM r2#(37)
p# = 100000000000001#
CLS
count = 1
DO
r2#(count) = p#
PRINT p#; count
p# = p# + 25000000000000#
a$ = INKEY$
IF a$ = "s" THEN STOP
count = count + 1
LOOP UNTIL p# >= 999999999999999#
r2#(count) = 999999999999999#



"C:\Users\Reactor1967\vcns\work\INSERTD.BAS"
```

```basic
v# = 1
r# = 1
CLS
count = 0
redo:
v# = v# + (v# - r# + 2)
COLOR 2, 0
PRINT , , v#
DO
v# = v# + (v# - r# + 1)
count = count + 1
COLOR 2, 0
PRINT , , v#
LOOP UNTIL v# >= 100
r# = 1 + v#
z = INT(RND * 2) + 0
IF z = 0 THEN v# = v# - 2
IF z = 1 THEN v# = v# - 3
COLOR 2, 0
PRINT , , "Data"; z; v#; count
REM a$ = INKEY$
INPUT a$
IF a$ = "s" THEN STOP
count = 0
v# = v# - (r# - v# + 2)
COLOR 2, 0
PRINT , , v#
DO
v# = v# - (r# - v# + 1)
count = count + 1
COLOR 2, 0
PRINT , , v#; count
LOOP UNTIL v# - (r# - v#) <= 0
REM a$ = INKEY$
INPUT a$
IF a$ = "s" THEN STOP
REM r# = v# + 1
r# = 1
GOTO redo:


"C:\Users\Reactor1967\vcns\work\INITR.BAS"
DIM r#(50)
r1# = 999999999999999#
count = 1
```

```
DO
r#(count) = r1#
PRINT r1#; count
r1# = INT(r1# / 2)
c = (r1# / 2) - INT(r1# / 2)
IF c = 0 THEN r1# = r1# + 1
INPUT a$
IF a$ = "s" THEN STOP
count = count + 1
LOOP UNTIL count = 51


"C:\Users\Reactor1967\vcns\work\INCREASE.BAS"
r1# = 1
v# = 1
CLS
RANDOMIZE TIMER
x = 0
count = 0
DO
z = INT(RND * 2) + 0
store# = v#
IF z = 0 THEN v# = v# + (v# - r1# + 1)
IF z = 1 THEN v# = v# + (v# - r1# + 2)
PRINT v#; r1#
IF x = 1 THEN r1# = (r1# * 2) + 1
IF x = 0 THEN x = 1 ELSE x = 0
REM INPUT a$
a$ = INKEY$
IF a$ = "s" THEN STOP
count = count + 1
LOOP UNTIL v# + (v# - r#) >= 999999999999999#


"C:\Users\Reactor1967\vcns\work\INCR3.BAS"
p# = 150000000000001#
r# = p# - 25000000000000#
CLS
RANDOMIZE TIMER
DO
PRINT p#; r#; p# - r#; z
c = INT(RND * 2) + 0
IF c = 0 THEN p# = p# + (p# - r# + 1)
```

```
IF c = 1 THEN p# = p# + (p# - r# + 2)
r# = r# + 25000000000000#
REM INPUT a$
a$ = INKEY$
IF a$ = "s" THEN STOP
LOOP UNTIL r# > 925000000000000#




"C:\Users\Reactor1967\vcns\work\INCR2.BAS"
p# = 125000000000001#
r1# = 100000000000001#
r2# = 100000000000000#
CLS
RANDOMIZE TIMER
v1# = 25000000000000#
redo:
DO
c = INT(RND * 2) + 0
IF (r1# / 2) - INT(r1# / 2) >= .5 THEN r# = r1#
IF (r2# / 2) - INT(r2# / 2) >= .5 THEN r# = r2#
PRINT p#; "pr"; (p# - r#); "pstr"; p# - store#; "vrte"; v1# + rate#
store# = p#
IF c = 0 THEN p# = p# + (p# - r# + 1)
IF c = 1 THEN p# = p# + (p# - r# + 2)
IF (r1# / 2) - INT(r1# / 2) >= .5 THEN r# = r1#
IF (r2# / 2) - INT(r2# / 2) >= .5 THEN r# = r2#
r1# = r1# + v1#
r2# = r2# + v2# mmmm 7/8899+6
INPUT a$
IF a$ = "s" THEN STOP
LOOP




"C:\Users\Reactor1967\vcns\work\GENISUS6.BAS"
REM Experimenting with using both v# & R# in equations. This
REM seems promising. L.B.
REM Go research some geometery equations. See if can apply something to
this.
v# = 3
r# = 1
CLS
RANDOMIZE TIMER
PRINT , , v#; r#
```

```
DO
z = INT(RND * 2) + 0
IF z = 0 THEN v# = v# + (v# - r# + 1)
IF z = 1 THEN v# = v# + (v# - r# + 2)
IF z = 0 THEN r# = (v# - r# - 2)
IF z = 1 THEN r# = (v# - r# - 1)
COLOR 2, 0
PRINT , v#; r#; v# - r#
a$ = INKEY$
REM INPUT a$
IF a$ = "s" THEN STOP
LOOP UNTIL v# + (v# - r#) >= 999999999999999#
```

```
"C:\Users\Reactor1967\vcns\work\GENISUS5.BAS"
v# = 1
r# = 1
CLS
RANDOMIZE TIMER
count = 0
DO
z = INT(RND * 2) + 0
IF z = 0 THEN v# = v# + (v# - r# + 1)
IF z = 1 THEN v# = v# + (v# - r# + 2)
IF z = 0 THEN r# = r# - (r# - v# + 1)
IF z = 1 THEN r# = r# - (r# - v# + 2)
COLOR 2, 0
REM LOCATE 13, 7
REM PRINT "Bit stored"; z; "|Vectors"; v#; "|Coordinates"; r#; "|Total bytes
stored"; INT(count / 4)
PRINT z; v#; r#
REM a$ = INKEY$
INPUT a$
IF a$ = "s" THEN STOP
count = count + 1
LOOP
```

```
"C:\Users\Reactor1967\vcns\work\GENISUS4.BAS"
v# = 50
b# = 1
CLS
```

```
RANDOMIZE TIMER
DO
z = INT(RND * 2) + 0
IF z = 0 THEN v# = v# + b# + 1
IF z = 1 THEN v# = v# + b# + 2
COLOR 2, 0
m$ = STR$(v#)
lth = LEN(m$) - 1
PRINT , , z; v#; b#; lth
a$ = INKEY$
REM INPUT a$
IF a$ = "s" THEN STOP
REM IF a$ = "d" THEN GOTO redo:
store# = b#
count = 1
DO
b# = b# + 25
t1 = (b# / 2) - INT(b# / 2)
t2 = (v# / 2) - INT(v# / 2)
test1 = (t1 = 0) AND (t2 = .5)
test2 = (t1 = .5) AND (t2 = 0)
IF test1 = -1 THEN EXIT DO
IF test2 = -1 THEN EXIT DO
stop$ = INKEY$
IF stop$ = "s" THEN STOP
count = count + 1
LOOP
dist# = b# - store#
LOOP


"C:\Users\Reactor1967\vcns\work\GENISUS3.BAS"
c# = 2
t# = 0
r# = 1
v# = 3
CLS
RANDOMIZE TIMER
DO
z = INT(RND * 2) + 0
IF z = 0 THEN v# = v# + ((v# - r#) * 1) + 1
IF z = 1 THEN v# = v# + ((v# - r#) * 1) + 2
COLOR 2, 0
msr$ = STR$(v#)
size# = LEN(msr$) - 1
```

```
PRINT , , size#; v#; r#; v# - r#
REM PRINT c#; t#; r#
REM INPUT a$
a$ = INKEY$
IF a$ = "s" THEN STOP
DO
store# = t#
IF c# - t# >= v# THEN EXIT DO
DO
t# = t# + 1
test = ((c# / 2) - INT(c# / 2)) <> ((t# / 2) - INT(t# / 2))
IF test = -1 THEN EXIT DO
LOOP
dist# = t# - store#
IF dist# = 1 THEN c# = c# + t# + 1
IF dist# = 2 THEN c# = c# + t# + 2
IF c# - t# >= v# THEN EXIT DO
r# = c# - t#
REM PRINT c#; t#; r#
REM INPUT a$
a$ = INKEY$
IF a$ = "s" THEN STOP
LOOP

LOOP


"C:\Users\Reactor1967\vcns\work\GENISUS1.BAS"
a# = 10
CLS
DIM matrix#(100)
FOR count = 1 TO 100
matrix#(count) = count
NEXT count
RANDOMIZE TIMER
CLS
p = 6
fp = 0
DO
PRINT , , z; a#; r#; x#
REM INPUT a$
a$ = INKEY$
IF a$ = "s" THEN STOP
IF a$ = "d" THEN GOTO decode:
test1# = (a# / 2) - INT(a# / 2)
```

```
DO
IF fp = 0 THEN p = p + 1
IF fp = 1 THEN p = p - 1
x# = matrix#(p)
test2# = (x# / 2) - INT(x# / 2)
IF test2# <> test1# THEN EXIT DO
IF p + 1 >= 85 THEN fp = 1
IF p - 1 <= 20 THEN fp = 0
LOOP
dist# = x#
r# = a# - x#
z = INT(RND * 2) + 0
dist2# = a# - r#
IF z = 0 THEN a# = a# + (a# - r# + 1)
IF z = 1 THEN a# = a# + (a# - r# + 2)
LOOP
decode:
PRINT "---------------"
store1# = a#
store2# = x#
test = (a# / 2) - INT(a# / 2)
IF test = 0 THEN a# = a# - 1
IF test = .5 THEN a# = a# - 2
a# = a# - x#
x# = store1# - a#
x# = store2# - (x# - store2#)
PRINT , , a#; x#




"C:\Users\Reactor1967\vcns\work\GENISES2.BAS"
REM Moving right along. Use a self coding chart to code R
REM for a dependent data chart. Much room for inprovement here.
a# = 2
b# = 0
v# = 5
r# = 1
CLS
RANDOMIZE TIMER
redo:
z = INT(RND * 2) + 0
store# = v#
IF z = 0 THEN v# = v# + (v# - r# + 1)
IF z = 1 THEN v# = v# + (v# - r# + 2)
COLOR 2, 0
PRINT , , v#; r#
```

```
REM INPUT a$
a$ = INKEY$
IF a$ = "s" THEN STOP
DO
test = (a# / 2) - INT(a# / 2)
store# = b#
DO
b# = b# + 1
IF (b# / 2) - INT(b# / 2) <> test THEN EXIT DO
stop$ = INKEY$
IF stop$ = "s" THEN STOP
LOOP
dist# = b# - store#
r# = a# - b#
IF dist# = 1 THEN a# = a# + b# + 1
IF dist# = 2 THEN a# = a# + b# + 2
REM PRINT a#; b#; r#
a$ = INKEY$
REM INPUT a$
IF a$ = "s" THEN STOP
LOOP UNTIL a# + (b# + 2) >= v#
IF r# >= v# THEN STOP
GOTO redo:


"C:\Users\Reactor1967\vcns\work\FRRANGE.BAS"
RANDOMIZE TIMER
store# = 999999999999999#
CLS
repeat:
p# = store#
a# = 1
PRINT p#
store# = p#
count = 1
DO
p# = p# - a#
count = count + 1
a# = a# * 2 + 1
LOOP UNTIL p# - a# <= 0
PRINT p#; count
PRINT store# - p#
INPUT a$
IF a$ = "s" THEN STOP
PRINT "-----------------------------"
```

```
p# = store#
a# = 1
PRINT p#
store# = p#
count = 1
DO
p# = p# - a#
count = count + 1
a# = a# * 2
LOOP UNTIL p# - a# <= 0
PRINT p#; count
PRINT store# - p#
INPUT a$
IF a$ = "s" THEN STOP




"C:\Users\Reactor1967\vcns\work\FRCT3.BAS"
v# = 1
r# = 1
d# = 2
CLS
RANDOMIZE TIMER
x = 0
DO
z = INT(RND * 2) + 0
IF z = 0 THEN v# = v# + (v# - r# + 1)
IF z = 1 THEN v# = v# + (v# - r# + 2)
COLOR 2, 0
PRINT , , v#; r#; v# - r#; d#
INPUT a$
REM a$ = INKEY$
IF a$ = "s" THEN STOP
IF a$ = "d" THEN GOTO decode:
test = (v# / 2) - INT(v# / 2)
IF test = 0 THEN r# = v# - (d# - 1)
IF test = .5 THEN r# = v# - d#
IF x = 0 THEN d# = d# + 2
REM IF x = 1 THEN d# = d# - 2
REM IF d# >= 100 THEN x = 1
REM IF d# <= 2 THEN x = 0
LOOP
decode:
DO
PRINT , , v#; r#; d#
INPUT a$
```

```
IF a$ = "s" THEN STOP
dist# = (v# - r#)
IF (dist# / 2) - INT(dist# / 2) = .5 THEN dist# = dist# - 1
dist# = dist# / 2
dist# = dist# + 1
v# = v# - dist#
d# = d# - 2
test = (v# / 2) - INT(v# / 2)
IF test = 0 THEN d2# = (d# * 2) + 1
IF test = .5 THEN d2# = (d# * 2) + 2
r# = v# - d2#
LOOP
```

```
"C:\Users\Reactor1967\vcns\work\FRCT2.BAS"
v# = 4
r# = 1
CLS
RANDOMIZE TIMER
DO
d# = v# - r#
IF d# / 2 - INT(d# / 2) = .5 THEN d# = d# - 1
d# = d# / 2
r# = v# - d#
IF (r# / 2) - INT(r# / 2) = 0 THEN r# = r# - 1
REM IF v# - r# > 100 THEN r# = r# + 100
z = INT(RND * 2) + 0
v2# = v#
IF z = 0 THEN v# = v# + (v# - r# + 1)
IF z = 1 THEN v# = v# + (v# - r# + 2)
COLOR 2, 0
PRINT , , v#; r#; v# - r#; v2# - r#
INPUT a$
REM a$ = INKEY$
IF a$ = "s" THEN STOP
LOOP
```

```
"C:\Users\Reactor1967\vcns\work\FRANG2.BAS"
p# = 1
r1# = 1
r2# = 999999999999999#
```

```
Genius# = 999999999999999#
CLS
store1# = 0
store2# = 0
store3# = 0
store4# = 0
add1# = 0
add2# = 0
RANDOMIZE TIMER
repeat:
store1# = p#
DO
c = INT(RND * 2) + 0
IF c = 0 THEN p# = p# + (p# - r1# + 1)
IF c = 1 THEN p# = p# + (p# - r1# + 2)
PRINT , p#; "+"; p# - store1#; Genius#
a$ = INKEY$
IF a$ = "s" THEN STOP
LOOP UNTIL p# + (p# - r1# + 2) > r2#
store2# = p#
PRINT store2# - store1#
IF (store2# - store1#) < Genius# THEN Genius# = (store2# - store1#)
a$ = INKEY$
IF a$ = "s" THEN STOP
store3# = p#
DO
c = INT(RND * 2) + 0
IF c = 0 THEN p# = p# - (r2# - p# + 1)
IF c = 1 THEN p# = p# - (r2# - p# + 2)
PRINT , p#; "-"; store3# - p#; Genius#
a$ = INKEY$
IF a$ = "s" THEN STOP
LOOP UNTIL p# - (r2# - p# + 2) < 0
store4# = p#
PRINT store3# - store4#
IF (store3# - store4#) < Genius# THEN Genius# = (store3# - store4#)
a$ = INKEY$
IF a$ = "s" THEN STOP
GOTO repeat:


"C:\Users\Reactor1967\vcns\work\FRAME3.BAS"
DIM reoranize(20)
p# = 1
r1# = 1
```

```
r2# = 999999999999999#
m# = 500000000000000#
maxp# = r1#
minp# = r2#
CLS
RANDOMIZE TIMER
repeat:
store# = p#
PRINT p#; ABS(p# - store#)
DO
store2# = p#
c = INT(RND * 2) + 0
IF c = 0 THEN p# = p# + (p# - r1# + 1)
IF c = 1 THEN p# = p# + (p# - r1# + 2)
PRINT p#; ABS(p# - store2#); ABS(p# - store#)
a$ = INKEY$
IF a$ = "s" THEN STOP
flip = 0
LOOP UNTIL p# + (p# - r1#) > r2#
REM INPUT a$
IF a$ = "s" THEN STOP
IF a$ = "d" THEN GOTO decode:
p# = p# - m#
flip = 1
GOTO repeat:
decode:
store# = p#
count = 20
```


```
"C:\Users\Reactor1967\vcns\work\FRAME2.BAS"
RANDOMIZE TIMER
r# = 1
r2# = 999999999999999#
minn# = r2#
maxn# = r1#
minp# = r2#
maxp# = r1#
RANDOMIZE TIMER
DO
IF ABS(p# - store#) > 250000000000000# THEN flip = 1 ELSE flip = 0
IF p# - store# > 0 THEN code$ = "+"
IF p# - store# < 0 THEN code$ = "-"
c = INT(RND * 2) + 0
test1 = (p# < 500000000000000#) AND (c = 0)
```

```basic
test2 = (p# < 500000000000000#) AND (c = 1)
test3 = (p# > 500000000000000#) AND (c = 0)
test4 = (p# > 500000000000000#) AND (c = 1)
store# = p#: REM using this to tell things
IF test1 = -1 THEN GOSUB test1:
IF test2 = -1 THEN GOSUB test2:
IF test3 = -1 THEN GOSUB test3:
IF test4 = -1 THEN GOSUB test4:
test5 = flip = 1 AND code$ = "+"
test6 = flip = 1 AND code$ = "-"
REM INPUT a$
a$ = INKEY$
IF a$ = "s" THEN STOP
LOOP
test1:
p# = p# + (p# - r# + 1)
RETURN
test2:
p# = p# + (p# - r# + 2)
RETURN
test3:
p# = p# - (r2# - p# + 1)
RETURN
test4:
p# = p# - (r2# - p# + 2)
RETURN




"C:\Users\Reactor1967\vcns\work\FRAME1.BAS"
RANDOMIZE TIMER
r# = 1
r2# = 999999999999999#
minn# = r2#
maxn# = r1#
minp# = r2#
maxp# = r1#
RANDOMIZE TIMER
DO
IF ABS(p# - store#) > 250000000000000# THEN flip = 1 ELSE flip = 0
IF p# - store# > 0 THEN code$ = "+"
IF p# - store# < 0 THEN code$ = "-"
c = INT(RND * 2) + 0
test1 = (p# < 500000000000000#) AND (c = 0)
test2 = (p# < 500000000000000#) AND (c = 1)
test3 = (p# > 500000000000000#) AND (c = 0)
```

```
test4 = (p# > 500000000000000#) AND (c = 1)
store# = p#: REM using this to tell things
IF test1 = -1 THEN GOSUB test1:
IF test2 = -1 THEN GOSUB test2:
IF test3 = -1 THEN GOSUB test3:
IF test4 = -1 THEN GOSUB test4:
test5 = flip = 1 AND code$ = "+"
test6 = flip = 1 AND code$ = "-"
test7 = (test5 = -1) AND p# > maxp#
test8 = (test5 = -1) AND p# < minp#
test9 = (test6 = -1) AND p# > maxn#
test10 = (test6 = -1) AND p# < minn#
IF test7 = -1 THEN maxp# = p#
IF test8 = -1 THEN minp# = p#
IF test9 = -1 THEN maxn# = p#
IF test10 = -1 THEN minn# = p#
PRINT "                  ", c; p#; code$; flip, b$
PRINT "MnP"; minp#; "MxP"; maxp#; "MnN"; minn#; "MxN"; maxn#
test11 = (flip = 1) AND (code$ = "+") AND ((p# < minp#) OR (p# > maxp#))
test12 = (flip = 1) AND (code$ = "-") AND ((p# < minn#) OR (p# > maxn#))
IF test11 = -1 THEN PRINT "++++++++++++++++++++++++++++++++++++++
+++++++++"
IF test12 = -1 THEN PRINT "-------------------------------------------"
REM INPUT a$
a$ = INKEY$
IF a$ = "s" THEN STOP
LOOP
test1:
p# = p# + (p# - r# + 1)
RETURN
test2:
p# = p# + (p# - r# + 2)
RETURN
test3:
p# = p# - (r2# - p# + 1)
RETURN
test4:
p# = p# - (r2# - p# + 2)
RETURN




"C:\Users\Reactor1967\vcns\work\FRACTION.BAS"
v# = 1
r# = 1
CLS
```

```
RANDOMIZE TIMER
DO
z = INT(RND * 2) + 0
IF z = 0 THEN v# = v# + (v# - r# + 1)
IF z = 1 THEN v# = v# + (v# - r# + 2)
ra1# = v# - r#
dist# = v# - r#
IF (dist# / 2) - INT(dist# / 2) = .5 THEN dist# = dist# - 1
dist# = dist# / 2
r# = v# - dist#
IF (r# / 2) - INT(r# / 2) = 0 THEN r# = r# - 1
ra2# = v# - r#
PRINT , , v#; r#; ra1#; ra2#
REM INPUT a$
a$ = INKEY$
IF a$ = "s" THEN STOP
LOOP




"C:\Users\Reactor1967\vcns\work\FOCUS2.BAS"
a# = 2
r# = 1
CLS
RANDOMIZE TIMER
x = 0
redo:
DO
dist# = ABS(a# - r#) - 0: REM you can minus this by 4 to make dist# go down.
z = INT(RND * 2) + 0
IF z = 0 THEN dist# = dist# * 2
IF z = 1 THEN dist# = (dist# * 2) + 1
a# = a# + dist#: REM you can minus the dist# too.
COLOR 2, 0
PRINT , , z; a#; r#; dist#
REM INPUT a$
a$ = INKEY$
IF a$ = "s" THEN STOP
IF a$ = "d" THEN EXIT DO
dist# = ABS(a# - r#) - 0: REM you can minus this by 4 to make dist# go down.
z = INT(RND * 2) + 0
IF z = 0 THEN dist# = dist# * 2
IF z = 1 THEN dist# = (dist# * 2) + 1
r# = r# + dist#: REM you can minus the dist# too.
PRINT , , z; a#; r#; dist#
```

```
REM INPUT a$
a$ = INKEY$
IF a$ = "s" THEN STOP
IF a$ = "d" THEN EXIT DO
LOOP
GOTO redo:
```

```
"C:\Users\Reactor1967\vcns\work\FOCUS.BAS"
REM this program teaches how to change the focus. On regular numbers
REM you can alway get a even number by multiplying it by two. Well, that
REM takes the number up in value but what about going down in value. And,
REM what about doing it constructivly. Here learning how to change the
REM focus on the difference between variables, where one variable
REM incleases 2:1 of another variable then switching that focus so that
REM the other variable increases 1:2 and so forth.
a# = 1
b# = 2
CLS
redo:
DO
PRINT a#; b#; "right"
a# = a# + 1
b# = b# + 2
REM INPUT a$
a$ = INKEY$
IF a$ = "s" THEN STOP
test = (a$ = "d") AND (a# / 2 - INT(a# / 2) = 0)
IF test = -1 THEN EXIT DO
LOOP
a$ = ""
DO
PRINT "left"; a#; b#
a# = a# + 2
b# = b# + 1
REM INPUT a$
a$ = INKEY$
IF a$ = "s" THEN STOP
test = (a$ = "d") AND (a# / 2 - INT(a# / 2) = 0)
IF test = -1 THEN EXIT DO
LOOP
a$ = ""
GOTO redo:
```

```
"C:\Users\Reactor1967\vcns\work\FNDRAGN.BAS"
RANDOMIZE TIMER
CLS
p# = 1
repeat:
r# = p#
DO
c = INT(RND * 2) + 0
IF c = 0 THEN p# = p# + (p# - r# + 1)
IF c = 1 THEN p# = p# + (p# - r# + 2)
PRINT p#
LOOP UNTIL p# + (p# - r# + 2) >= 999999999999999#
PRINT "-------------------------"
PRINT p# - r#
INPUT a$
IF a$ = "s" THEN STOP
DO
c = INT(RND * 2) + 0
IF c = 0 THEN p# = p# - (r# - p# + 1)
IF c = 1 THEN p# = p# - (r# - p# + 2)
PRINT p#
LOOP UNTIL p# - (r# - p# + 2) <= 1
PRINT "-------------------------"
PRINT r# - p#
INPUT a$
IF a$ = "s" THEN STOP
GOTO repeat:
```


```
"C:\Users\Reactor1967\vcns\work\FNCYVC.BAS"
DIM r#(255)
DECLARE SUB Delay (a AS SINGLE)
DECLARE SUB DisplayBox (X AS INTEGER, Y AS INTEGER,
FrameColourA AS INTEGER, FrameColourB AS INTEGER, TextColourA AS
INTEGER, TextColourB AS INTEGER, FrameColour AS INTEGER,
FrameLabel AS STRING, Text AS STRING, Size AS INTEGER, Speed AS
SINGLE, _
WaitKey AS INTEGER)
DECLARE SUB CloseBox (X AS INTEGER, Y AS INTEGER, Size AS
INTEGER, Lines AS INTEGER, Speed AS SINGLE)

DIM SHARED OnScreen(25, 80) AS INTEGER
DIM SHARED OnColour(25, 80) AS INTEGER
DIM SHARED Lines AS INTEGER
```

```
CLS
COLOR 7, 0
FOR a = 1 TO 25
  FOR B = 1 TO 80
    PRINT "�";
  NEXT B
NEXT a
PRINT "THIS PROGRAM IS POWERED BY A RANDOM NUMBER
GENERATOR FOR DEMONSTRATION!"
PRINT "THE RIGHT SIDE OF THE SCREEN CODES DATA UP FROM A
LOW VECTOR NUMBER"
PRINT "THE LEFT SIDE OF THE SCREEN CODES DATA DOWN FROM
A HIGH VECTOR NUMBER"
PRINT "THIS IS A CODING MONSTER THAN CAN STORE AN
UNLIMITED AMOUNT OF BINARY"
PRINT "DATA AS 1 FIFTEEN DIGIT
NUMBER!!!!!!!!!!!!!!!!!!!!!!!!!!!!!!!!!!!!!!!!!!"
' Demonstration of DisplayBox
'
' Display Header, Text, Delay, No Wait
CALL DisplayBox(1, 1, 9, 1, 7, 1, 15, "VECTOR COORDINATE
NUMERICAL SYSTEMS", "THIS PROGRAM CAN CODE AN
UNLIMITED AMOUNT OF BINARY DATA!!!", 39, .05, 0)
' Display Text, Delay, Wait
CALL DisplayBox(4, 25, 7, 0, 3, 0, 0, "", "PUSH ANY KEY!!!", 24, .05, 1)
CALL CloseBox(4, 25, 24, Lines, .05)
' Display Text, Delay, No Wait
CALL DisplayBox(4, 25, 7, 0, 3, 0, 0, "", "POWERING UP!!!", 14, .05, 0)
' Display Text, Delay, No Wait
CALL DisplayBox(10, 15, 15, 7, 14, 7, 0, "", "VCNS READY", 6, .05, 0)
' Display Text, No Delay, No Wait
CALL DisplayBox(12, 40, 9, 1, 15, 1, 0, "", "READY TO CODE UNLIMITED
DATA", 31, 0, 0)
DO
a$ = INKEY$
IF a$ > "" THEN EXIT DO
IF a$ = "s" THEN STOP
LOOP
REM GOT TO GET R TO SYNCHRONIZE WITH P
REM THERE IS A GAP BETWEEN 1 AND THE NEXT R#
REM FILL IN THAT GAP OR ELIMANATE THAT RANGE ALL
TOGETHER.
REM ----------------------Initilize R#-------------------------------------
CLS
r#(1) = 1
```

```
REM PRINT r#(1)
pt1# = 5
rf# = 100
count = 2
a# = 0
replaced1:
DO
a# = a# + pt1#
test = (a# / 2) - INT(a# / 2)
r#(count) = a#
IF (r#(count) / 2) - INT(r#(count) / 2) = 0 THEN r#(count) = r#(count) + 1
REM PRINT r#(count); count
count = count + 1
REM INPUT a$
a$ = INKEY$
IF a$ = "s" THEN STOP
IF count >= 254 THEN EXIT DO
LOOP UNTIL a# = rf#
pt1# = pt1# * 10
rf# = rf# * 10
IF count >= 254 THEN GOTO replaced2:
IF a# + pt1# > 999999999999999# THEN GOTO replaced2:
GOTO replaced1:
replaced2:
r#(254) = 950000000000001#
r#(255) = 999999999999999#
REM ---------------------Code P# up.--------------------------------------
range = 1
p# = 1
RANDOMIZE TIMER
repeat:
DO
bt = (p# / 2) - INT(p# / 2)
IF bt >= .5 THEN bt = 1
COLOR 2, 0
PRINT , , "     "; p#; bt; r#(range)
a$ = INKEY$
REM INPUT a$
REM IF a$ = "s" THEN STOP
IF p# > 1000 THEN c = INT(RND * 2) + 0 ELSE c = 0
IF c = 0 THEN p# = p# + (p# - (r#(range))) + 1
IF c = 1 THEN p# = p# + (p# - (r#(range))) + 2
IF a$ = "s" THEN STOP
IF p# > (r#(range + 1)) THEN range = range + 1
LOOP UNTIL p# + (p# - (r#(range))) + 2 >= 999999999999999#
bt = (p# / 2) - INT(p# / 2)
```

```
IF bt >= .5 THEN bt = 1
PRINT ; "     "; p#; bt; r#(range)
REM INPUT a$
REM IF a$ = "s" THEN STOP
REM --------------------Reduce P# down.--------------------------------
range = range + 1
count = 0
DO
bt = (p# / 2) - INT(p# / 2)
IF bt >= .5 THEN bt = 1
COLOR 2, 0
PRINT ; "     "; p#; bt; r#(range)
a$ = INKEY$
REM INPUT a$
IF a$ = "s" THEN STOP
c = INT(RND * 2) + 0
IF c = 0 THEN p# = p# - ((r#(range)) - p# + 1)
IF c = 1 THEN p# = p# - ((r#(range)) - p# + 2)
IF p# < (r#(range - 1)) THEN range = range - 1
count = count + 1
LOOP UNTIL p# < 150000000000000#
bt = (p# / 2) - INT(p# / 2)
IF bt >= .5 THEN bt = 1
PRINT ; "     "; p#; bt; r#(range)
REM INPUT a$
IF a$ = "s" THEN STOP
range = range - 1
GOTO repeat:

SUB CloseBox (X AS INTEGER, Y AS INTEGER, Size AS INTEGER, Lines
AS INTEGER, Speed AS SINGLE) STATIC

' WARNING: X and Y positions MUST be the same as well as the size!
'         if different values are passed, the box will not be cleared or
'         an error will occur.

' *** Command Syntax ***

' CALL CloseBox(X, Y, Size, Lines, Speed)
' X/Y = X/Y Exactly that of the Box Drawn!
' Size = Size Exactly that of the Box Drawn!
' Lines = You dont need to touch this!
' Speed = The speed to close

Size = Size + 10
```

```
IF Speed > 0 THEN
  ' Descending Soud
  FOR I = 1500 TO 500 STEP -(35 - (Lines * 5))
    SOUND I, I / 20000
  NEXT
END IF

' A is stated to signify the X position... while B is stated to
' signify columns... Purpose: Select the line X and draw all
'                  text of that line (Y to Length)
FOR a = (Lines + X) TO (X - 1) STEP -1
  ' Delay between line draws
  CALL Delay(Speed)
  LOCATE a + 2, Y
  FOR E = Y TO (Y + Size - 4)
    COLOR 8, 0
    IF a = X - 1 THEN
      ' Redraw top line which would not be redrawn
      ' normally because of my crappy programming.
      Foreground = OnColour(a + 2, E) AND 15
      Background = OnColour(a + 2, E) \ 16
      COLOR Foreground, Background
      PRINT CHR$(OnScreen(a + 2, E));
    ELSE
      ' Draw dimmed bottom line to create moving
      ' shade effect
      COLOR 8, 0
      LOCATE ((a - 1) + Lines), E
      PRINT CHR$(OnScreen(((a - 1) + Lines), E));
      ' Fix first two Y co-ordinates after dimming
      LOCATE (a - 1 + Lines), Y
      FOR F = Y TO Y + 1
        Foreground = OnColour(a + 1, F) AND 15
        Background = OnColour(a + 1, F) \ 16
        COLOR Foreground, Background
        PRINT CHR$(OnScreen(a + 1, F));
      NEXT F
      ' Draw Full Coloured Pulled Line
      LOCATE a + 2, E
      Foreground = OnColour(a + 2, E) AND 15
      Background = OnColour(a + 2, E) \ 16
      COLOR Foreground, Background
      PRINT CHR$(OnScreen(a + 2, E));
    END IF
  NEXT E
NEXT a
```

```
' Clean up very top line of Box Frame
LOCATE X, Y
FOR F = Y TO (Y + Size - 4)
    ' Draw Full Coloured Pulled Line
    Foreground = OnColour(X, F) AND 15
    Background = OnColour(X, F) \ 16
    COLOR Foreground, Background
    PRINT CHR$(OnScreen(X, F));
NEXT F

' Re-Initialize these values! VERY VERY IMPORTANT!
ERASE OnColour
ERASE OnScreen
Lines = 0

END SUB

SUB Delay (a AS SINGLE) STATIC

Start! = TIMER
DO
LOOP UNTIL TIMER - Start! >= a

END SUB

SUB DisplayBox (X AS INTEGER, Y AS INTEGER, FrameColourA AS
INTEGER, FrameColourB AS INTEGER, TextColourA AS INTEGER,
TextColourB AS INTEGER, FrameColour AS INTEGER, FrameLabel AS
STRING, Text AS STRING, Size AS INTEGER, Speed AS SINGLE, WaitKey
AS _
INTEGER) STATIC

' Warning: DO NOT specify Size > 74 or < a Word in the String!!!

' *** Command Syntax ***

' CALL
DisplayBox(X,Y,FrameColourA,FrameColourB,TextColourA,TextColourB,Fra
meColour, "FrameLabel","Text",Max Columns, Speed of Roll, Pause Toggle)

' X/Y Position - Draw Box at Location (Row, Column)
' FrameColourA/B - Colour of Frame (Foreground, Background)
' TextColourA/B - Colour of Text (Foreground, Background)
' FrameColour - Foreground Colour of Frame Header
' FrameLabel - The Frame Header
```

```
' "" - Your Text to Display Here
' Size - Maximum Display Columns
' Delay - Delay of Roll Up/Down
' WaitKey - 0 = No Pause, 1 = Pause

' Initialize all variables
CharProg = 1
OrgX = X
NewX = X
Lines = LEN(Text) / Size
Lines = INT(Lines + .5)

' Determine if Sound is on... if so: Ascending Sound
IF WaitKey = 1 OR Speed > 0 THEN
FOR I = 500 TO 1500 STEP (35 - Lines * 5)
   SOUND I, I / 20000
NEXT
END IF

' Capture Text to be Over-Written by Box and Shadow
FOR a = Y TO (Y + Size + 6)
   FOR B = X TO (X + Lines + 2)
      OnScreen(B, a) = SCREEN(B, a)
      OnColour(B, a) = SCREEN(B, a, 1)
   NEXT B
NEXT a

' Draw Box
LOCATE NewX, Y: COLOR FrameColourA, FrameColourB: PRINT " ��";

' Display Frame Header
IF LEN(FrameLabel) < Size THEN
   COLOR FrameColour
   PRINT FrameLabel;
   COLOR FrameColourA, FrameColourB
   FOR a = 1 TO (Size - LEN(FrameLabel) - 1)
      PRINT "�";
   NEXT a
ELSE

' Draw Top Text Border
FOR a = 1 TO Size - 1
   PRINT "�";
NEXT a
END IF
```

```
PRINT "Ŀ "

FOR a = 1 TO Lines
   NewX = NewX + 1
   COLOR FrameColourA, FrameColourB
   LOCATE NewX, Y
   PRINT " ��";
   FOR c = 1 TO Size - 1
      PRINT "�";
   NEXT c
   PRINT "�� "

   COLOR 8, 0
   ' Draw dimmed bottom
   FOR B = Y TO (Y + Size + 6)
      LOCATE ((NewX - 1) + Lines), B
      PRINT CHR$(OnScreen(((NewX - 1) + Lines), B));
   NEXT B

   CALL Delay(Speed / 2)

   COLOR FrameColourA, FrameColourB
   LOCATE NewX, Y
   PRINT " � ";
   COLOR TextColourA, TextColourB
   PRINT SPACE$(Size - 1);
   COLOR FrameColourA, FrameColourB
   PRINT " � ";

   ' Draw dimmed edges
   COLOR 8, 0:
   FOR B = (Y + Size + 5) TO (Y + Size + 6)
      PRINT CHR$(OnScreen(NewX, B));
   NEXT B
NEXT a

NewX = NewX + 1
COLOR FrameColourA, FrameColourB
LOCATE NewX, Y
PRINT " ��";
FOR c = 1 TO Size - 1
      PRINT "�";
NEXT c
PRINT "�� ";

' Draw final dimmed edge
```

```
COLOR 8, 0
FOR B = (Y + Size + 5) TO (Y + Size + 6)
   PRINT CHR$(OnScreen(NewX, B));
NEXT B

' Draw dimmed bottom
FOR B = Y + 2 TO (Y + Size + 6)
   LOCATE (X + Lines + 2), B
   PRINT CHR$(OnScreen((X + Lines + 2), B));
NEXT B

' Display Text
FOR D = 1 TO Lines
   COLOR TextColourA, TextColourB
   DO
      Temp$ = MID$(Text, CharProg, Size)
      IF LEN(Text) - CharProg <= Size THEN
         X = X + 1
         LOCATE X, Y + 3
         PRINT MID$(Text, CharProg, Size)
         CharProg = CharProg + LEN(Temp$)
      ELSE
         X = X + 1
         LOCATE X, Y + 3
         FOR Ccnt = LEN(Temp$) TO 1 STEP -1
            IF MID$(Temp$, Ccnt, 1) = " " THEN EXIT FOR
         NEXT Ccnt
         PRINT LEFT$(Temp$, Ccnt)
         CharProg = CharProg + Ccnt
      END IF
   LOOP UNTIL CharProg >= LEN(Text)
NEXT D

IF WaitKey = 1 THEN
   DO WHILE INKEY$ = ""
   LOOP
END IF

END SUB




"C:\Users\Reactor1967\vcns\work\FLVCNS.BAS"
INPUT "the name of your program"; file$
OPEN file$ FOR BINARY ACCESS READ AS #1
REM GET [#]filenumber%[,[recordnumber&][,variable]]
```

```
filenum = 1
DO
GET #1, filenum, vcns%
PRINT vcns%
INPUT a$
REM a$ = INKEY$
IF a$ = "s" THEN CLOSE #1
IF a$ = "s" THEN STOP
filenum = filenum + 1
LOOP UNTIL (EOF(1))
PRINT "Your done"
CLOSE #1
```

```
"C:\Users\Reactor1967\vcns\work\FLTRTST2.BAS"
DECLARE SUB rangedown (p#, r#, reference#())
DECLARE SUB rangeup (p#, r#, reference#())
DIM reference#(502)
count = 1
p# = 1
r# = 1
CLS
a# = 15
reference#(1) = 1
FOR count = 2 TO 502
a# = a# + INT(a# / 15)
reference#(count) = a#
NEXT count
count = 1
r# = 1
p# = 1
DO
PRINT p#
c = INT(RND * 2) + 0
IF c = 0 THEN p# = p# + (p# - r# + 1)
IF c = 1 THEN p# = p# + (p# - r# + 2)
PRINT p#
test = p# < reference#(count) AND p# < reference#(count + 1)
IF test = -1 THEN r# = reference#(count + 1)
IF p# > reference#(count + 1) THEN count = count + 1
```

```
INPUT a$
IF a$ = "s" THEN STOP
LOOP


SUB rangedown (p#, r#, reference#())
count = 1
DO
IF reference#(count) > p# THEN r# = reference#(count)
IF reference#(count) > p# THEN EXIT DO
count = count + 1
IF count >= 503 THEN EXIT DO
LOOP
IF count = 503 THEN r# = 999999999999999#
PRINT "                         "; count
END SUB


SUB rangeup (p#, r#, reference#())
count = 1
DO
IF reference#(count + 1) > p# THEN r# = reference#(count)
IF reference#(count + 1) > p# THEN EXIT DO
count = count + 1
LOOP
PRINT "                         "; count
END SUB




"C:\Users\Reactor1967\vcns\work\FLTRTST.BAS"
DECLARE SUB rangedown (p#, r#, reference#())
DECLARE SUB rangeup (p#, r#, reference#())
DIM reference#(502)
count = 1
p# = 1
r# = 1
CLS
a# = 15
reference#(1) = 1
FOR count = 2 TO 502
a# = a# + INT(a# / 15)
reference#(count) = a#
NEXT count
file$ = "output.txt"
OPEN file$ FOR APPEND AS #1
repeat:
store# = r#
```

```
CALL rangeup(p#, r#, reference#())
PRINT #1, store#; "="; r#
DO
c = INT(RND * 2) + 0
IF c = 0 THEN p# = p# + (p# - r# + 1)
IF c = 1 THEN p# = p# + (p# - r# + 2)
PRINT p#
INPUT a$
REM a$ = INKEY$
IF a$ = "s" THEN STOP
LOOP UNTIL p# + (p# - r# + 2) >= 999999999999999#
store# = r#
CALL rangedown(p#, r#, reference#())
PRINT #1, store#; "="; r#
PRINT "-------------------------------------------"
DO
c = INT(RND * 2) + 0
IF c = 0 THEN p# = p# - (r# - p# + 1)
IF c = 1 THEN p# = p# - (r# - p# + 2)
PRINT p#
INPUT a$
REM a$ = INKEY$
IF a$ = "s" THEN STOP
LOOP UNTIL p# - (r# - p# + 2) <= 0
PRINT "-------------------------------------------"
GOTO repeat:

SUB rangedown (p#, r#, reference#())
count = 1
DO
IF reference#(count) > p# THEN r# = reference#(count)
IF reference#(count) > p# THEN EXIT DO
count = count + 1
IF count >= 503 THEN EXIT DO
LOOP
IF count = 503 THEN r# = 999999999999999#
PRINT "                            "; count
END SUB

SUB rangeup (p#, r#, reference#())
count = 1
DO
IF reference#(count + 1) > p# THEN r# = reference#(count)
IF reference#(count + 1) > p# THEN EXIT DO
count = count + 1
LOOP
```

```
PRINT "                                  "; count
END SUB



"C:\Users\Reactor1967\vcns\work\FLOATRP.BAS"
r1# = 1
r2# = 2
p# = 1
store# = 0
CLS
RANDOMIZE TIMER
DO
c = INT(RND * 2) + 0
IF c = 0 THEN store# = r1#
IF c = 1 THEN store# = (r1# + 1)
p# = store# + p#
r1# = r1# + store#
PRINT p#; r1#
INPUT a$
IF a$ = "s" THEN STOP
test = r1# + r1# >= 999999999999999# OR p# + r1# >= 999999999999999#
LOOP UNTIL test = -1
PRINT "-----------------------------------"
repeat:
DO
c = INT(RND * 2) + 0
IF c = 0 THEN store# = r2#
IF c = 1 THEN store# = r2# - 1
p# = p# - store#
r1# = r1# - store# - 9999999999991#
r2# = r2# + store#
PRINT p#; r2#, r1#
INPUT a$
IF a$ = "s" THEN STOP
test = r1# - r2# <= -1
LOOP UNTIL test = -1
PRINT "-------------------------------------"
DO
c = INT(RND * 2) + 0
IF c = 0 THEN store# = r1#
IF c = 1 THEN store# = r1# + 1
p# = store# + p#
r1# = r1# + store#
r2# = r2# - store# - 9999999999991#
PRINT p#; r1#, r2#
```

```
INPUT a$
IF a$ = "s" THEN STOP
test = r2# - r1# <= -1
LOOP UNTIL test = -1
PRINT "-------------------------------------"
GOTO repeat:
```

```
"C:\Users\Reactor1967\vcns\work\FLIPTOG.BAS"
REM program outline
REM Section one -----------------------------------------------------------
REM comparing and testing vector coordinates
REM Section two -----------------------------------------------------------
REM make decision whether to
REM A. compare and find highest vector number
REM B. test one and both vectors to see flip state
REM    1. Either can flip only one vector
REM    2. Or can flip both vectors
REM 1. flip and record
REM    A. both vectors or just one vector
REM 2. code data and record.
REM Section three -----------------------------------------------------------
```

```
"C:\Users\Reactor1967\vcns\work\FLIPRP.BAS"
r# = 1
p# = 1
r2# = 999999999999999#
really# = 999999999999999#
CLS
RANDOMIZE TIMER
repeat:
store# = p#
DO
c = INT(RND * 2) + 0
IF c = 0 THEN p# = p# + (p# - r# + 1)
```

```
IF c = 1 THEN p# = p# + (p# - r# + 2)
PRINT p#; c
REM test = ((p# - store#) >= 25000000000000#) AND (p# >=
500000000000000#)
a$ = INKEY$
IF a$ = "s" THEN STOP
LOOP UNTIL p# >= 500000000000000#
PRINT "                "; p# - store#
add# = p# - store#
IF add# < really# THEN really# = add#
store# = p#
DO
c = INT(RND * 2) + 0
IF c = 0 THEN p# = p# - (r2# - p# + 1)
IF c = 1 THEN p# = p# - (r2# - p# + 2)
PRINT p#; c
REM test = ((store# - p#) >= 250000000000000#) AND (p# <=
500000000000000#)
a$ = INKEY$
IF a$ = "s" THEN STOP
IF a$ = "d" THEN GOTO decode:
LOOP UNTIL p# <= 500000000000000#
PRINT "                "; store# - p#
add# = store# - p#
IF add# < really# THEN really# = add#
GOTO repeat:
decode:
STOP
store# = p#
DO
c = p# / 2 - INT(p# / 2)
IF c = 0 THEN p# = p# + ((r2# - p# + 1) / 2)
IF c >= .5 THEN p# = p# + ((r2# - p# + 2) / 2)
PRINT p#
a$ = INKEY$
IF a$ = "s" THEN STOP
test = ((p# - store#) >= 250000000000000#) AND (p# > 500000000000000#)
LOOP UNTIL test = -1
store# = p#
DO
c = p# / 2 - INT(p# / 2)
IF c = 0 THEN p# = p# - ((p# - r# + 1) / 2)
IF c >= .5 THEN p# = p# - ((p# - r# + 2) / 2)
PRINT p#
a$ = INKEY$
IF a$ = "s" THEN STOP
```

```
test = ((p# - store#) >= 250000000000000#) AND (p# < 500000000000000#)
LOOP UNTIL test = -1




"C:\Users\Reactor1967\vcns\work\FLIPB3.BAS"
a# = 1
b# = 3
RANDOMIZE TIMER
CLS
carry# = a#
redo:
FOR count2 = 1 TO 51
IF carry# = a# THEN d = INT(RND * 3) + 0 ELSE d = 99
FOR count = 1 TO 3
PRINT a#; b#
b# = b# + 2
test1 = d = 0 AND count = 1
test2 = d = 1 AND count = 2
test3 = d = 2 AND count = 3
IF test1 = -1 THEN carry# = b#
IF test2 = -1 THEN carry# = b#
IF test3 = -1 THEN carry# = b#
NEXT count
PRINT a#; b#; carry#
a# = a# + 2
a$ = INKEY$
INPUT a$
IF a$ = "s" THEN STOP
IF carry# > 101 THEN EXIT FOR
NEXT count2
REM --------------------------------
PRINT a#; b#; carry#; d
INPUT "Ready for next chart"; a$
IF a$ = "s" THEN STOP
a# = 307
b# = 305
FOR count2 = 1 TO 51
IF carry# = a# THEN d = INT(RND * 3) + 0 ELSE d = 99
FOR count = 1 TO 3
PRINT a#; b#
b# = b# - 2
test1 = d = 0 AND count = 1
test2 = d = 1 AND count = 2
test3 = d = 2 AND count = 3
IF test1 = -1 THEN carry# = b#
```

```
IF test2 = -1 THEN carry# = b#
IF test3 = -1 THEN carry# = b#
NEXT count
PRINT a#; b#; carry#
a# = a# - 2
a$ = INKEY$
INPUT a$
IF a$ = "s" THEN STOP
IF carry# < 101 THEN EXIT FOR
NEXT count2


STOP
GOTO redo:


"C:\Users\Reactor1967\vcns\work\FINDR.BAS"
p1# = 1
p2# = 999999999999999#
r1# = 1
r2# = 999999999999999#
CLS
RANDOMIZE TIMER
repeat:
DO
c = INT(RND * 2) + 0
IF c = 0 THEN p1# = p1# + (p1# - r1# + 1)
IF c = 0 THEN p2# = p2# - (r2# - p2# + 1)
IF c = 1 THEN p1# = p1# + (p1# - r1# + 2)
IF c = 1 THEN p2# = p2# - (r2# - p2# + 2)
PRINT p1#; p2#; r1#; r2#
REM a$ = INKEY$
IF a$ = "s" THEN STOP
LOOP UNTIL p2# - (r2# - p2# + 2) <= 1 OR p1# + (p1# - r1# + 2) >=
999999999999999#
REM INPUT a$
a$ = INKEY$
xIF a$ = "s" THEN STOP
store1# = p1#
store2# = p2#
p1# = store2#
p2# = store1#
r1# = p1#
r2# = p2#
GOTO repeat:
```

```
"C:\Users\Reactor1967\vcns\work\FINALLY.BAS"
p# = 1
r1# = 1
r2# = 999999999999999#
RANDOMIZE TIMER
CLS
repeat:
store# = p#
DO
test1 = ((p# - store#) < 250000000000000#) AND (p# + (p# - r1# + 2) <
500000000000000#)
test2 = ((p# - store#) >= 250000000000000#) AND (p# + (p# - r1#) >
500000000000000#)
test3 = ((p# - store#) >= 250000000000000#) AND (p# + (p# - r1# + 2) <
500000000000000#)
IF test1 = -1 THEN GOTO test1:
IF test2 = -1 THEN GOTO test2:
IF test3 = -1 THEN GOTO test3:
returnpoint:
PRINT p#; p# - store#
REM INPUT a$
a$ = INKEY$
IF a$ = "s" THEN STOP
LOOP
test1:
c = INT(RND * 2) + 0
IF c = 0 THEN p# = p# + (p# - r1# + 1)
IF c = 1 THEN p# = p# + (p# - r1# + 2)
GOTO returnpoint:
test2:
p# = p# + (p# - r1# + 2)
GOTO flip:
test3:
p# = p# + (p# - r1# + 1)
GOTO returnpoint:
flip:
PRINT p#
PRINT "Well, You made it."
INPUT x$
IF x$ = "s" THEN STOP
store# = p#
DO
c = (p# / 2) - INT(p# / 2)
```

```
IF c = 0 THEN p# = p# - ((p# - r1# + 1) / 2)
IF c >= .5 THEN p# = p# - ((p# - r1# + 2) / 2)
range# = ABS(store# - p#)
PRINT p#
test = (range# >= 250000000000000#) AND (c >= .5)
IF test = -1 THEN EXIT DO
a$ = INKEY$
IF a$ = "s" THEN STOP
LOOP
```

```
"C:\Users\Reactor1967\vcns\work\FEEDBACK.BAS"
v8# = 0
v7# = 0
v6# = 0
v5# = 0
v4# = 0
v3# = 0
v2# = 0
v1# = 1
RANDOMIZE TIMER
CLS
count = 0
DO
COLOR 2, 0
PRINT z; v8#; v7#; v6#; v5#; v4#; v3#; v2#; v1#; count
count = count + 1
a$ = INKEY$
REM INPUT a$
IF a$ = "s" THEN STOP
z = INT(RND * 2)
v1# = v1# + v1# + z
IF v1# > 999999 THEN carry = 1 ELSE carry = 0
IF carry = 1 THEN v1# = v1# - 1000000
v2# = v2# + v2# + carry
IF v2# > 999999 THEN carry = 1 ELSE carry = 0
IF carry = 1 THEN v2# = v2# - 1000000
v3# = v3# + v3# + carry
IF v3# > 999999 THEN carry = 1 ELSE carry = 0
IF carry = 1 THEN v3# = v3# - 1000000
v4# = v4# + v4# + carry
IF v4# > 999999 THEN carry = 1 ELSE carry = 0
IF carry = 1 THEN v4# = v4# - 1000000
```

```
v5# = v5# + v5# + carry
IF v5# > 999999 THEN carry = 1 ELSE carry = 0
IF carry = 1 THEN v5# = v5# - 1000000
v6# = v6# + v6# + carry
IF v6# > 999999 THEN carry = 1 ELSE carry = 0
IF carry = 1 THEN v6# = v6# - 1000000
v7# = v7# + v7# + carry
IF v7# > 999999 THEN carry = 1 ELSE carry = 0
IF carry = 1 THEN v7# = v7# - 1000000
v8# = v8# + v8# + carry
IF v8# > 999999 THEN carry = 1 ELSE carry = 0
IF carry = 1 THEN v8# = v8# - 1000000
v1# = v1# + carry
LOOP


SUB vector (v4#, v3#, v2#, v1#, c$)
v1# = v1# + 1
IF v1# = 1000 THEN v2# = v2# + 1
IF v2# = 1000 THEN v3# = v3# + 1
IF v3# = 1000 THEN v4# = v4# + 1
IF v1# = 1000 THEN v1# = 0
IF v2# = 1000 THEN v2# = 0
IF v3# = 1000 THEN v3# = 0
IF v4# > 1000 THEN STOP
c$ = ""
c$ = STR$(v4#) + STR$(v3#) + STR$(v2#) + STR$(v1#)
END SUB


"C:\Users\Reactor1967\vcns\work\FCHART2.BAS"
r# = 7
v1# = 12
v2# = v1# + ((v1# - r#) * 2) + 1
v3# = v1# + ((v1# - r#) * 2) + 2
v4# = v1# + ((v1# - r#) * 2) + 3
CLS
RANDOMIZE TIMER
DO
PRINT v1#; v2#; v3#; v4#; r#
REM PRINT (v2# / 3) - INT(v2# / 3)
count = 0
z = INT(RND * 3) + 0
IF z = 0 THEN v1# = v2#
IF z = 1 THEN v1# = v3#
```

```
IF z = 2 THEN v1# = v4#
d# = INT(v1# / 6)
r# = (d# * 6) + 1
r# = r# - 6
v2# = v1# + ((v1# - r#) * 2) + 1
v3# = v1# + ((v1# - r#) * 2) + 2
v4# = v1# + ((v1# - r#) * 2) + 3
REM INPUT a$
a$ = INKEY$
IF a$ = "s" THEN STOP
LOOP
```

```
"C:\Users\Reactor1967\vcns\work\FCHART.BAS"
v# = 1
CLS
RANDOMIZE TIMER
DO
PRINT v# * 2; (v# * 2) + 1, v#
v# = v# + 1
INPUT a$
IF a$ = "s" THEN STOP
LOOP
```

```
"C:\Users\Reactor1967\vcns\work\EVNCHRT.BAS"
a# = 2
b# = 4
c# = 6
CLS
DO
PRINT a#; b#; c#
a# = a# + 2
b# = b# + 4
c# = c# + 4
INPUT a$
IF a$ = "s" THEN STOP
LOOP
```

```
"C:\Users\Reactor1967\vcns\work\DSOLVED.BAS"
a# = 1
```

```
b# = 3
c# = 5
d# = 7
carry# = 1
RANDOMIZE TIMER
CLS
carry# = 31
redo:
x = 0
DO
REM LOCATE 15, 25
PRINT "1 to 49 +"; a#; b#; c#; d#; carry#
REM PRINT carry#
test = a# <= 49
IF test = -1 THEN code = 1 ELSE code = 0
IF code# = 1 THEN dt = INT(RND * 2) + 0
test1 = (code = 1) AND dt = 0 AND (carry# = a#)
test2 = (code = 1) AND dt = 1 AND (carry# = a#)
IF test1 = -1 THEN carry# = b#
IF test2 = -1 THEN carry# = c#
test3 = (test1 = -1) OR (test2 = -1)
test4 = (test3 = -1) AND x = 0
IF test4 = -1 THEN carry# = d#
x = 1
a# = a# + 2
b# = b# + 4
c# = c# + 4
d# = d# + 4
INPUT a$
REM a$ = INKEY$
IF a$ = "s" THEN STOP
LOOP UNTIL a# = 101
a# = 101
b# = 97
c# = 95
d# = 93
x = 0
DO
REM LOCATE 15, 25
PRINT "101 to 55 -"; a#; b#; c#; d#; carry#
REM PRINT carry#
test = a# >= 55
IF test = -1 THEN code = 1 ELSE code = 0
IF code# = 1 THEN dt = INT(RND * 2) + 0
test1 = (code = 1) AND dt = 0 AND (carry# = a#)
test2 = (code = 1) AND dt = 1 AND (carry# = a#)
```

```
IF test1 = -1 THEN carry# = b#
IF test2 = -1 THEN carry# = c#
test3 = (test1 = -1) OR (test2 = -1)
test4 = (test3 = -1) AND x = 0
IF test4 = -1 THEN carry# = d#
x = 1
a# = a# - 2
b# = b# - 4
c# = c# - 4
d# = d# - 4
INPUT a$
REM a$ = INKEY$
IF a$ = "s" THEN STOP
IF a$ = "d" THEN GOTO decode:
LOOP UNTIL a# = 1
a# = 1
b# = 1
c# = 3
d# = 5
GOTO redo:
decode:




"C:\Users\Reactor1967\vcns\work\DOWN.BAS"
REM: This is intended to be a subroutine for vcframe4.bas: This sub if works
REM: should finish out the vcns project for commerical status for my benifet.
REM                        COPYRIGHT C 2002
REM                    Lloyd Dudley Burris
REM ----------------------Initilize R#-------------------------------------
CLS
DIM r#(255)
r#(1) = 1
REM PRINT r#(1):rem Output controlled here when needed for this sub.
pt1# = 5
rf# = 100
count = 2
a# = 0
replaced1:
DO
a# = a# + pt1#
test = (a# / 2) - INT(a# / 2)
r#(count) = a#
IF (r#(count) / 2) - INT(r#(count) / 2) = 0 THEN r#(count) = r#(count) + 1
REM PRINT r#(count); count
count = count + 1
```

```
IF a$ = "s" THEN STOP
IF count >= 254 THEN EXIT DO
LOOP UNTIL a# = rf#
pt1# = pt1# * 10
rf# = rf# * 10
IF count >= 254 THEN GOTO replaced2:
IF a# + pt1# > 999999999999999# THEN GOTO replaced2:
GOTO replaced1:
replaced2:
r#(254) = 950000000000001#
r#(255) = 999999999999999#
REM ---------------------Code P# up.--------------------------------------
RANDOMIZE TIMER
DO
p# = INT(RND * 999999999999999#) + 0
a$ = INKEY$
IF a$ = "s" THEN STOP
test = p# >= 975000000000000#
LOOP UNTIL test = -1
p# = p# - 25000000000000#
PRINT p#
range = 254
range = range + 1
count = 0
DO
bt = (p# / 2) - INT(p# / 2)
IF bt >= .5 THEN bt = 1
COLOR 2, 0
PRINT p#; "-"; bt; r#(range); count
a$ = INKEY$
IF a$ = "s" THEN STOP
IF (p# - (r#(range) - p# + 1)) < (r#(range - 1)) THEN c = 1 ELSE c = 0
IF c = 0 THEN p# = p# - ((r#(range)) - p# + 1)
IF c = 1 THEN p# = p# - ((r#(range)) - p# + 2)
IF p# < (r#(range - 1)) THEN range = range - 1
count = count + 1
LOOP UNTIL p# <= 475000000000000#
bt = (p# / 2) - INT(p# / 2)
IF bt >= .5 THEN bt = 1
IF bt = 0 THEN p# = p# + 1
bt = (p# / 2) - INT(p# / 2)
IF bt >= .5 THEN bt = 1
PRINT p#; "-"; bt; r#(range); count
a$ = INKEY$
REM INPUT a$
IF a$ = "s" THEN SYSTEM
```

```
REM IF a$ = "d" THEN GOTO decode2:
range = range - 1
```

```
"C:\Users\Reactor1967\vcns\work\DOUBLE.BAS"
REM v# = v# + ((v# - r#) * base) + n:for more than 1 state
REM Having binary 2 with 4 states for coding data and system commands.
REM Here each odd equals a  binary 1
REM Here each even equals a binary 0
REM but each even number has two states for coding system commands
REM and each odd  number has two states for coding system commands
v# = 2
r1# = 4
r2# = 6
CLS
DO
PRINT v#; r1#; r2#; "="; r1# / 2; r2# / 2
r1# = r1# + 4
r2# = r2# + 4
v# = v# + 2
INPUT a$
IF a$ = "s" THEN STOP
IF a$ = "d" THEN EXIT DO
LOOP
v# = 1
r1# = 3
r2# = 5
CLS
DO
PRINT v#; r1#; r2#; "="; r1# / 2; r2# / 2
r1# = r1# + 4
r2# = r2# + 4
v# = v# + 2
INPUT a$
IF a$ = "s" THEN STOP
LOOP
```

```
"C:\Users\Reactor1967\vcns\work\DISTCALC.BAS"
REM Distance Calculator
REM If the distance equals one of the two numbers
REM on the right side then the distance to the next
REM vector equals the left side of the chart.
d# = 1
```

```
CLS
DO
PRINT d#; (d# * 2) + 1; (d# * 2) + 2
INPUT a$
IF a$ = "s" THEN STOP
d# = d# + 1
LOOP
```

```
"C:\Users\Reactor1967\vcns\work\DISTCAL.BAS"
REM this is a distance calculator for figuring distance before
REM coding a vector and distance after coding a vector
REM you can start off with anything and add anything
REM If should closely go by the vector equations you are using
REM v# = v# + ((v# - r#) * base-1) + n
REM v# = v# - ((r# - v#) * base-1) + n
REM the plus and minus signs are interchangable depending on what
REM you are trying to do. So is the v# and r#
dist# = 2
CLS
DO
PRINT , , dist#; (dist# + 2) + dist#; (dist# + 4) + dist#
dist# = dist# + 2
a$ = INKEY$
INPUT a$
IF a$ = "s" THEN STOP
LOOP
```

```
"C:\Users\Reactor1967\vcns\work\DISTANCE.BAS"
a# = 1
b# = 2
c# = 3
CLS
RANDOMIZE TIMER
distance# = ABS(a# - b#)
redo:
DO
d = INT(RND * 2) + 0
IF d = 0 THEN st# = b#
IF d = 1 THEN st# = c#
a# = st#
```

```
DO
distance# = distance# + 1
test = ((a# + distance#) / 2) - (INT((a# + distance#) / 2)) = 0
LOOP UNTIL test = -1
b# = a# + distance#
c# = a# + distance# + 1
PRINT a#; b#; c#; distance#
INPUT a$
REM a$ = INKEY$
IF a$ = "s" THEN STOP
LOOP UNTIL distance# > 200
DO
d = INT(RND * 2) + 0
IF d = 0 THEN st# = b#
IF d = 1 THEN st# = c#
a# = st#
DO
distance# = distance# - 1
test = ((a# + distance#) / 2) - (INT((a# + distance#) / 2)) = 0
LOOP UNTIL test = -1
b# = a# + distance#
c# = a# + distance# + 1
PRINT a#; b#; c#; distance#
INPUT a$
REM a$ = INKEY$
IF a$ = "s" THEN STOP
LOOP UNTIL distance# < 100
GOTO redo:


"C:\Users\Reactor1967\vcns\work\DFROLL.BAS"
p# = 1
r# = 1
CLS
DO
flip = 0
c = INT(RND * 2) + 0
IF c = 0 THEN p# = p# + (p# - r# + 1)
IF c = 1 THEN p# = p# + (p# - r# + 2)
PRINT , , p#
IF p# + (p# - r#) > 100000000000000# THEN flip = 1
IF p# + (p# - r#) > 200000000000000# THEN flip = 2
IF p# + (p# - r#) > 300000000000000# THEN flip = 3
IF p# + (p# - r#) > 400000000000000# THEN flip = 4
IF p# + (p# - r#) > 500000000000000# THEN flip = 5
```

```
IF p# + (p# - r#) > 600000000000000# THEN flip = 6
IF p# + (p# - r#) > 700000000000000# THEN flip = 7
IF p# + (p# - r#) > 800000000000000# THEN flip = 8
IF p# + (p# - r#) > 900000000000000# THEN flip = 9
IF p# + (p# - r#) < 100000000000000# THEN p# = p# + (p# - r# + 1)
IF p# + (p# - r#) > 100000000000000# THEN p# = p# + (p# - r# + 2)
IF flip = 1 THEN p# = p# - 100000000000000#
IF flip = 2 THEN p# = p# - 200000000000000#
IF flip = 3 THEN p# = p# - 300000000000000#
IF flip = 4 THEN p# = p# - 400000000000000#
IF flip = 5 THEN p# = p# - 500000000000000#
IF flip = 6 THEN p# = p# - 600000000000000#
IF flip = 7 THEN p# = p# - 700000000000000#
IF flip = 8 THEN p# = p# - 800000000000000#
IF flip = 9 THEN p# = p# - 900000000000000#
COLOR 2, 1
PRINT , , p#
REM INPUT a$
a$ = INKEY$
IF a$ = "s" THEN STOP
PRINT " "
LOOP UNTIL p# > 999999999999999#




"C:\Users\Reactor1967\vcns\work\DEL2.BAS"
v# = 101
v1# = 101
r# = 101
CLS
RANDOMIZE TIMER
DO
z = INT(RND * 2) + 0
IF z = 0 THEN v# = v# + ABS(v# - v1#) + 1
IF z = 1 THEN v# = v# + ABS(v# - v1#) + 2
IF v1# + (v1# - r#) >= (r# + 50) THEN result = 1 ELSE result = 0
IF result = 0 THEN v1# = v1# + (v1# - r# + 1)
IF result = 1 THEN v1# = v1# + (v1# - r# + 2)
IF result = 1 THEN r# = r# + 52
COLOR 2, 0
PRINT , , v1#, r#; v#
a$ = INKEY$
REM INPUT a$
IF a$ = "s" THEN STOP
```

LOOP

```
"C:\Users\Reactor1967\vcns\work\DEL1.BAS"
v# = 1
r# = 1
high# = 0
CLS
RANDOMIZE TIMER
carry# = 1
DO
FOR count = 1 TO 10
z = INT(RND * 2) + 0
v1# = v# + ((v# - r#) * 1) + 1
v2# = v# + ((v# - r#) * 1) + 2
IF v1# > high# THEN high# = v1#
DO
IF v1# < high# THEN r# = r# + 2
v1# = v# + ((v# - r#) * 1) + 1
v2# = v# + ((v# - r#) * 1) + 2
a$ = INKEY$
IF a$ = "s" THEN STOP
IF v1# >= high# THEN EXIT DO
LOOP
IF v2# > high# THEN high# = v2#
IF v1# > high# THEN high# = v1#
test1 = (carry# = v#) AND (z = 0)
test2 = (carry# = v#) AND (z = 1)
IF test1 = -1 THEN carry# = v1#
IF test2 = -1 THEN carry# = v2#
PRINT r#; v#; "="; v1#; v2#; carry#
NEXT count
r# = r# + 10
LOOP
```

```
"C:\Users\Reactor1967\vcns\work\DCODE2.BAS"
a# = 102
d# = 1
CLS
RANDOMIZE TIMER
redo:
DO
```

```
z = INT(RND * 2) + 0
IF (d# * 2) >= 500 THEN z = 2
IF z = 0 THEN a# = a# + ((a# - r#) * 2) + 1
IF z = 1 THEN a# = a# + ((a# - r#) * 2) + 2
IF z = 2 THEN a# = a# + ((a# - r#) * 2) + 3
IF z = 0 THEN d# = (d# * 3) + 1
IF z = 1 THEN d# = (d# * 3) + 2
IF z = 2 THEN d# = d# * 2
PRINT z; a#; d#; a# / 3
IF z = 2 THEN d# = d# - 500
r# = a# - d#
INPUT a$
IF a$ = "s" THEN STOP
LOOP


"C:\Users\Reactor1967\vcns\work\DCODE.BAS"
a# = 2
r# = 1
d# = 1
RANDOMIZE TIMER
CLS
redo:
DO
z = INT(RND * 2) + 0
IF z = 0 THEN a# = a# + (a# - r# + 1)
IF z = 1 THEN a# = a# + (a# - r# + 2)
test = (a# / 2) - (INT(a# / 2))
IF test = 0 THEN d# = (d# * 2) + 1
IF test >= .5 THEN d# = (d# * 2)
PRINT ; , , a#; d#
IF d# >= 500 THEN d# = d# - 500
r# = a# - d#
REM INPUT a$
a$ = INKEY$
IF a$ = "s" THEN STOP
LOOP UNTIL a# >= 9000
r# = a# + d#
DO
z = INT(RND * 2) + 0
IF z = 0 THEN a# = a# - (r# - a# + 1)
IF z = 1 THEN a# = a# - (r# - a# + 2)
test = (a# / 2) - (INT(a# / 2))
IF test = 0 THEN d# = (d# * 2) + 1
IF test >= .5 THEN d# = (d# * 2)
```

```
PRINT ; , , a#; d#
IF d# >= 500 THEN d# = d# - 500
r# = a# + d#
REM INPUT a$
a$ = INKEY$
IF a$ = "s" THEN STOP
LOOP UNTIL a# <= 2000
r# = a# - d#
GOTO redo:


"C:\Users\Reactor1967\vcns\work\CTEST.BAS"
p# = 1
r1# = 1
r2# = 499999999999999#
RANDOMIZE TIMER
CLS
z = 0
repeat:
DO
IF p# + (p# - r1# + 1) < 250000000000000# THEN c = 0
IF p# + (p# - r1# + 1) >= 250000000000000# THEN c = 1
IF c = 0 THEN p# = p# + (p# - r1# + 1)
IF c = 1 THEN p# = p# + (p# - r1# + 2)
PRINT , , p#; c
REM INPUT a$
a$ = INKEY$
IF a$ = "s" THEN STOP
LOOP UNTIL p# >= 250000000000000#
IF z = 0 THEN z = 1 ELSE z = 0
IF z = 0 THEN p# = p# - 250000000000000#
IF z = 0 THEN GOTO repeat:
REM p# = p# + (p# - r1# + 2)
DO
IF p# - (r2# - p# + 1) > 250000000000000# THEN c = 0
IF p# - (r2# - p# + 1) < 250000000000000# THEN c = 1
IF c = 0 THEN p# = p# - (r2# - p# + 1)
IF c = 1 THEN p# = p# - (r2# - p# + 2)
PRINT , , p#; c
REM INPUT a$
a$ = INKEY$
IF a$ = "s" THEN STOP
LOOP UNTIL p# - (r2# - p# + 2) <= 0
GOTO repeat:
```

```
"C:\Users\Reactor1967\vcns\work\CRSSOVR.BAS"
a# = 1
b# = 2
CLS
DO
PRINT a# - 1; a#; b#; b# + 1
a# = a# + 1
b# = b# + 2
INPUT a$
IF a$ = "s" THEN STOP
LOOP




"C:\Users\Reactor1967\vcns\work\CRDGENB3.BAS"
DIM up1#(500)
DIM up2#(500)
DIM up3#(500)
DIM dn1#(500)
DIM dn2#(500)
DIM dn3#(500)
a# = 1
b# = 1
c# = 3
CLS
count = 1
RANDOMIZE TIMER
DO
REM PRINT a#; b#; c#
up1#(count) = a#
up2#(count) = b#
up3#(count) = c#
a# = a# + 2
b# = b# + 4
c# = c# + 4
REM INPUT a$
a$ = INKEY$
IF a$ = "s" THEN STOP
count = count + 1
LOOP UNTIL count = 501
REM PRINT a#; b#; c#
count = 1
a# = 2003
b# = 2001
```

```
c# = 2003
PRINT "---------"
DO
REM PRINT a#; b#; c#; count
dn1#(count) = a#
dn2#(count) = b#
dn3#(count) = c#
a# = a# - 2
b# = b# - 4
c# = c# - 4
REM INPUT a$
a$ = INKEY$
IF a$ = "s" THEN STOP
count = count + 1
LOOP UNTIL count = 501
REM PRINT a#; b#; c#
REM -------------------------------------------------------------
count = 1
CLS
redo:
DO
PRINT "+"; up1#(count); up2#(count); up3#(count)
z = INT(RND * 2) + 0
IF z = 0 THEN a# = up2#(count)
IF z = 1 THEN a# = up3#(count)
IF a# >= 1000 THEN EXIT DO
count = 1
DO
IF a# = up1#(count) THEN EXIT DO
a$ = INKEY$
IF a$ = "s" THEN STOP
count = count + 1
LOOP
REM change point 1.............
REM INPUT a$
a$ = INKEY$
IF a$ = "s" THEN STOP
LOOP
count = 1
DO
IF dn1#(count) = a# THEN EXIT DO
count = count + 1
a$ = INKEY$
IF a$ = "s" THEN STOP
LOOP
PRINT "-----------------------------------"
```

```
REM change point 2 .............
REM INPUT a$
a$ = INKEY$
IF a$ = "s" THEN STOP
DO
PRINT "-"; dn1#(count); dn2#(count); dn3#(count)
z = INT(RND * 2) + 0
IF z = 0 THEN a# = dn2#(count)
IF z = 1 THEN a# = dn3#(count)
IF a# <= 1000 THEN EXIT DO
count = 1
DO
count = count + 1
IF dn1#(count) = a# THEN EXIT DO
a$ = INKEY$
IF a$ = "s" THEN STOP
LOOP
REM change point 3 -----------
a$ = INKEY$
REM INPUT a$
IF a$ = "s" THEN STOP
LOOP
count = 1
DO
IF up1#(count) = a# THEN EXIT DO
count = count + 1
a$ = INKEY$
IF a$ = "s" THEN STOP
LOOP
PRINT "----------------------------------"
REM change point 4 .............
REM INPUT a$
a$ = INKEY$
IF a$ = "s" THEN STOP
GOTO redo:


"C:\Users\Reactor1967\vcns\work\COVCNS.BAS"
DECLARE SUB coding (c, p#, code$, store#, b$, outp)
CLS
p# = 1
PRINT "This will code any file to vcns for demonstration of the power of vcns"
INPUT "the name of your program"; file$
INPUT "Do you want to see the output y/n (faster n)"; a$
IF a$ = "Y" THEN outp = 1
```

```
IF a$ = "y" THEN outp = 1
IF a$ = "n" THEN outp = 0
IF a$ = "N" THEN outp = 0
OPEN file$ FOR BINARY ACCESS READ AS #1
filenum = 1
DO
GET #1, filenum, vcns%
a$ = INKEY$
IF a$ = "s" THEN CLOSE #1
IF a$ = "s" THEN SYSTEM
filenum = filenum + 1
b$ = STR$(vcns%)
GOSUB coding:
LOOP UNTIL (EOF(1))
CLOSE #1
CLS
LOCATE 20, 30
PRINT "Your done"
PRINT c1; p#; code$
INPUT "Enter the name of the file to save under."; file$
OPEN file$ FOR OUTPUT AS #1
PRINT #1, p#, code$
CLOSE #1
SYSTEM
coding:
b$ = LTRIM$(b$)
b$ = RTRIM$(b$)
length = LEN(b$)
FOR count1 = 1 TO length
b$ = LTRIM$(b$)
b$ = RTRIM$(b$)
z = ASC(MID$(b$, count1, 1))
b2$ = STR$(z)
b2$ = LTRIM$(b2$)
b2$ = RTRIM$(b2$)
length2 = LEN(b2$)
FOR count2 = 1 TO length2
b2$ = LTRIM$(b2$)
b2$ = RTRIM$(b2$)
bc$ = MID$(b2$, count2, 1)
IF bc$ = "0" THEN o$ = "0000"
IF bc$ = "1" THEN o$ = "0001"
IF bc$ = "2" THEN o$ = "0010"
IF bc$ = "3" THEN o$ = "0011"
IF bc$ = "4" THEN o$ = "0100"
IF bc$ = "5" THEN o$ = "0101"
```

```
IF bc$ = "6" THEN o$ = "0110"
IF bc$ = "7" THEN o$ = "0111"
IF bc$ = "8" THEN o$ = "1000"
IF bc$ = "9" THEN o$ = "1001"
FOR count3 = 1 TO 4
c = VAL(MID$(o$, count3, 1))
CALL coding(c, p#, code$, store#, b$, outp)
NEXT count3
NEXT count2
NEXT count1
RETURN

SUB coding (c, p#, code$, store#, b$, outp)
r# = 1
r2# = 999999999999999#
RANDOMIZE TIMER
start:
IF ABS(p# - store#) > 250000000000000# THEN flip = 1 ELSE flip = 0
IF p# - store# > 0 THEN code$ = "+"
IF p# - store# < 0 THEN code$ = "-"
REM We took our random number generator out here.
REM substituting real data for the first time
test1 = (p# < 500000000000000#) AND (c = 0)
test2 = (p# < 500000000000000#) AND (c = 1)
test3 = (p# > 500000000000000#) AND (c = 0)
test4 = (p# > 500000000000000#) AND (c = 1)
store# = p#: REM using this to tell things
IF test1 = -1 THEN GOSUB test1:
IF test2 = -1 THEN GOSUB test2:
IF test3 = -1 THEN GOSUB test3:
IF test4 = -1 THEN GOSUB test4:
c1 = p# / 2 - INT(p# / 2)
IF c1 >= .5 THEN c1 = 1
IF outp = 1 THEN PRINT "                    ", c1; p#; code$; flip, b$
IF outp = 0 THEN LOCATE 20, 30
IF outp = 0 THEN PRINT "Working......";
EXIT SUB
test1:
p# = p# + (p# - r# + 1)
RETURN
test2:
p# = p# + (p# - r# + 2)
RETURN
test3:
p# = p# - (r2# - p# + 1)
RETURN
```

```
test4:
p# = p# - (r2# - p# + 2)
RETURN


END SUB



"C:\Users\Reactor1967\vcns\work\COUNTR.BAS"
r# = 1
CLS
PRINT "Working........"
repeat:
p# = 999999999999999#
count = 1
DO
PRINT p#; count
t# = p#
t# = p# - r#
IF t# / 2 - INT(t# / 2) >= .5 THEN t# = t# - 1
t# = t# / 2
p# = p# - t#
count = count + 1
INPUT a$
REM a$ = INKEY$
IF a$ = "s" THEN STOP
LOOP UNTIL p# <= r# + 2
PRINT r#; count; INT(r# / 2)
IF count < 50 THEN PRINT p#; r#; count
IF count < 50 THEN STOP
r# = r# + 2
a$ = INKEY$
IF a$ = "s" THEN STOP
GOTO repeat:



"C:\Users\Reactor1967\vcns\work\CONTROLV.BAS"
CLS
p# = 125678909876543#
redo:
r# = p# - 25000000000000#
t# = (r# / 2) - INT(r# / 2)
IF t# = 0 THEN r# = r# + 1
RANDOMIZE TIMER
```

```
count = 1
DO
PRINT , , p#; c; r#
c = INT(RND * 2) + 0
IF c = 0 THEN p# = p# + (p# - r# + 1)
IF c = 1 THEN p# = p# + (p# - r# + 2)
r# = p# - 25000000000000#
t# = (r# / 2) - INT(r# / 2)
IF t# = 0 THEN r# = r# + 1
a$ = INKEY$
IF a$ = "s" THEN STOP
count = count + 1
LOOP UNTIL p# + (p# - r#) > 925000000000000#
r# = p# + 25000000000000#
t# = (r# / 2) - INT(r# / 2)
IF t# = 0 THEN r# = r# + 1
DO
PRINT , , p#; c; r#
c = INT(RND * 2) + 0
IF c = 0 THEN p# = p# - (r# - p# + 1)
IF c = 1 THEN p# = p# - (r# - p# + 2)
r# = p# + 25000000000000#
t# = (r# / 2) - INT(r# / 2)
IF t# = 0 THEN r# = r# + 1
a$ = INKEY$
IF a$ = "s" THEN STOP
count = count + 1
LOOP UNTIL p# - (r# - p#) < 150000000000000#
GOTO redo:




"C:\Users\Reactor1967\vcns\work\CONTROLR.BAS"
v# = 1
CLS
RANDOMIZE TIMER
redo:
DO
z = INT(RND * 2) + 0
a# = v# - r#
r# = v#
REM r# = 1
IF (r# / 2) - INT(r# / 2) = 0 THEN r# = r# - 1
DO
```

```
r# = r# - 2
IF z = 0 THEN v1# = v# + (v# - r# + 1)
IF z = 1 THEN v1# = v# + (v# - r# + 2)
REM IF v1# - r# < a# THEN EXIT DO
IF v1# - r# > a# THEN EXIT DO
a$ = INKEY$
IF a$ = "s" THEN STOP
LOOP
IF z = 0 THEN v# = v# + (v# - r# + 1)
IF z = 1 THEN v# = v# + (v# - r# + 2)
COLOR 2, 0
PRINT , , r#; v#; v# - r#
IF v# - r# = 84 THEN STOP
a$ = INKEY$
REM INPUT a$
IF a$ = "s" THEN STOP
LOOP UNTIL v# - r# > 100
DO
z = INT(RND * 2) + 0
a# = v# - r#
REM r# = v#
r# = v# - a#
IF (r# / 2) - INT(r# / 2) = 0 THEN r# = r# - 1
DO
r# = r# + 2
IF z = 0 THEN v1# = v# + (v# - r# + 1)
IF z = 1 THEN v1# = v# + (v# - r# + 2)
IF v1# - r# < a# THEN EXIT DO
REM IF v1# - r# > a# THEN EXIT DO
a$ = INKEY$
IF a$ = "s" THEN STOP
LOOP
IF z = 0 THEN v# = v# + (v# - r# + 1)
IF z = 1 THEN v# = v# + (v# - r# + 2)
COLOR 2, 0
PRINT , , r#; v#; v# - r#
a$ = INKEY$
REM INPUT a$
IF a$ = "s" THEN STOP
LOOP UNTIL v# - r# < 10
GOTO redo:


"C:\Users\Reactor1967\vcns\work\CONTROL5.BAS"
r# = 100000000000000#
```

```
r1# = 100000000000000#
p# = 125000000000000#
t1# = 125000000000000#
RANDOMIZE TIMER
CLS
redo:
DO
PRINT , p#; c; r# + 1; p# - r#
c = INT(RND * 2) + 0
IF c = 0 THEN p# = p# + (p# - (r# + 1) + 1)
IF c = 1 THEN p# = p# + (p# - (r# + 1) + 2)
t2# = p# - t1#
r# = r1# + t2#
REM INPUT a$
a$ = INKEY$
IF a$ = "s" THEN STOP
LOOP UNTIL p# + (p# - (r# + 1)) >= 900000000000000#
PRINT "----------------------------------------"
t2# = p# - t1#
r# = r1# + t2# + 50000000000000#
DO
PRINT , p#; c; r# + 1; r# - p#
c = INT(RND * 2) + 0
IF c = 0 THEN p# = p# - ((r# + 1) - p# + 1)
IF c = 1 THEN p# = p# - ((r# + 1) - p# + 2)
t2# = p# - t1#
r# = r1# + t2# + 50000000000000#
a$ = INKEY$
REM INPUT a$
IF a$ = "s" THEN STOP
LOOP UNTIL p# + ((r# + 1) - p#) <= 200000000000000#
r# = r# - 50000000000000#
PRINT "----------------------------------------"
GOTO redo:


"C:\Users\Reactor1967\vcns\work\CONTROL4.BAS"
DIM vectors#(1000)
p# = 123456789098765#
CLS
count = 0
t = 1
DO
test# = (p# / 2) - INT(p# / 2)
```

```
IF test# >= .5 THEN add# = 25000000000001#
IF test# = 0 THEN add# = 25000000000000#
PRINT p#; (p# + add#); (p# + add# + 1)
count = count + 1
vectors#(count) = (p# + add#)
count = count + 1
vectors#(count) = (p# + add# + 1)
REM INPUT a$
a$ = INKEY$
IF a$ = "s" THEN STOP
p# = vectors#(t)
t = t + 1
IF count >= 1000 THEN EXIT DO
LOOP
t = 1
CLS
PRINT "Working...."
rpt# = 1
DO
p# = vectors#(t)
FOR count = (t + 1) TO 1000
IF p# = vectors#(count) THEN PRINT rpt#
IF p# = vectors#(count) THEN rpt# = rpt# + 1
IF p# = vectors#(count) THEN EXIT FOR
a$ = INKEY$                         s
IF a$ = "s" THEN STOP
NEXT count
t = t + 1
a$ = INKEY$
IF a$ = "s" THEN STOP
LOOP UNTIL t = 1000


"C:\Users\Reactor1967\vcns\work\CONTROL3.BAS"
DIM vectors#(7999)
p# = 112345678909871#
r1# = 100000000000000#
r2# = 150000000000000#
count = 0
CLS
t# = 1
DO
PRINT p#; "="; p# + (p# - (r1# + 1) + 1); p# + (p# - (r1# + 1) + 2)
count = count + 1
vectors#(count) = p# + (p# - (r1# + 1) + 1)
```

```
count = count + 1
vectors#(count) = p# + (p# - (r1# + 1) + 2)
IF p# >= r2# THEN r1# = r1# + 50000000000000#
IF p# >= r2# THEN r2# = r2# + 50000000000000#
INPUT a$
IF a$ = "s" THEN STOP
p# = vectors#(t#)
t# = t# + 1
LOOP UNTIL count >= 7998


"C:\Users\Reactor1967\vcns\work\CONTROL2.BAS"
DIM vectors#(999)
p# = 125098765432123#
count = 0
redo:
CLS
t# = 1
man = 1
DO
r# = p# - 25000000000000#
test# = (r# / 2) - INT(r# / 2)
IF test# = 0 THEN r# = r# + 1
PRINT p#; (p# + (p# - r# + 1)); (p# + (p# - r# + 2)); man
count = count + 1
vectors#(count) = p# + (p# - r# + 1)
count = count + 1
vectors#(count) = p# + (p# - r# + 2)
REM a$ = INKEY$
INPUT a$
IF a$ = "s" THEN STOP
p# = vectors#(t#)
t# = t# + 1
IF count >= 998 THEN EXIT DO
man = man + 1
LOOP


"C:\Users\Reactor1967\vcns\work\CONFRM.BAS"
RANDOMIZE TIMER
DO
p# = INT(RND * 999999999999999#) + 0
LOOP UNTIL p# >= 500000000000000#
```

```
PRINT p#
r# = p#
count = 0
DO
count = count + 1
c = INT(RND * 2) + 0
IF c = 0 THEN p# = p# - ((r# - p#) + 1)
IF c = 1 THEN p# = p# - ((r# - p#) + 2)
PRINT p#; count
a$ = INKEY$
IF a$ = "s" THEN STOP
LOOP UNTIL p# - (r# - p#) <= 0
a# = 1
DO
count = count - 1
p# = p# - a#
a# = (a# * 2) + 1
test1 = p# - r# <= 50
test2 = r# - p# <= 50
IF test1 = -1 THEN STOP
IF test2 = -1 THEN STOP
LOOP
```

```
"C:\Users\Reactor1967\vcns\work\CODEVR.BAS"
x1# = r#
x2# = 0
x3# = 0
CLS
```

```
DO
x1# = x1# + 1
x2# = x1# + (x1# - r# + 2)
x3# = x1# + (x1# - r# + 4)
PRINT , , x1#; x2#; x3#
IF x1# = v# THEN STOP
a$ = INKEY$
IF a$ = "s" THEN STOP
IF x1# > v# THEN PRINT "OUT OF RANGE ERROR!!!"
IF x1# > v# THEN STOP
LOOP
IF z = 0 THEN v# = x1#
IF z = 1 THEN v# = x2#
```


"C:\Users\Reactor1967\vcns\work\CODE3.BAS"
```
a# = 2
r# = 1
CLS
RANDOMIZE TIMER
redo:
DO
PRINT z; a#; r#; a# / 3
REM a$ = INKEY$
INPUT a$
IF a$ = "s" THEN STOP
z = INT(RND * 3) + 0
IF z = 0 THEN a# = a# + r# + 1
IF z = 1 THEN a# = a# + r# + 2
IF z = 2 THEN a# = a# + r# + 3
r# = r# * 3
IF z = 0 THEN r# = r# + 1
IF z = 1 THEN r# = r# + 2
IF z = 2 THEN r# = r# + 3
LOOP UNTIL a# >= 500000000000000#
PRINT "------------------------------"
a# = a# - 500000000000000#
r# = r# - 500000000000000#
GOTO redo:
```


"C:\Users\Reactor1967\vcns\work\CODE1.BAS"
```
REM PRINT v4#; v3#; v2#; v1#, v8#; v7#; v6#; v5#, v12#; v11#; v10#; v9#,
```

```
v16#; v15#; v14#; v13#
REM PRINT pos$, vec$, c1$, c2$
REM WRITE #1, v16#, v15#, v14#, v13#, v12#, v11#, v10#, v9#, v8#, v7#,
v6#, v5#, v4#, v3#, v2#, v1#
REM WRITE #1, pos$, vec$, c1$, c2$
RANDOMIZE TIMER
CLS
test$ = "chart1"
OPEN test$ FOR INPUT AS #1
INPUT #1, v16#, v15#, v14#, v13#, v12#, v11#, v10#, v9#, v8#, v7#, v6#, v5#,
v4#, v3#, v2#, v1#
INPUT #1, pos$, vec$, c1$, c2$
DO
z = INT(RND * 2) + 0
IF z = 0 THEN carry$ = c1$
IF z = 1 THEN carry$ = c2$
PRINT pos$, vec$, c1$, c2$, carry$; z
a$ = INKEY$
REM INPUT a$
IF a$ = "s" THEN CLOSE #1
IF a$ = "s" THEN STOP
DO
INPUT #1, v16#, v15#, v14#, v13#, v12#, v11#, v10#, v9#, v8#, v7#, v6#, v5#,
v4#, v3#, v2#, v1#
INPUT #1, pos$, vec$, c1$, c2$
stop$ = INKEY$
IF stop$ = "s" THEN CLOSE #1
IF stop$ = "s" THEN STOP
LOOP UNTIL carry$ = vec$
LOOP


"C:\Users\Reactor1967\vcns\work\CNTRLR3.BAS"
v# = 100
r# = 99
a# = v# - r#
CLS
RANDOMIZE TIMER
DO
z = INT(RND * 2) + 0
r# = v# - a#
IF (r# / 2) - INT(r# / 2) = 0 THEN r# = r# - 1
DO
r# = r# - 2
IF z = 0 THEN v1# = v# + (v# - r# + 1)
```

```
IF z = 1 THEN v1# = v# + (v# - r# + 2)
a$ = INKEY$
IF a$ = "s" THEN STOP
LOOP UNTIL v1# - r# > a#
a# = v# - r#
IF z = 0 THEN v# = v# + (v# - r# + 1)
IF z = 1 THEN v# = v# + (v# - r# + 2)
COLOR 2, 0
PRINT , , r#; v#; v# - r#
REM INPUT a$
a$ = INKEY$
IF a$ = "s" THEN STOP
LOOP
```

```
"C:\Users\Reactor1967\vcns\work\CMBNCHRT.BAS"
DECLARE SUB crd2 (v13#, v14#, v15#, v16#, c2$)
DECLARE SUB vector (v5#, v6#, v7#, v8#, vec$)
DECLARE SUB position (v1#, v2#, v3#, v4#, pos$)
DECLARE SUB crd1 (v9#, v10#, v11#, v12#, c1$)
v1# = 0
v2# = 0
v3# = 0
v4# = 0
v5# = 0
v6# = 0
v7# = 0
v8# = 0
v9# = 0
v10# = 0
v11# = 0
v12# = 0
v13# = 1
v14# = 0
v15# = 0
v16# = 0
CLS
chart$ = "chart1"
OPEN chart$ FOR OUTPUT AS #1
DO
track# = track# + 1
CALL position(v1#, v2#, v3#, v4#, pos$)
CALL vector(v5#, v6#, v7#, v8#, vec$)
FOR count = 1 TO 2
CALL crd1(v9#, v10#, v11#, v12#, c1$)
```

```
CALL crd2(v13#, v14#, v15#, v16#, c2$)
NEXT count
COLOR 2, 0
PRINT track#, pos$, vec$, c1$, c2$
WRITE #1, v16#, v15#, v14#, v13#, v12#, v11#, v10#, v9#, v8#, v7#, v6#,
v5#, v4#, v3#, v2#, v1#
WRITE #1, pos$, vec$, c1$, c2$
a$ = INKEY$
REM INPUT a$
IF a$ = "s" THEN CLOSE #1
IF a$ = "s" THEN STOP
REM IF a$ = "s" THEN EXIT DO
LOOP UNTIL v6# >= 999
CLOSE #1
STOP
OPEN chart$ FOR INPUT AS #1
DO
INPUT #1, v16#, v15#, v14#, v13#, v12#, v11#, v10#, v9#, v8#, v7#, v6#, v5#,
v4#, v3#, v2#, v1#
INPUT #1, pos$, vec$, c1$, c2$
PRINT v4#; v3#; v2#; v1#, v8#; v7#; v6#; v5#, v12#; v11#; v10#; v9#, v16#;
v15#; v14#; v13#
PRINT pos$, vec$, c1$, c2$
INPUT a$
IF a$ = "s" THEN STOP
LOOP
CLOSE #1

SUB crd1 (v9#, v10#, v11#, v12#, c1$)
v9# = v9# + 1
IF v9# = 1000 THEN v10# = v10# + 1
IF v10# = 1000 THEN v11# = v11# + 1
IF v11# = 1000 THEN v12# = v12# + 1
IF v9# = 1000 THEN v9# = 0
IF v10# = 1000 THEN v10# = 0
IF v11# = 1000 THEN v11# = 0
IF v12# >= 1000 THEN STOP
c1$ = ""
c1$ = c1$ + STR$(v12#) + STR$(v11#) + STR$(v10#) + STR$(v9#)
END SUB

SUB crd2 (v13#, v14#, v15#, v16#, c2$)
v13# = v13# + 1
IF v13# = 1000 THEN v14# = v14# + 1
IF v14# = 1000 THEN v15# = v15# + 1
IF v15# = 1000 THEN v16# = v16# + 1
```

```
IF v13# = 1000 THEN v13# = 0
IF v14# = 1000 THEN v14# = 0
IF v15# = 1000 THEN v15# = 0
IF v16# >= 1000 THEN STOP
c2$ = ""
c2$ = c2$ + STR$(v16#) + STR$(v15#) + STR$(v14#) + STR$(v13#)
END SUB


SUB position (v1#, v2#, v3#, v4#, pos$)
v1# = v1# + 1
IF v1# = 1000 THEN v2# = v2# + 1
IF v2# = 1000 THEN v3# = v3# + 1
IF v3# = 1000 THEN v4# = v4# + 1
IF v1# = 1000 THEN v1# = 0
IF v2# = 1000 THEN v2# = 0
IF v3# = 1000 THEN v3# = 0
IF v4# >= 1000 THEN STOP
pos$ = ""
pos$ = STR$(v4#) + STR$(v3#) + STR$(v2#) + STR$(v1#)
END SUB


SUB vector (v5#, v6#, v7#, v8#, vec$)
v5# = v5# + 1
IF v5# = 1000 THEN v6# = v6# + 1
IF v6# = 1000 THEN v7# = v7# + 1
IF v7# = 1000 THEN v8# = v8# + 1
IF v5# = 1000 THEN v5# = 0
IF v6# = 1000 THEN v6# = 0
IF v7# = 1000 THEN v7# = 0
IF v8# >= 1000 THEN STOP
vec$ = ""
vec$ = vec$ + STR$(v8#) + STR$(v7#) + STR$(v6#) + STR$(v5#)
END SUB




"C:\Users\Reactor1967\vcns\work\CHRTB2.BAS"
CLS
v# = 1000
r# = 1001
dist# = 50
count1# = 1
count2# = 1
count3# = 1
test1$ = "bank"
REM put instructions here for restarting. comment out other lines.
```

```
OPEN test1$ FOR OUTPUT AS #1
WRITE #1, count1#, r#, v#
CLOSE #1
DO
test1$ = "bank"
OPEN test1$ FOR INPUT AS #1
DO
INPUT #1, a#, b#, c#
stop$ = INKEY$
IF stop$ = "s" THEN STOP
LOOP UNTIL a# = count1#
CLOSE #1
count1# = count1# + 1
v# = c#
rs# = b#
r# = 1001
redo:
DO
v1# = v# - (r# - v# + 1)
v2# = v# - (r# - v# + 2)
test = (r# - v1#) <= dist#
r# = r# - 2
IF test = -1 THEN EXIT DO
stop$ = INKEY$
IF stop$ = "s" THEN STOP
LOOP
IF r# < v# THEN dist# = dist# + 1
IF r# < v# THEN r# = 1001
IF r# < v# THEN GOTO redo:
test1$ = "bank"
OPEN test1$ FOR INPUT AS #1
DO
INPUT #1, a#, b#, c#
test = ((v1# = c#) AND (r# = b#)) OR ((v2# = c#) AND (r# = b#))
IF test = -1 THEN EXIT DO
stop$ = INKEY$
IF stop$ = "s" THEN STOP
LOOP UNTIL a# = count2#
CLOSE #1
IF test = -1 THEN GOTO redo:
COLOR 2, 0
PRINT , , v#; rs#; v1#; v2#; r#; r# - v#; r# - v1#
a$ = INKEY$
REM INPUT a$
IF a$ = "s" THEN STOP
count2# = count2# + 1
```

```
test1$ = "bank"
OPEN test1$ FOR APPEND AS #1
WRITE #1, count2#, r#, v1#
count2# = count2# + 1
WRITE #1, count2#, r#, v1#
CLOSE #1
test1$ = "chartu"
OPEN test1$ FOR APPEND AS #1
WRITE #1, count3#, rs#, v#, v1#, v2#, r#
CLOSE #1#
count3# = count3# + 1
test1$ = "vcns"
OPEN test1$ FOR OUTPUT AS #1
WRITE #1, count1#, count2#, dist#
CLOSE #1
LOOP UNTIL v# < 100


"C:\Users\Reactor1967\vcns\work\CHRTB1.BAS"
CLS
v# = 100
r# = 1
dist# = 50
count1# = 1
count2# = 1
count3# = 1
test1$ = "bank"
REM put instructions here for restarting. comment out other lines.
OPEN test1$ FOR OUTPUT AS #1
WRITE #1, count1#, r#, v#
CLOSE #1
DO
test1$ = "bank"
OPEN test1$ FOR INPUT AS #1
DO
INPUT #1, a#, b#, c#
stop$ = INKEY$
IF stop$ = "s" THEN STOP
LOOP UNTIL a# = count1#
CLOSE #1
count1# = count1# + 1
v# = c#
rs# = b#
REM IF (r# / 2) - INT(r# / 2) = 0 THEN r# = r# - 1
r# = 1
```

```
redo:
DO
v1# = v# + (v# - r# + 1)
v2# = v# + (v# - r# + 2)
test = (v1# - r#) <= dist#
r# = r# + 2
IF test = -1 THEN EXIT DO
stop$ = INKEY$
IF stop$ = "s" THEN STOP
LOOP
IF r# > v# THEN dist# = dist# + 1
IF r# > v# THEN r# = 1
IF r# > v# THEN GOTO redo:
test1$ = "bank"
OPEN test1$ FOR INPUT AS #1
DO
INPUT #1, a#, b#, c#
test = ((v1# = c#) AND (r# = b#)) OR ((v2# = c#) AND (r# = b#))
IF test = -1 THEN EXIT DO
stop$ = INKEY$
IF stop$ = "s" THEN STOP
LOOP UNTIL a# = count2#
CLOSE #1
IF test = -1 THEN GOTO redo:
COLOR 2, 0
PRINT , , v#; rs#; v1#; v2#; r#; v# - r#; v1# - r#
a$ = INKEY$
REM INPUT a$
IF a$ = "s" THEN STOP
count2# = count2# + 1
test1$ = "bank"
OPEN test1$ FOR APPEND AS #1
WRITE #1, count2#, r#, v1#
count2# = count2# + 1
WRITE #1, count2#, r#, v1#
CLOSE #1
test1$ = "chartu"
OPEN test1$ FOR APPEND AS #1
WRITE #1, count3#, rs#, v#, v1#, v2#, r#
CLOSE #1#
count3# = count3# + 1
test1$ = "vcns"
OPEN test1$ FOR OUTPUT AS #1
WRITE #1, count1#, count2#, dist#
CLOSE #1
LOOP UNTIL v# > 999
```

```
"C:\Users\Reactor1967\vcns\work\CHART.BAS"
a# = 500
b# = 250
CLS
FOR count = 1 TO 125
PRINT a#; b#
a# = a# - 1
b# = b# - 1
INPUT a$
IF a$ = "s" THEN STOP
NEXT count
PRINT "-------------"
a# = 250
FOR count = 1 TO 62
PRINT a#; b#
a# = a# - 1
b# = b# - 1
INPUT a$
IF a$ = "s" THEN STOP
NEXT count
PRINT "-------------"
a# = 125
FOR count = 1 TO 31
PRINT a#; b#
a# = a# - 1
b# = b# - 1
INPUT a$
IF a$ = "s" THEN STOP
NEXT count
PRINT "-------------"
a# = 62
FOR count = 1 TO 15
PRINT a#; b#
a# = a# - 1
b# = b# - 1
INPUT a$
IF a$ = "s" THEN STOP
NEXT count
PRINT "-------------"
a# = 31
FOR count = 1 TO 7
PRINT a#; b#
a# = a# - 1
```

```
b# = b# - 1
INPUT a$
IF a$ = "s" THEN STOP
NEXT count
PRINT "-------------"
a# = 15
FOR count = 1 TO 6
PRINT a#; b#
a# = a# - 1
b# = b# - 1
INPUT a$
IF a$ = "s" THEN STOP
NEXT count
PRINT "-------------"
a# = 7
FOR count = 1 TO 3
PRINT a#; b#
a# = a# - 1
b# = b# - 1
INPUT a$
IF a$ = "s" THEN STOP
NEXT count
PRINT "-------------"
a# = 6
PRINT a#; b#
a# = a# - 1
b# = b# - 1
INPUT "All done!"; a$
```

```
"C:\Users\Reactor1967\vcns\work\CHANGER.BAS"
REM A LITTLE HINT. TO KEEP THE SAME SYMETERY(FRACTIONAL
VALUES r = r + (base *2)
r# = 1
a# = 1
b# = a# + ((a# - r#) * 3) + 1
c# = a# + ((a# - r#) * 3) + 2
d# = a# + ((a# - r#) * 3) + 3
e# = a# + ((a# - r#) * 3) + 4
CLS
```

```
DO
PRINT a#; b#; c#; d#; e#; r#
PRINT a#; (b# / 4) - INT(b# / 4); (c# / 4) - INT(c# / 4); (d# / 4) - INT(d# / 4);
(e# / 4) - INT(e# / 4)
PRINT b# - a#
a# = a# + 1
b# = a# + ((a# - r#) * 3) + 1
c# = a# + ((a# - r#) * 3) + 2
d# = a# + ((a# - r#) * 3) + 3
e# = a# + ((a# - r#) * 3) + 4
INPUT a$
IF a$ = "s" THEN STOP
IF a$ = "d" THEN INPUT "Please Enter your New r value."; r#
LOOP




"C:\Users\Reactor1967\vcns\work\CGENB3.BAS"
a# = 1
b# = 1
carry# = 1
CLS
RANDOMIZE TIMER
redo:
DO
IF carry# = a# THEN d = INT(RND * 3) + 0 ELSE d = 99
FOR count = 1 TO 3
PRINT a#; b#; carry#
IF b# = 1 THEN x = 0
IF b# = 101 THEN x = 1
IF x = 0 THEN b# = b# + 2
IF x = 1 THEN b# = b# - 2
test1 = (d = 0) AND (count = 1)
test2 = (d = 1) AND (count = 2)
test3 = (d = 2) AND (count = 3)
IF test1 = -1 THEN carry# = b#
IF test2 = -1 THEN carry# = b#
IF test3 = -1 THEN carry# = b#
NEXT count
IF a# = 1 THEN z = 0
IF a# = 101 THEN z = 1
IF z = 0 THEN a# = a# + 2
IF z = 1 THEN a# = a# - 2
REM a$ = INKEY$
INPUT a$
IF a$ = "s" THEN STOP
```

```
LOOP
REM LOOP UNTIL a# = 101
PRINT a#; b#
GOTO redo:
decode:
DO
PRINT a#; b#; carry#
FOR count = 1 TO 3
IF b# = 1 THEN x = 0
IF b# = 101 THEN x = 1
IF x = 0 THEN b# = b# - 2
IF x = 1 THEN b# = b# + 2
IF carry# = b# THEN carry# = a#
NEXT count
IF a# = 1 THEN z = 0
IF a# = 101 THEN z = 1
IF z = 0 THEN a# = a# - 2
IF z = 1 THEN a# = a# + 2
REM a$ = INKEY$
INPUT a$
IF a$ = "s" THEN STOP
LOOP
REM LOOP UNTIL a# = 101
PRINT a#; b#
```

```
"C:\Users\Reactor1967\vcns\work\BIN.BAS"
CLS
RANDOMIZE TIMER
count = 0
DO
c = INT(RND * 2) + 0
COLOR 2, 0
PRINT c;
FOR count = 1 TO 4000
NEXT count
LOOP
```

```
"C:\Users\Reactor1967\vcns\work\BFCODE.BAS"
p# = 1
r# = 1
r2# = 999999999999999#
CLS
```

```
RANDOMIZE TIMER
count = 0
DO
count = count + 1
IF ABS(p# - store#) > 250000000000000# THEN flip = 1 ELSE flip = 0
c = p# / 2 - INT(p# / 2)
IF c >= .5 THEN c = 1
IF p# - store# > 0 THEN code$ = "+"
IF p# - store# < 0 THEN code$ = "-"
PRINT c; p#; code$; flip
c = INT(RND * 2) + 0
test1 = (p# < 500000000000000#) AND (c = 0)
test2 = (p# < 500000000000000#) AND (c = 1)
test3 = (p# > 500000000000000#) AND (c = 0)
test4 = (p# > 500000000000000#) AND (c = 1)
store# = p#
IF test1 = -1 THEN GOSUB test1:
IF test2 = -1 THEN GOSUB test2:
IF test3 = -1 THEN GOSUB test3:
IF test4 = -1 THEN GOSUB test4:
REM INPUT a$
a$ = INKEY$
IF a$ = "s" THEN STOP
LOOP
test1:
p# = p# + (p# - r# + 1)
RETURN
test2:
p# = p# + (p# - r# + 1)
RETURN
test3:
p# = p# - (r2# - p# + 1)
RETURN
test4:
p# = p# - (r2# - p# + 2)
RETURN


"C:\Users\Reactor1967\vcns\work\BETA1.BAS"
v# = 100
r# = 93
CLS
RANDOMIZE TIMER
d# = 1
DO
```

```
z = INT(RND * 2) + 0
REM IF z = 1 THEN z = .5
d# = 1
DO
v1# = v# + d#
IF (v1# / 2) - INT(v1# / 2) = z THEN v1# = v1# + 0
IF (v1# / 2) - INT(v1# / 2) <> z THEN v1# = v1# + 1
dist# = v1# - v#
dist# = (dist# * 2) - 1
r1# = v1# - dist#
REM PRINT z; v1# - v#; r1# - r#
IF v1# - v# = r1# - r# THEN EXIT DO
d# = d# + 1
REM INPUT stop$
stop$ = INKEY$
IF stop$ = "s" THEN STOP
LOOP
v# = v1#
r# = r1#
COLOR 2, 0
PRINT , , z; v#; r#
a$ = INKEY$
REM INPUT a$
IF a$ = "s" THEN STOP
IF a$ = "d" THEN EXIT DO
LOOP
PRINT "ready for decode"
INPUT a$
REM a$ = INKEY$
IF a$ = "s" THEN STOP
decode:
DO
dist# = v# - r#
store# = v#
IF (dist# / 2) - INT(dist# / 2) = .5 THEN dist# = dist# - 1
dist# = dist# / 2
dist# = dist# + 1
v# = v# - dist#
dist# = store# - v#
r# = r# - dist#
PRINT , , v#; r#
IF v# = 100 THEN PRINT "Decode Sucessful!!!"
IF v# = 100 THEN SYSTEM
REM INPUT a$
a$ = INKEY$
IF a$ = "s" THEN STOP
```

```
"C:\Users\Reactor1967\vcns\work\BCHART.BAS"
DECLARE SUB rangef (r#(), p#, range!)
REM                    COPYRIGHT C 2002
REM                    Lloyd Dudley Burris
REM ----------------------Initilize R#-------------------------------------
CLS
DIM va#(2998)
DIM vb#(2998)
DIM ra#(2998)
DIM rb#(2998)
DIM vp#(2998)
DIM rp#(2998)
DIM ln#(2998)
DIM r#(255)
r#(1) = 1
REM PRINT r#(1)
pt1# = 5
rf# = 100
count = 2
a# = 0
point1:
DO
a# = a# + pt1#
test = (a# / 2) - INT(a# / 2)
r#(count) = a#
IF (r#(count) / 2) - INT(r#(count) / 2) = 0 THEN r#(count) = r#(count) + 1
REM PRINT r#(count); count
count = count + 1
REM INPUT a$
a$ = INKEY$
IF a$ = "s" THEN STOP
IF count >= 254 THEN EXIT DO
LOOP UNTIL a# = rf#
pt1# = pt1# * 10
rf# = rf# * 10
IF count >= 254 THEN GOTO POINT2:
IF a# + pt1# > 999999999999999# THEN GOTO POINT2:
GOTO point1:
POINT2:
r#(254) = 950000000000001#
r#(255) = 999999999999999#
REM ----------------------init eqf-------------------------------------
```

```
REM put a location line pointer for all three variables so program can use
REM this chart to code actual data. Do small dry run. Then generate this
REM to a big file to use for actual data incoding and decoding. L.B.
REM va#()
REM vb#()
REM ra#()
REM rb#()
REM vp#()
REM rp#()
pcount = 0
count = 1
p# = 1
CALL rangef(r#(), p#, range)
vp#(count) = p#
rp#(count) = range
REM---
CALL rangef(r#(), p#, range)
p1# = p# + (p# - r#(range) + 1)
va#(count) = p1#
store# = p#
p# = p1#
CALL rangef(r#(), p#, range)
ra#(count) = range
p# = store#
REM ---
CALL rangef(r#(), p#, range)
p2# = p# + (p# - r#(range) + 2)
vb#(count) = p2#
store# = p#
p# = p2#
CALL rangef(r#(), p#, range)
rb#(count) = range
REM -----------------------------------
REM use pcount to get new vectors
REM use count to save new vectors
pcount = 0
DO
pcount = pcount + 1
count = count + 1
IF count = 4998 THEN EXIT DO
IF pcount = 4998 THEN EXIT DO
p# = va#(pcount)
rp#(count) = ra#(pcount)
ln#(count) = pcount
PRINT p#;
CALL rangef(r#(), p#, range)
```

```
vp#(count) = p#
p1# = p# + (p# - r#(range) + 1)
p2# = p# + (p# - r#(range) + 2)
p# = p1#
CALL rangef(r#(), p#, range)
va#(count) = p#
ra#(count) = range
p# = p2#
CALL rangef(r#(), p#, range)
vb#(count) = p#
rb#(count) = range
REM ----------------------------------
count = count + 1
IF count >= 2998 THEN EXIT DO
p# = vb#(pcount)
rp#(count) = rb#(pcount)
ln#(count) = pcount
PRINT p#;
CALL rangef(r#(), p#, range)
IF p# + (p# - r#(range)) >= 999999999999999# THEN EXIT DO
vp#(count) = p#
p1# = p# + (p# - r#(range) + 1)
p2# = p# + (p# - r#(range) + 2)
p# = p1#
CALL rangef(r#(), p#, range)
va#(count) = p#
ra#(count) = range
p# = p2#
CALL rangef(r#(), p#, range)
vb#(count) = p#
rb#(count) = range
a$ = INKEY$
LOOP
count = 1
DO
PRINT vp#(count); va#(count); vb#(count)
PRINT rp#(count); ra#(count); rb#(count)
PRINT ln#(count)
PRINT "              "
REM a$ = INKEY$
INPUT a$
IF a$ = "s" THEN STOP
count = count + 1
LOOP UNTIL count = 2998

SUB rangef (r#(), p#, range)
```

```
range = 255
DO
test = r#(range) <= p#
IF test = -1 THEN EXIT DO
range = range - 1
a$ = INKEY$
IF a$ = "s" THEN STOP
LOOP UNTIL range = 0
IF range = 0 THEN PRINT "ERROR rangef failed"
IF range = 0 THEN SYSTEM
END SUB
```

```
"C:\Users\Reactor1967\vcns\work\BALANCE.BAS"
v# = 100
r# = 51
CLS
RANDOMIZE TIMER
PRINT v#; r#
DO
z = INT(RND * 2) + 0
vs# = v#
IF z = 0 THEN v# = v# + (v# - r# + 1)
IF z = 1 THEN v# = v# + (v# - r# + 2)
COLOR 2, 0
rs# = r#: cs# = cr#
r# = r# + (v# - vs#)
dist# = r# - rs#
dist# = (dist# * 2) - 1
cr# = r# - dist#
PRINT , , z; v# - vs#; v#; r#; v# - r#; r# - rs#; cr#; cr# - cs#
REM a$ = INKEY$
INPUT a$
IF a$ = "s" THEN STOP
LOOP
```

```
"C:\Users\Reactor1967\vcns\work\BAK1.BAS"
a# = 1
b# = 2
c# = 3
d# = 4
CLS
RANDOMIZE TIMER
```

```
DO
z = INT(RND * 2) + 0
IF z = 0 THEN st# = b#
IF z = 1 THEN st# = c#
IF flip = 1 THEN st# = d#
distance# = st# - a#
a# = a# + distance#
b# = b# + (distance# * 3)
c# = c# + (distance# * 3)
d# = d# + (distance# * 3)
PRINT a#; b#; c#; d#; flip
IF b# - a# >= 4 THEN flip = 1 ELSE flip = 0
IF b# - a# >= 4 THEN a# = a# + 6
IF b# - a# >= 4 THEN a# = a# + 6
REM INPUT a$
a$ = INKEY$
IF a$ = "s" THEN STOP
IF a$ = "d" THEN GOTO decode:
LOOP UNTIL b# >= 999999999999999#
decode:
test1# = INT(((a# / 3) - INT(a# / 3)) * 10) = 6: REM b#
test2# = INT(((a# / 3) - INT(a# / 3)) * 10) = 0: REM c#
test3# = INT(((a# / 3) - INT(a# / 3)) * 10) = 3: REM d#
PRINT "b#"; test1#
PRINT "c#"; test2#
PRINT "d#"; test3#




"C:\Users\Reactor1967\vcns\work\BABYSTEP.BAS"
v# = 100
CLS
RANDOMIZE TIMER
DO
test = (v# / 2) - INT(v# / 2)
IF test = 0 THEN r# = v# - 9
IF test = .5 THEN r# = v# - 10
store1# = v#
z = INT(RND * 2) + 0
IF z = 0 THEN v# = v# + (v# - r# + 1)
IF z = 1 THEN v# = v# + (v# - r# + 2)
dist# = v# - store1#
dist# = (dist# * 2) - 1
store2# = cr#
cr# = v# - dist#
COLOR 2, 0
```

```
PRINT , , z; v# - store1#; v#; v# - cr#; cr#; cr# - store2#
REM PRINT , , z; cr#; cr# - store2#
INPUT a$
IF a$ = "s" THEN STOP
LOOP




"C:\Users\Reactor1967\vcns\work\BABY2.BAS"
v# = 100
cr# = 0
CLS
RANDOMIZE TIMER
DO
d# = 1
store2# = cr#
z = INT(RND * 2) + 0
store1# = v#
DO
r# = v# - d#
REM IF (r# / 2) - INT(r# / 2) = 0 THEN d# = d# + 1
REM IF (r# / 2) - INT(r# / 2) = 0 THEN r# = v# - d#
IF z = 0 THEN v1# = v# + (v# - r# + 1)
IF z = 1 THEN v1# = v# + (v# - r# + 2)
dist# = v1# - v#
dist# = (dist# * 2) - 1
cr# = v1# - dist#
REM test = ABS((v1# - v#) - (cr# - store2#)) = 1
test2 = ABS((v1# - v#) - (cr# - store2#)) = 2
REM PRINT z; v1# - v#; cr# - store2#
REM IF test = -1 THEN EXIT DO
IF test2 = -1 THEN EXIT DO
d# = d# + 1
REM INPUT a$
a$ = INKEY$
IF a$ = "s" THEN STOP
LOOP
v# = v1#
COLOR 2, 0
PRINT , , v# - store1#; z; v#; cr#; cr# - store2#
INPUT a$
REM a$ = INKEY$
IF a$ = "s" THEN STOP
LOOP
```

```
"C:\Users\Reactor1967\vcns\work\B3FLIP.BAS"
RANDOMIZE TIMER
CLS
high# = 0
r# = 1
v# = 1
redo:
r# = v# - 1
v1# = v# + ((v# - r#) * 2) + 1
v2# = v# + ((v# - r#) * 2) + 2
v3# = v# + ((v# - r#) * 2) + 3
DO
COLOR 2, 0
PRINT , , z; v#, v1#; v2#; v3#; high#
a$ = INKEY$
REM INPUT a$
IF a$ = "s" THEN STOP
z = INT(RND * 2) + 0
IF z = 0 THEN v# = v1#
IF z = 1 THEN v# = v2#
IF z = 2 THEN v# = v3#
v1# = v# + ((v# - r#) * 2) + 1
v2# = v# + ((v# - r#) * 2) + 2
v3# = v# + ((v# - r#) * 2) + 3
LOOP UNTIL v# + ((v# - r#) * 2) >= 1000000
IF v# > high# THEN high# = v#
r# = v# + 1
v1# = v# - ((r# - v#) * 2) - 1
v2# = v# - ((r# - v#) * 2) - 2
v3# = v# - ((r# - v#) * 2) - 3
DO
COLOR 2, 0
PRINT , , z; v#, v1#; v2#; v3#; high#
a$ = INKEY$
REM INPUT a$
IF a$ = "s" THEN STOP
z = INT(RND * 2) + 0
IF z = 0 THEN v# = v1#
IF z = 1 THEN v# = v2#
IF z = 2 THEN v# = v3#
v1# = v# - ((r# - v#) * 2) - 1
v2# = v# - ((r# - v#) * 2) - 2
v3# = v# - ((r# - v#) * 2) - 3
```

```
LOOP UNTIL v# - ((r# - v#) * 2) <= 0
GOTO redo:




"C:\Users\Reactor1967\vcns\work\ATTEMPT.BAS"
a# = 1
b# = 2
c# = 3
CLS
DO
PRINT , , a#; "="; b#; c#
IF xp2 = 0 THEN a# = a# + 1
IF xp = 0 THEN b# = b# + 2
IF xp = 0 THEN c# = c# + 2
IF xp2 = 1 THEN a# = a# - 1
IF xp = 1 THEN b# = b# - 2
IF xp = 1 THEN c# = c# - 2
REM a$ = INKEY$
INPUT a$
IF a$ = "s" THEN STOP
IF b# >= 98 THEN xp = 1
IF b# <= 2 THEN xp = 0
IF a# >= 99 THEN xp2 = 1
IF a# <= 1 THEN xp2 = 0
LOOP




"C:\Users\Reactor1967\vcns\work\ALPHA10A.BAS"
a# = 999999999999996#
r# = 999999999999997#
c# = 999999999999999#
d# = 999999999999998#
e# = 999999999999997#
CLS
RANDOMIZE TIMER
DO
z = INT(RND * 2) + 0
IF z = 0 THEN a# = a# - (r# - a# + 1)
IF z = 1 THEN a# = a# - (r# - a# + 2)
PRINT "data"; r#; a#; r# - a#
redo:
IF r# >= a# THEN store# = r# - 1
IF r# <= a# THEN GOTO ot:
dist# = c# - r#
```

```
dist# = dist# * 2
dist# = d# - dist#
r# = dist#
GOTO redo:
ot:
r# = store#
IF test = 0 THEN a# = a# - (r# - a# + 2)
IF test = -1 THEN a# = a# - (r# - a# + 1)
PRINT "code"; r#; a#
REM INPUT a$
a$ = INKEY$
IF a$ = "s" THEN STOP
LOOP UNTIL a# <= 500000000000000#


"C:\Users\Reactor1967\vcns\work\ALPHA10.BAS"
DECLARE SUB coderd (r#, v#)
DECLARE SUB coderu (r#, v#)
CLS
RANDOMIZE TIMER
r# = 1
v# = 1
redo:
DO
REM data:
PRINT "code"; r#; v#
n1# = v# + (v# - r# + 1)
n2# = v# + (v# - r# + 2)
z = INT(RND * 2) + 0
IF z = 0 THEN v# = n1#
IF z = 1 THEN v# = n2#
PRINT "data"; r#; v#
REM read code bit:
test = (r# * 2) + 1 > v#
IF test = 0 THEN CALL coderu(r#, v#)
n1# = v# + (v# - r# + 1)
n2# = v# + (v# - r# + 2)
IF test = 0 THEN v# = n2#
IF test = -1 THEN v# = n1#
REM INPUT a$
a$ = INKEY$
IF a$ = "s" THEN STOP
LOOP UNTIL (v# * 2) >= 500000000000000#
test = (r# * 2) + 1 > v#
IF test = 0 THEN CALL coderu(r#, v#)
```

```
DO
IF v# - 50000000000000# <= 0 THEN EXIT DO
IF r# - 50000000000000# <= 0 THEN EXIT DO
v# = v# - 50000000000000#
r# = r# - 50000000000000#
a$ = INKEY$
IF a$ = "s" THEN STOP
LOOP
REM PRINT v#; r#
REM STOP
test = (r# * 2) + 1 > v#
IF test = 0 THEN CALL coderu(r#, v#)
GOTO redo:

SUB coderu (r#, v#)
DO
IF (r# * 2) >= v# THEN EXIT DO
r# = r# * 2
LOOP
r# = r# + 1
END SUB




"C:\Users\Reactor1967\vcns\work\ALPHA9.BAS"
DECLARE SUB deflate (r#, v#, n1#, n2#, c#, p#)
r# = 1
v# = 1
n1# = v# + (v# - r# + 1)
n2# = v# + (v# - r# + 2)
c# = 0
p# = 1
CLS
RANDOMIZE TIMER
redo:
DO
FOR count = 1 TO 10
z = INT(RND * 2) + 0
test1 = (p# = v#) AND (z = 0)
test2 = (p# = v#) AND (z = 1)
IF test1 = -1 THEN p# = n1#
IF test2 = -1 THEN p# = n2#
dt = (p# / 2) - INT(p# / 2)
IF dt >= .5 THEN dt = 1
PRINT "r="; r#; "data="; dt; "v="; v#; "n1="; n1#; "n2="; n2#; "c="; c#; "p=";
p#; "n1-r="; n1# - r#
```

```
v# = v# + 1
n1# = v# + (v# - r# + 1)
n2# = v# + (v# - r# + 2)
round# = ((v# - r#) / 2) - (INT((v# - r#) / 2))
IF round# >= .5 THEN round = 1 ELSE round = 0
c# = (v# - r#) / 2 + round#
add# = c#
c# = r# + c# - 1
REM INPUT a$
REM IF a$ = "s" THEN STOP
NEXT count
r# = r# + 10
n1# = v# + (v# - r# + 1)
n2# = v# + (v# - r# + 2)
round# = ((v# - r#) / 2) - (INT((v# - r#) / 2))
IF round# >= .5 THEN round = 1 ELSE round = 0
c# = (v# - r#) / 2 + round#
c# = r# + c# - 1
a$ = INKEY$
REM INPUT a$
IF a$ = "s" THEN STOP
LOOP UNTIL v# >= 99100
CALL deflate(r#, v#, n1#, n2#, c#, p#)
GOTO redo:

SUB deflate (r#, v#, n1#, n2#, c#, p#)
DO
dt = (v# / 2) - INT(v# / 2)
IF dt >= .5 THEN dt = 1
PRINT "r="; r#; "data="; dt; "v="; v#; "n1="; n1#; "n2="; n2#; "c="; c#; "p=";
p#; "n1-r="; n1# - r#
r# = r# - 2
v# = v# - 2
n1# = n1# - 2
n2# = n2# - 2
c# = c# - 2
p# = p# - 2
a$ = INKEY$
IF a$ = "s" THEN STOP
LOOP UNTIL v# <= 100

END SUB
```

"C:\Users\Reactor1967\vcns\work\ALPHA8.BAS"

```
DECLARE SUB ot (r1#(), p#(), r2#(), crd#())
DECLARE SUB Parsing (r1#(), p#(), r2#(), crd#())
REM this program creats a chart that shows what r that p came
REM from and what r p goes to. It can be used for coding or
REM decoding p. Its the latest and greatest in my work on vcns.
DECLARE SUB rcalc (a#, rt#)
c# = 1
r# = 1
rt# = 1
a# = 1
DIM r1#(1000)
DIM p#(1000)
DIM r2#(1000)
DIM crd#(1000)
RANDOMIZE TIMER
CLS
PRINT "Initilizing VCNS......"
x = 1
FOR count = 1 TO 500
a# = c#
CALL rcalc(a#, rt#)
r# = rt#
r1#(x) = r#
a# = a# + (a# - r# + 1)
p#(x) = a#
CALL rcalc(a#, rt#)
r# = rt#
r2#(x) = r#
x = x + 1
a# = c#
CALL rcalc(a#, rt#)
r# = rt#
r1#(x) = r#
a# = a# + (a# - r# + 2)
p#(x) = a#
CALL rcalc(a#, rt#)
r# = rt#
r2#(x) = r#
c# = c# + 1
x = x + 1
NEXT count
PRINT "Parsing VCNS...."
CALL Parsing(r1#(), p#(), r2#(), crd#())
CALL ot(r1#(), p#(), r2#(), crd#())
PRINT "VCNS IS POWERED UP!!!!!!!"
INPUT a$
```

```
IF a$ = "s" THEN STOP
v = 1
DO
a# = p#(v)
r# = r2#(v)
z = INT(RND * 2) + 0
IF z = 0 THEN a# = a# + (a# - r# + 1)
IF z = 1 THEN a# = a# + (a# - r# + 2)
count = v
DO
count = count + 1
IF count = 1001 THEN GOTO decode:
test = (p#(count) = a#) AND (r1#(count) = r#) AND (crd#(count) = v)
p$ = INKEY$
IF p$ = "s" THEN STOP
LOOP UNTIL test = -1
v = count
PRINT r1#(v); p#(v); r2#(v); crd#(v)
INPUT a$
IF a$ = "s" THEN STOP
LOOP
decode:
PRINT "Decoding...."
redo2:
PRINT r1#(v); p#(v); r2#(v); crd#(v)
v = crd#(v)
INPUT a$
IF a$ = "s" THEN STOP
IF p#(v) = 0 THEN SYSTEM
GOTO redo2:

SUB ot (r1#(), p#(), r2#(), crd#())
FOR count = 1 TO 1000
PRINT r1#(count); p#(count); r2#(count), crd#(count)
INPUT a$
IF a$ = "s" THEN STOP
NEXT count
END SUB

SUB Parsing (r1#(), p#(), r2#(), crd#())
position1 = 1
redo:
a# = p#(position1)
r# = r2#(position1)
a# = a# + (a# - r# + 1)
count = position1
```

```
DO
count = count + 1
test = (p#(count) = a#) AND r1#(count) = r#
IF test = -1 THEN EXIT DO
IF count = 1000 THEN EXIT SUB
LOOP
crd#(count) = position1
a# = p#(position1)
r# = r2#(position1)
a# = a# + (a# - r# + 2)
count = position1
DO
count = count + 1
test = (p#(count) = a#) AND r1#(count) = r#
IF test = -1 THEN EXIT DO
IF count = 1000 THEN EXIT SUB
LOOP
crd#(count) = position1
position1 = position1 + 1
IF position1 = 1000 THEN EXIT SUB
a$ = INKEY$
IF a$ = "s" THEN STOP
GOTO redo:
END SUB

SUB rcalc (a#, rt#)
t# = 0
rt# = 1
DO
FOR count = 1 TO 10
IF t# = a# THEN EXIT FOR
t# = t# + 1
NEXT count
IF t# = a# THEN EXIT DO
rt# = rt# + 10
LOOP
END SUB




"C:\Users\Reactor1967\vcns\work\ALPHA7.BAS"
DECLARE SUB down (a1#, a2#, b#, c#, p#)
a2# = 1
a1# = 0
b# = 0
c# = 0
```

```
p# = 1
RANDOMIZE TIMER
CLS
redo:
FOR count = 1 TO 10
a1# = a1# + 1
b# = b# + 2
c# = c# + 2
z = INT(RND * 2) + 0
test1 = (z = 0) AND (p# = a1#)
test2 = (z = 1) AND (p# = a1#)
IF test1 = -1 THEN p# = b#
IF test2 = -1 THEN p# = c#
PRINT "|a1"; a1#; "|b"; b#; "|c"; c#; "|a2"; a2#; "|ba"; b# - a1#; "|ba2"; b# - a2#;
"|Binary Data"; p#; "|"; z; "|"
stop$ = INKEY$
IF stop$ = "s" THEN STOP
NEXT count
DO
IF a2# + 2# >= p# THEN EXIT DO
a2# = a2# + 2
stop$ = INKEY$
IF stop$ = "s" THEN STOP
LOOP
z = INT(RND * 2) + 0
test1 = (z = 0) AND (p# = a2#)
test2 = (z = 1) AND (p# = a2#)
IF test1 = -1 THEN p# = b#
IF test2 = -1 THEN p# = c#
PRINT "|a1"; a1#; "|b"; b#; "|c"; c#; "|a2"; a2#; "|ba"; b# - a1#; "|ba2"; b# - a2#;
"|Binary Data"; p#; "|"; z; "|"
a$ = INKEY$
REM INPUT a$
IF a$ = "s" THEN STOP
FOR count = 1 TO 10
a2# = a2# + 1
b# = b# + 2
c# = c# + 2
z = INT(RND * 2) + 0
test1 = (z = 0) AND (p# = a2#)
test2 = (z = 1) AND (p# = a2#)
IF test1 = -1 THEN p# = b#
IF test2 = -1 THEN p# = c#
PRINT "|a1"; a1#; "|b"; b#; "|c"; c#; "|a2"; a2#; "|ba"; b# - a1#; "|ba2"; b# - a2#;
"|Binary Data"; p#; "|"; z; "|"
stop$ = INKEY$
```

```
IF stop$ = "s" THEN STOP
NEXT count
DO
IF a1# + 2# >= p# THEN EXIT DO
a1# = a1# + 2
stop$ = INKEY$
IF stop$ = "s" THEN STOP
LOOP
z = INT(RND * 2) + 0
test1 = (z = 0) AND (p# = a1#)
test2 = (z = 1) AND (p# = a1#)
IF test1 = -1 THEN p# = b#
IF test2 = -1 THEN p# = c#
PRINT "|a1"; a1#; "|b"; b#; "|c"; c#; "|a2"; a2#; "|ba"; b# - a1#; "|ba2"; b# - a2#;
"|Binary Data"; p#; "|"; z; "|"
IF p# >= 99800 THEN CALL down(a1#, a2#, b#, c#, p#)
a$ = INKEY$
REM INPUT a$
IF a$ = "s" THEN STOP
GOTO redo:
decode1:

SUB down (a1#, a2#, b#, c#, p#)
DO
a1# = a1# - 10: b# = b# - 10: c# = c# - 10: a2# = a2# - 10: p# = p# - 10
PRINT "|a1"; a1#; "|b"; b#; "|c"; c#; "|a2"; a2#; "|ba"; b# - a1#; "|ba2"; b# - a2#;
"|Binary Data"; p#; "|"; z; "|"
a$ = INKEY$
IF a$ = "s" THEN STOP
LOOP UNTIL b# <= 10100

END SUB




"C:\Users\Reactor1967\vcns\work\ALPHA6.BAS"
REM goal of this program is not to take a# or a2 past p#
REM In past programs I discovered P# can not repeat itself on
REM b or c before appearing in a. Not taking a or a2 past p
REM solves this problem and also I still can sink my chart.
REM Problem now is when to take a and a2 down and how much
REM and keep them in alignment with p and the b and c. And,
REM do this in a code-able and de-code-able fashion.
a# = 1
a2# = 1
b# = 2
```

```
c# = 3
p# = 1
CLS
RANDOMIZE TIMER
overredo:
FOR count = 1 TO 10
z = INT(RND * 2) + 0
test1 = (p# = a#) AND (z = 0)
test2 = (p# = a#) AND (z = 1)
IF test1 = -1 THEN p# = b#
IF test2 = -1 THEN p# = c#
PRINT "a"; a#; "b"; b#; "c"; c#; "a2"; a2#; "ba"; b# - a#; "ba2"; b# - a2#; "|
Binary Data"; p#; "|"
a# = a# + 1
b# = b# + 2
c# = c# + 2
IF ((a2# + 4) <= p#) THEN a2# = a2# + 4
NEXT count
PRINT "a"; a#; "b"; b#; "c"; c#; "a2"; a2#; "ba"; b# - a#; "ba2"; b# - a2#; "|
Binary Data"; p#; "|"
REM INPUT a$
a$ = INKEY$
IF a$ = "s" THEN STOP
redo:
FOR count = 1 TO 10
z = INT(RND * 2) + 0
test1 = (p# = a2#) AND (z = 0)
test2 = (p# = a2#) AND (z = 1)
IF test1 = -1 THEN p# = b#
IF test2 = -1 THEN p# = c#
PRINT "a"; a#; "b"; b#; "c"; c#; "a2"; a2#; "ba"; b# - a#; "ba2"; b# - a2#; "|
Binary Data"; p#; "|"
IF (a# + 4 <= p#) THEN a# = a# + 4
b# = b# + 2
c# = c# + 2
a2# = a2# + 1
stop$ = INKEY$
IF stop$ = "s" THEN STOP
NEXT count
PRINT "a"; a#; "b"; b#; "c"; c#; "a2"; a2#; "ba"; b# - a#; "ba2"; b# - a2#; "|
Binary Data"; p#; "|"
a$ = INKEY$
REM INPUT a$
IF a$ = "s" THEN STOP
FOR count = 1 TO 10
z = INT(RND * 2) + 0
```

```
test1 = (p# = a#) AND (z = 0)
test2 = (p# = a#) AND (z = 1)
IF test1 = -1 THEN p# = b#
IF test2 = -1 THEN p# = c#
PRINT "a"; a#; "b"; b#; "c"; c#; "a2"; a2#; "ba"; b# - a#; "ba2"; b# - a2#; "|
Binary Data"; p#; "|"
a# = a# + 1
b# = b# + 2
c# = c# + 2
IF ((a2# + 4) <= p#) THEN a2# = a2# + 4
stop$ = INKEY$
IF stop$ = "s" THEN STOP
NEXT count
PRINT "a"; a#; "b"; b#; "c"; c#; "a2"; a2#; "ba"; b# - a#; "ba2"; b# - a2#; "|
Binary Data"; p#; "|"
REM INPUT a$
a$ = INKEY$
IF a$ = "s" THEN STOP
GOTO overredo:




"C:\Users\Reactor1967\vcns\work\ALPHA5.BAS"
a# = 1
b1# = 2
c1# = 3
b2# = 12
c2# = 13
carry# = 1
RANDOMIZE TIMER
CLS
redo:
FOR count = 1 TO 10
PRINT a#; b1#; c1#; carry#
z = INT(RND * 2) + 0
test1 = (carry# = a#) AND z = 0
test2 = (carry# = a#) AND z = 1
IF test1 = -1 THEN carry# = b1#
IF test2 = -1 THEN carry# = c1#
a# = a# + 1
b1# = b1# + 2
c1# = c1# + 2
NEXT count
PRINT a#; b1#; c1#; carry#
REM a$ = INKEY$
INPUT a$
```

```
REM IF a$ = "s" THEN STOP
FOR count = 1 TO 10
PRINT a#; b2#; c2#; carry#
z = INT(RND * 2) + 0
test1 = (carry# = a#) AND z = 0
test2 = (carry# = a#) AND z = 1
IF test1 = -1 THEN carry# = b2#
IF test2 = -1 THEN carry# = c2#
a# = a# + 1
b2# = b2# + 2
c2# = c2# + 2
NEXT count
PRINT a#; b2#; c2#; carry#
INPUT a$
REM a$ = INKEY$
IF a$ = "s" THEN STOP
IF a$ = "d" THEN GOTO decode:
GOTO redo:
decode:


"C:\Users\Reactor1967\vcns\work\ALPHA4.BAS"
a# = 1
b# = 2
c# = 3
d# = 4
RANDOMIZE TIMER
carry# = 1
CLS
redo:
x = 1
FOR count = 1 TO 10
test1 = (carry# = a#) AND (x = 1)
IF test1 = -1 THEN carry# = d#
IF test1 = -1 THEN x = 0
z = INT(RND * 2) + 0
test2 = (x = 0) AND (carry# = a#) AND (z = 0)
test3 = (x = 0) AND (carry# = a#) AND (z = 1)
IF test2 = -1 THEN carry# = b#
IF test3 = -1 THEN carry# = c#
PRINT a#; b#; c#; d#; carry#
a# = a# + 1
b# = b# + 3
c# = c# + 3
d# = d# + 3
```

```
NEXT count
PRINT a#; b#; c#; d#; carry#
a# = a# + 20
carry# = carry# + 20
a$ = INKEY$
REM INPUT a$
IF a$ = "s" THEN STOP
GOTO redo:


"C:\Users\Reactor1967\vcns\work\ALPHA3.BAS"
a# = 1
b# = 2
c# = 3
a2# = 11
b2# = 12
c2# = 13
carry# = 1
CLS
RANDOMIZE TIMER
redo:
FOR count = 1 TO 10
z = INT(RND * 2) + 0
test1 = (carry# = a#) AND (z = 0)
test2 = (carry# = a#) AND (z = 1)
IF test1 = -1 THEN carry# = b#
IF test2 = -1 THEN carry# = c#
PRINT a#; b#; c#; carry#
a# = a# + 1
b# = b# + 2
c# = c# + 2
NEXT count
z = INT(RND * 2) + 0
test1 = (carry# = a#) AND (z = 0)
test2 = (carry# = a#) AND (z = 1)
IF test1 = -1 THEN carry# = b#
IF test2 = -1 THEN carry# = c#
PRINT a#; b#; c#; carry#

FOR count = 1 TO 10
a# = a# + 1
NEXT count
INPUT a$
IF a$ = "s" THEN STOP
FOR count = 1 TO 10
```

```
z = INT(RND * 2) + 0
test3 = (carry# = a2#) AND (z = 0)
test4 = (carry# = a2#) AND (z = 1)
IF test3 = -1 THEN carry# = b2#
IF test4 = -1 THEN carry# = c2#
PRINT a2#; b2#; c2#; carry#
a2# = a2# + 1
b2# = b2# + 2
c2# = c2# + 2
NEXT count
z = INT(RND * 2) + 0
test3 = (carry# = a2#) AND (z = 0)
test4 = (carry# = a2#) AND (z = 1)
IF test3 = -1 THEN carry# = b2#
IF test4 = -1 THEN carry# = c2#

PRINT a2#; b2#; c2#; carry#
FOR count = 1 TO 10
a2# = a2# + 1
NEXT count
INPUT a$
IF a$ = "s" THEN STOP
GOTO redo:
```

```
"C:\Users\Reactor1967\vcns\work\ALPHA2.BAS"
a# = 1
b# = 2
c# = 3
d# = 4
c# = 1
```

```
"C:\Users\Reactor1967\vcns\work\ALMOST.BAS"
REM attempt to cycle self coding and self decoding v#;r# with second
REM v2#. Both v# and v2# use the same r#. Key here is getting v2# timed
REM with v# and r#. I do this by monitoring the distance between v2# & r#.
REM This program seems almost to decode. Its the most sucessful I,ve been
REM in a long long time. If the cycle/timing and be corrected this program
REM will fly.
v# = 1
r# = 1
v2# = 100
CLS
```

```
count = 1
DO
test = (v2# - r#) <= 100
IF test = -1 THEN x = INT(RND * 2) + 0
test3 = (test = -1) AND (x = 0)
test4 = (test = -1) AND (x = 1)
IF test3 = -1 THEN v2# = v2# + (v2# - r#) + 1
IF test4 = -1 THEN v2# = v2# + (v2# - r#) + 2
IF (v# + (v# - r#)) >= (r# + 50) THEN z = 1
IF (v# + (v# - r#)) < (r# + 50) THEN z = 0
IF z = 0 THEN v# = v# + ((v# - r#) * 1) + 1
IF z = 1 THEN v# = v# + ((v# - r#) * 1) + 2
COLOR 2, 0
PRINT , , z; v#; r#; v2#
a$ = INKEY$
REM INPUT a$
IF a$ = "s" THEN STOP
IF a$ = "d" THEN EXIT DO
IF z = 1 THEN r# = r# + 52
LOOP UNTIL v2# + (v2# - r#) >= 999999999999999#
DO
test = (v2# - r#) >= 100
IF test = -1 THEN GOSUB decode:
dist# = (v# - r#)
IF (dist# / 2) - INT(dist# / 2) = .5 THEN dist# = dist# - 1
dist# = dist# / 2
dist# = dist# + 1
v# = v# - dist#
test = (v# / 2) - INT(v# / 2)
IF test = .5 THEN r# = r# - 52
COLOR 2, 0
PRINT , , v#; r#; v2#
IF v2# = 100 THEN STOP
IF v# = 1 THEN STOP
INPUT a$
REM a$ = INKEY$
IF a$ = "s" THEN STOP
LOOP
STOP
decode:
dist# = (v2# - r#)
IF (dist# / 2) - INT(dist# / 2) = .5 THEN dist# = dist# - 1
dist# = dist# / 2
dist# = dist# + 1
v2# = v2# - dist#
RETURN
```

```
"C:\Users\Reactor1967\vcns\work\AKEEPER.BAS"
v# = 1
r# = 1
CLS
RANDOMIZE TIMER
DO
z = INT(RND * 2) + 0
REM (v# - r#) Filter**You have to reverse this when decoding*********
dist# = v# - r#
test# = (v# - r#) / 3
test# = test# - INT(test#)
test# = test# * 10
test# = INT(test#)
IF test# = 6 THEN dist# = dist# - 1
IF test# = 3 THEN dist# = dist# + 1
REM End of (v# - r#)
filter******************************************
testx = v# + dist# + 3 >= (r# + 100)
IF testx = -1 THEN z = 2
IF z = 0 THEN v# = v# + dist# + 1
IF z = 1 THEN v# = v# + dist# + 2
IF z = 2 THEN v# = v# + dist# + 3
COLOR 2, 0
PRINT , , z; v#; r#; v# / 3
IF testx = -1 THEN r# = r# + 102
a$ = INKEY$
REM INPUT a$
IF a$ = "s" THEN STOP
LOOP




"C:\Users\Reactor1967\vcns\work\AGAIN.BAS"
a# = 1
b# = 1
c# = 2
CLS
DO
PRINT a#; b#; c#; "="; a# + b#; a# + c#
INPUT a$
IF a$ = "s" THEN STOP
a# = a# + 1
b# = b# + 1
```

```
c# = c# + 1
LOOP
```

"C:\Users\Reactor1967\vcns\work\ADVANCE1.BAS"
```
REM CREATE EQUATIONS FOR THEN NEW BREAK
THROUGH!!!!!!!!!!!!!!!!!!!!!!
REM Now Create chart for range of even and odd values to subtract and
REM add for encoding and decoding at different ranges.
CLS
RANDOMIZE TIMER
DO
p# = INT(RND * 99999999999999#) + 0
test2# = p# / 2 - INT(p# / 2)
IF test2# >= .5 THEN test2# = 1
c# = p#
REM ----------------------------- Encode
d = INT(RND * 2) + 0
c2# = c# - d: REM made changes here
c# = c# + c2#
REM PRINT "p#"; p#
test# = c# / 2 - INT(c# / 2)
REM IF test# = 0 THEN STOP: REM made changes here
IF test# >= .5 THEN test# = 1
PRINT , ; "p#"; p#; "c#"; c#; "Data"; test#
REM x# = (p# - r#) Equations
REM p# = p# + x# Equations
REM ----------------------------------- Decode
REM x# = p# - r# Equations
REM p# = (((p# - x#) / 2) + x#) Equations
c# = c# - d: REM made changes here
c# = c# / 2
c# = c# + d: REM made changes here
REM PRINT c#
REM PRINT p#
REM IF p# = c# THEN PRINT "GREAT!!!"
IF p# <> c# THEN STOP
REM PRINT "-------------------------------"
a$ = INKEY$
IF a$ = "s" THEN STOP
LOOP
```

"C:\Users\Reactor1967\vcns\work\ADDR.BAS"

```
REM NOTE: TRY CODING YOUR DIFFERECE BETWEEN P & R UP AND
DOWN.
REM calculate p-r as a difference then code it up or down and
REM use it to find the next r1# and r2#
REM start with diff = 25*10^12 + 1 * 2 + c minus or + p for r
REM Keeping up with diff to calculate R should work.
p# = 125000000000001#
r1# = 100000000000001#
r2# = 100000000000000#
RANDOMIZE TIMER
CLS
redo:
count = 1
DO
c = INT(RND * 2) + 0
store2# = r#
IF (r1# / 2) - INT(r1# / 2) >= .5 THEN r# = r1#
IF (r2# / 2) - INT(r2# / 2) >= .5 THEN r# = r2#
PRINT p#; c; r1#; r2#; p# - r#; k
store# = p#
IF c = 0 THEN p# = p# + (p# - r# + 1)
IF c = 1 THEN p# = p# + (p# - r# + 2)
IF c = 0 THEN k = 1
IF c = 1 THEN k = 2
add# = 25000000000000# + k
REM add# = p# - store#
r1# = r1# + add#
r2# = r2# + add#
INPUT a$
REM a$ = INKEY$
IF a$ = "s" THEN STOP
IF a$ = "d" THEN GOTO decode:
count = count + 1
LOOP UNTIL p# >= 925000000000000#
IF (r1# / 2) - INT(r1# / 2) >= .5 THEN r# = r1#
IF (r2# / 2) - INT(r2# / 2) >= .5 THEN r# = r2#
r1# = p# + p# - r#
r2# = p# + p# - r#
PRINT "--------------------------------------------------------"
DO
c = INT(RND * 2) + 0
IF (r1# / 2) - INT(r1# / 2) >= .5 THEN r# = r1#
IF (r2# / 2) - INT(r2# / 2) >= .5 THEN r# = r2#
PRINT p#; c; r1#; r2#; r# - p#; k
store# = p#
IF c = 0 THEN p# = p# - (r# - p# + 1)
```

```
IF c = 1 THEN p# = p# - (r# - p# + 2)
k# = (r# - p#)
IF (k# / 2) - INT(k# / 2) >= .5 THEN k# = k# - 1
subtract# = p# - k#
r1# = r1# - subtract#
r2# = r2# - subtract#
INPUT a$
REM a$ = INKEY$
IF a$ = "s" THEN STOP
LOOP UNTIL p# <= 150000000000000#
r1# = p# - (r# - p#)
r2# = p# - (r# - p#)
PRINT "----------------------------------------------------------"
GOTO redo:
decode:
c = (p# / 2) - INT(p# / 2)
IF c >= .5 THEN c = 1
IF c = 0 THEN k = 1
IF c = 1 THEN k = 2
subtract# = 25000000000000# + k
r1# = r1# - subtract#
r2# = r2# - subtract#
IF (r1# / 2) - INT(r1# / 2) >= .5 THEN r# = r1#
IF (r2# / 2) - INT(r2# / 2) >= .5 THEN r# = r2#
subtract# = p# - r#
IF (subtract# / 2) - INT(subtract# / 2) >= .5 THEN subtract# = subtract# - 1
p# = p# - (((subtract#) / 2) + 1)
PRINT p#; c; r1#; r2#; p# - r#; k
INPUT a$
IF a$ = "s" THEN STOP
IF a$ = "f" THEN GOTO decode2:
GOTO decode:
decode2:
PRINT r# - p#



"C:\Users\Reactor1967\vcns\work\821023.BAS"
count = 0
v# = 1
CLS
RANDOMIZE TIMER
count = 1
DO
z = INT(RND * 2) + 0
IF z = 0 THEN v# = v# + ((count + 1) * 1) + 1
```

```
IF z = 1 THEN v# = v# + ((count + 1) * 1) + 2
PRINT count; v#; z
test1 = (count / 2) - INT(count / 2)
test2 = (v# / 2) - INT(v# / 2)
test3 = (test1 = 0) AND (test2 = 0)
test4 = (test1 = 0) AND (test2 = .5)
test5 = (test1 = .5) AND (test2 = 0)
test6 = (test1 = .5) AND (test2 = .5)
IF test3 = -1 THEN count = count + 2
IF test4 = -1 THEN count = count + 1
IF test5 = -1 THEN count = count + 1
IF test6 = -1 THEN count = count + 2
a$ = INKEY$
IF a$ = "s" THEN STOP
LOOP


"C:\Users\Reactor1967\vcns\work\122902C.BAS"
v# = 1
r# = 1
dist# = 0
CLS
DO
z = INT(RND * 2) + 0
IF z = 0 THEN v# = v# + (v# - r# + 2)
IF z = 1 THEN v# = v# + (v# - r# + 4)
COLOR 2, 0
PRINT z; v#; r#; dist#
r# = r# + dist#
IF z = 0 THEN dist# = dist# + 2
IF z = 1 THEN dist# = dist# + 4
a$ = INKEY$
IF a$ = "s" THEN STOP
LOOP


"C:\Users\Reactor1967\vcns\work\122902B.BAS"
v# = 1
r# = 1
dist# = 0
CLS
DO
z = INT(RND * 2) + 0
IF z = 0 THEN v# = v# + (v# - r# + 2)
```

```
IF z = 1 THEN v# = v# + (v# - r# + 4)
COLOR 2, 0
PRINT z; v#; r#; dist#
r# = r# + dist#
IF z = 0 THEN dist# = dist# + 2
IF z = 1 THEN dist# = dist# + 4
a$ = INKEY$
IF a$ = "s" THEN STOP
LOOP


"C:\Users\Reactor1967\vcns\work\122902A.BAS"
v# = 1
r# = 1
dist# = 0
CLS
DO
z = INT(RND * 2) + 0
dist# = v# - r#
IF z = 0 THEN v# = v# + (v# - r# + 2)
IF z = 1 THEN v# = v# + (v# - r# + 4)
COLOR 2, 0
PRINT z; v#; r#; dist#
r# = r# + dist#
a$ = INKEY$
IF a$ = "s" THEN STOP
LOOP


"C:\Users\Reactor1967\vcns\work\122802B.BAS"
v# = 1
r# = 1
CLS
RANDOMIZE TIMER
redo:
DO
z = INT(RND * 2) + 0
test = (r# * 2) + 1 <= (v# + (v# - r# + 1))
IF z = 0 THEN v# = v# + (v# - r# + 1)
IF z = 1 THEN v# = v# + (v# - r# + 2)
COLOR 2, 0
PRINT , v#; r#; v# - r#
IF test = -1 THEN r# = (r# * 2) + 1
a$ = INKEY$
```

```
REM INPUT a$
IF a$ = "s" THEN STOP
LOOP UNTIL v# + (v# - r#) >= 999999999999999#
test = (r# - 1) / 2 <= (v# - (v# - r# + 1))
IF test = -1 THEN r# = (r# - 1) / 2
DO
z = INT(RND * 2) + 0
test = (r# - 1) / 2 <= (v# - (v# - r# + 1))
IF z = 0 THEN v# = v# - (v# - r# + 1)
IF z = 1 THEN v# = v# - (v# - r# + 2)
COLOR 2, 0
IF test = -1 THEN r# = (r# - 1) / 2
PRINT , v#; r#; v# - r#
a$ = INKEY$
REM INPUT a$
IF a$ = "s" THEN STOP
LOOP UNTIL v# - (v# - r#) <= 100000000000000#
GOTO redo:




"C:\Users\Reactor1967\vcns\work\122802.BAS"
v# = 1
r# = 1
CLS
RANDOMIZE TIMER
redo:
DO
z = INT(RND * 2) + 0
test = (r# * 2) + 1 <= (v# + (v# - r# + 1))
IF z = 0 THEN v# = v# + (v# - r# + 1)
IF z = 1 THEN v# = v# + (v# - r# + 2)
COLOR 2, 0
PRINT , v#; r#; v# - r#
IF test = -1 THEN r# = (r# * 2) + 1
a$ = INKEY$
REM INPUT a$
IF a$ = "s" THEN STOP
LOOP UNTIL v# + (v# - r#) >= 999999999999999#
test = (r# - 1) / 2 <= (v# - (v# - r# + 1))
IF test = -1 THEN r# = (r# - 1) / 2
DO
z = INT(RND * 2) + 0
test = (r# - 1) / 2 <= (v# - (v# - r# + 1))
```

```
IF z = 0 THEN v# = v# - (v# - r# + 1)
IF z = 1 THEN v# = v# - (v# - r# + 2)
COLOR 2, 0
PRINT , v#; r#; v# - r#
IF test = -1 THEN r# = (r# - 1) / 2
a$ = INKEY$
REM INPUT a$
IF a$ = "s" THEN STOP
LOOP UNTIL v# - (v# - r#) <= 1
GOTO redo:




"C:\Users\Reactor1967\vcns\work\111802.BAS"
REM the value itself is a binary 1 or 0
REM the value devided by 2 if 0 means
REM do not derement r. If 1 then that
REM means decrement r with the equation
REM below.
REM -------------------------------
REM for decoding try
REM dist# = v# - r#
REM r# = r# - ((dist# - N + 1)/ 3)
REM --------------------------------
REM r# = r# - ((((v# - r#) - n) + 1) / base-1) decoding
REM r# = r# + (((v# - r#) * (base - 1)) - 1) coding
CLS
count = 0
v# = 1
r# = 1
RANDOMIZE TIMER
COLOR 2, 0
PRINT , , z1; z2; v#; r#; v# - r#
dist# = 0
DO
z1 = INT(RND * 2) + 0
z2 = INT(RND * 2) + 0
test1 = (z1 = 0) AND (z2 = 0)
test2 = (z1 = 0) AND (z2 = 1)
test3 = (z1 = 1) AND (z2 = 0)
test4 = (z1 = 1) AND (z2 = 1)
IF (v# - r#) >= 14 THEN dist# = (((v# - r#) * 3) - 1) ELSE dist# = 0
IF test1 = -1 THEN v# = v# + ((v# - r#) * 2) + 1
IF test2 = -1 THEN v# = v# + ((v# - r#) * 2) + 2
```

```
IF test3 = -1 THEN v# = v# + ((v# - r#) * 2) + 3
IF test4 = -1 THEN v# = v# + ((v# - r#) * 2) + 4
COLOR 2, 0
REM PRINT , , z1; z2; v#; r#; v# - vs#; ((v# - vs#) * 3) - 1
PRINT , , z1; z2; v#; r#; v# - r#
r# = r# + dist#
a$ = INKEY$
INPUT a$
IF a$ = "s" THEN STOP
LOOP




"C:\Users\Reactor1967\vcns\work\102502.BAS"
v# = 1
r# = 1
CLS
RANDOMIZE TIMER
DO
z = INT(RND * 2) + 0
test = (v# + (v# - r#)) >= (r# + 4)
IF z = 0 THEN v# = v# + (v# - r# + 1)
IF z = 1 THEN v# = v# + (v# - r# + 2)
COLOR 2, 0
PRINT , , z; v#; r#; r# - r1#
r1# = r#
v1# = v#
IF test = -1 THEN r# = r# + 6
a$ = INKEY$
REM INPUT a$
IF a$ = "s" THEN STOP
LOOP




"C:\Users\Reactor1967\vcns\work\0100603G.BAS"
v# = 1
r# = 1
dist# = 0
CLS
RANDOMIZE TIMER
DO
z = INT(RND * 2) + 0
test = (v# / 2) - INT(v# / 2)
test1 = (test = 0) AND (z = 0)
test2 = (test = 0) AND (z = 1)
```

```
test3 = (test = .5) AND (z = 0)
test4 = (test = .5) AND (z = 1)
IF test1 = -1 THEN dist# = dist# + 2
IF test2 = -1 THEN dist# = dist# + 1
IF test3 = -1 THEN dist# = dist# + 1
IF test4 = -1 THEN dist# = dist# + 2
x# = dist#
IF z = 0 THEN x# = x# - 1
IF z = 1 THEN x# = x# - 2
x# = INT(x# / 2)
r# = v# - x#
IF z = 0 THEN v# = v# + (v# - r# + 1)
IF z = 1 THEN v# = v# + (v# - r# + 2)
PRINT , , z; v#; r#; dist#; v# - r#
a$ = INKEY$
INPUT a$
IF a$ = "s" THEN STOP
LOOP




"C:\Users\Reactor1967\vcns\work\92702A.BAS"
v# = 1
r# = 1
CLS
RANDOMIZE TIMER
c1# = 1
c2# = 2
DO
z = INT(RND * 2) + 0
test = (v# / 2) - INT(v# / 2)
test1 = (z = 0) AND (test = 0)
test2 = (z = 1) AND (test = 0)
test3 = (z = 0) AND (test = .5)
test4 = (z = 1) AND (test = .5)
IF test1 = -1 THEN v# = v# + c1# + 1
IF test2 = -1 THEN v# = v# + c1# + 2
IF test3 = -1 THEN v# = v# + c2# + 1
IF test4 = -1 THEN v# = v# + c2# + 2
COLOR 2, 0
dist# = (v# - vs#)
dist# = dist# - 1
dist# = dist# * 2
r# = v# - dist#
PRINT , , z; v#; r#; r# - rs#
rs# = r#
```

```
vs# = v#
ds# = (v# - r#)
a$ = INKEY$
IF a$ = "s" THEN STOP
c1# = c1# + 2
c2# = c2# + 2
LOOP




"C:\Users\Reactor1967\vcns\work\92602B.BAS"
v# = 1
r# = 1
d# = 100
CLS
RANDOMIZE TIMER
DO
z = INT(RND * 2) + 0
IF z = 0 THEN v# = v# + (v# - r# + 1)
IF z = 1 THEN v# = v# + (v# - r# + 2)
COLOR 2, 0
PRINT , , d#; z; v#; r#; v# - r#
a$ = INKEY$
IF a$ = "s" THEN STOP
IF a$ = "d" THEN EXIT DO
IF v# >= d# THEN r# = r# + 100
IF v# >= d# THEN d# = d# + 100
LOOP
DO
COLOR 2, 0
PRINT , , d#; z; v#; r#; v# - r#
a$ = INKEY$
INPUT a$
IF a$ = "s" THEN STOP
dist# = v# - r#
IF (dist# / 2) - INT(dist# / 2) = .5 THEN dist# = dist# - 1
dist# = dist# / 2
dist# = dist# + 1
v# = v# - dist#
IF v# < d# THEN r# = r# - 100
IF v# < d# THEN d# = d# - 100
LOOP
```

```
"C:\Users\Reactor1967\vcns\work\92602.BAS"
v# = 4
r# = 1
booster = 100#
CLS
RANDOMIZE TIMER
DO
z = INT(RND * 2) + 0
IF v# >= (r# + 100) THEN r# = r# + 98
IF z = 0 THEN v# = v# + (v# - r# - 1)
IF z = 1 THEN v# = v# + (v# - r# - 2)
COLOR 2, 0
PRINT , , v#; r#; v# - r#
a$ = INKEY$
IF a$ = "s" THEN STOP
IF a$ = "d" THEN EXIT DO
LOOP
REM when decoding if result is even subtract nothing
REM if result is odd minus 1.
REM this is after v# - r# is divided by two
ref# = v#
DO
IF ref# - v# >= 98 THEN r# = r# - 98
dist# = v# - r#
test = (v# / 2) - INT(v# / 2)
IF (dist# / 2) - INT(dist# / 2) = .5 THEN dist# = dist# - 1
dist# = dist# / 2
IF test = .5 THEN dist# = dist# - 1
v# = v# - dist#
COLOR 2, 0
PRINT , , v#; r#
a$ = INKEY$
INPUT a$
IF a$ = "s" THEN STOP
LOOP


"C:\Users\Reactor1967\vcns\work\92402C.BAS"
v# = 1
r# = 1
r2# = r#
vcarry# = 1
rcarry# = 1
CLS
RANDOMIZE TIMER
```

```
DO
test = v# + (v# - r#) >= (r# + 100)
v1# = v# + (v# - r# + 1)
v2# = v# + (v# - r# + 2)
PRINT v#; v1#; v2#; r#; r2#; vcarry#; rcarry#
IF test = -1 THEN r2# = r# + 100
test2 = (vcarry# = v#) AND (rcarry# = r#)
IF test2 = -1 THEN z = INT(RND * 2) + 0
test3 = (test2 = -1) AND (z = 0)
test4 = (test2 = -1) AND (z = 1)
IF test3 = -1 THEN vcarry# = v1#
IF test4 = -1 THEN vcarry# = v2#
IF test3 = -1 THEN rcarry# = r2#
IF test4 = -1 THEN rcarry# = r2#
v# = v# + 1
IF v# >= (r# + 100) THEN r# = r# + 100
a$ = INKEY$
IF a$ = "s" THEN STOP
IF a$ = "d" THEN EXIT DO
IF v# > vcarry# THEN STOP
LOOP
DO
test = v# + (v# - r#) >= (r# + 100)
v1# = v# + (v# - r# + 1)
v2# = v# + (v# - r# + 2)
PRINT v#; v1#; v2#; r#; r2#; vcarry#; rcarry#
test2 = ((vcarry# = v1#) OR (vcarry# = v2#))
test3 = (test2 = -1) AND (rcarry# = r2#)
IF test3 = -1 THEN vcarry# = v#
IF test3 = -1 THEN rcarry# = r#
IF test = -1 THEN r2# = r# + 100
v# = v# - 1
IF v# < r# THEN r# = r# - 100
a$ = INKEY$
REM IF v# = 110 THEN STOP
INPUT a$
IF a$ = "s" THEN STOP
LOOP UNTIL v# = 1




"C:\Users\Reactor1967\vcns\work\92402B.BAS"
r# = 1
v# = 1
r2# = 1
```

```
CLS
DO
z = INT(RND * 2) + 0
IF z = 1 THEN r2# = r# + (v# - r# + 2)
IF z = 0 THEN r2# = r# + (v# - r# + 1)
IF z = 0 THEN v# = v# + (v# - r# + 3)
IF z = 1 THEN v# = v# + (v# - r# + 4)
r# = r2#
COLOR 2, 0
vr# = v# - vs#
vr# = vr# - 1
vr# = vr# * 2
vr# = v# - vr#
rr# = r# - rs#
rr# = rr# - 1
rr# = rr# * 2
rr# = r# - rr#
PRINT , , vr# - ds# - 2; vr#; v#; r#; rr#; rr# - ds2#
ds# = vr#
ds2# = rr#
vs# = v#
rs# = r#
a$ = INKEY$
IF a$ = "s" THEN STOP
LOOP




"C:\Users\Reactor1967\vcns\work\92402A.BAS"
d# = 1
v# = 4
r# = (v# - d#)
CLS
DO
FOR count = 1 TO 10
z = 0
v# = v# + (v# - r# + 1)
COLOR 2, 0
PRINT , , z; v#; r#; d#; count
d# = d# + 2
r# = (v# - d#)
a$ = INKEY$
IF a$ = "s" THEN STOP
NEXT count
z = INT(RND * 2) + 0
```

```
IF z = 0 THEN v# = v# + (v# - r# + 1)
IF z = 1 THEN v# = v# + (v# - r# + 2)
COLOR 2, 0
PRINT , , z; v#; r#; d#; count; "DATA---------------"
d# = d# + 2
r# = (v# - d#)
a$ = INKEY$
IF a$ = "s" THEN STOP
LOOP




"C:\Users\Reactor1967\vcns\work\92102C.BAS"
v# = 1
r# = 1
d1# = 0
CLS
RANDOMIZE TIMER
DO
z = INT(RND * 2) + 0
ds1# = (v# - r#)
IF z = 0 THEN v# = v# + (v# - r# + 1)
IF z = 1 THEN v# = v# + (v# - r# + 2)
ds2# = v# - r#
COLOR 2, 0
PRINT , , v#; r#; ds1#; ds2#; d1#
a$ = INKEY$
REM INPUT a$
IF a$ = "s" THEN STOP
DO
d1# = d1# + 1
r# = v# - d1#
test = (r# / 2) - INT(r# / 2) = .5
a$ = INKEY$
IF a$ = "s" THEN STOP
LOOP UNTIL test = -1
LOOP




"C:\Users\Reactor1967\vcns\work\92002A.BAS"
v# = 1
r# = 1
dt# = 1000
RANDOMIZE TIMER
```

```
CLS
redo:
DO
test = (v# + (v# - r#) >= (r# + 100))
IF test = 0 THEN z = 0
IF test = -1 THEN z = 1
REM z = INT(RND * 2) + 0
IF z = 0 THEN v# = v# + (v# - r# + 1)
IF z = 1 THEN v# = v# + (v# - r# + 2)
COLOR 2, 0
PRINT , , z; v#; r#; v# - r#
a$ = INKEY$
REM INPUT a$
IF a$ = "s" THEN STOP
IF test = -1 THEN r# = r# + 102
LOOP UNTIL (v# - r# <= 25) AND (((v# + (v# - r# + 1)) - r#) >= 25)
COLOR 2, 0
PRINT , , z; v#; r#; v# - r#
z = INT(RND * 2) + 0
IF z = 0 THEN v# = v# + (v# - r# + 1)
IF z = 1 THEN v# = v# + (v# - r# + 2)
COLOR 2, 0
PRINT , , z; v#; r#; v# - r#; "---------"; dt#
a$ = INKEY$
REM INPUT a$
IF a$ = "s" THEN STOP
GOTO redo:


"C:\Users\Reactor1967\vcns\work\91902B.BAS"
v# = 1
r# = 1
CLS
RANDOMIZE TIMER
DO
z = INT(RND * 2) + 0
REM (v# - r#) Filter**You have to reverse this when decoding*********
dist# = v# - r#
test# = (v# - r#) / 3
test# = test# - INT(test#)
test# = test# * 10
test# = INT(test#)
IF test# = 6 THEN dist# = dist# - 1
IF test# = 3 THEN dist# = dist# + 1
REM End of (v# - r#)
```

```
filter*******************************************
test = v# + dist# + 3 >= (r# + 100)
IF test = -1 THEN z = 2
IF z = 0 THEN v# = v# + dist# + 1
IF z = 1 THEN v# = v# + dist# + 2
IF z = 2 THEN v# = v# + dist# + 3
COLOR 2, 0
PRINT , , z; v#; r#; v# / 3
IF test = -1 THEN r# = r# + 102
a$ = INKEY$
REM INPUT a$
IF a$ = "s" THEN STOP
LOOP




"C:\Users\Reactor1967\vcns\work\91902A.BAS"
REM 0 .666 then v# - r# = .3333333333
REM 2 .333 then v# - r# = 0
REM 1 0.00 then v# - r# = .666666
v# = 1
r# = 1
CLS
DO
s = z
z = INT(RND * 3) + 0
test = v# + ((v# - r#) * 2) >= (r# + 100)
IF z = 0 THEN v# = v# + ((v# - r#) * 2) + 1
IF z = 1 THEN v# = v# + ((v# - r#) * 2) + 2
IF z = 2 THEN v# = v# + ((v# - r#) * 2) + 3
COLOR 2, 0
PRINT , , z; v# / 3
a$ = INKEY$
INPUT a$
IF a$ = "s" THEN STOP
LOOP




"C:\Users\Reactor1967\vcns\work\91802B.BAS"
REM trying to find a way to add 3 states to regular incode by
REM changing the constants in the equations
v# = 1
r# = 1
CLS
RANDOMIZE TIMER
```

```
DO
z = INT(RND * 2) + 0
redo:
dist# = v# - r#
dist# = dist# + 1
DO
v1# = v# + dist#
test = (v1# / 3)
test = test - INT(test)
test = test * 10
test = INT(test)
test1 = (z = 0) AND (test = 0)
test2 = (z = 1) AND (test = 3)
test3 = (z = 2) AND (test = 6)
test4 = (test1 = -1) OR (test2 = -1) OR (test3 = -1)
IF test4 = -1 THEN EXIT DO
dist# = dist# + 1
a$ = INKEY$
IF a$ = "s" THEN STOP
LOOP
IF v# - r# >= 100 THEN z = 2
IF v# - r# >= 100 THEN GOTO redo:
v# = v1#
COLOR 2, 0
PRINT , , z; v#; v# / 3
a$ = INKEY$
IF a$ = "s" THEN STOP
IF z = 2 THEN r# = r# + 100
LOOP




"C:\Users\Reactor1967\vcns\work\91802A.BAS"
v# = 1
r# = 1
CLS
RANDOMIZE TIMER
DO
z = INT(RND * 2) + 0
test = v# + (v# - r# + 4) >= (r# + 102)
IF test = -1 THEN z = 2
IF z = 0 THEN v# = v# + (v# - r# + 1)
IF z = 1 THEN v# = v# + (v# - r# + 3)
IF z = 2 THEN v# = v# + (v# - r# + 4)
COLOR 2, 0
PRINT , , z; v#; r#; v# / 2
```

```
a$ = INKEY$
REM input a$
IF a$ = "s" THEN STOP
IF test = -1 THEN r# = r# + 102
LOOP
```

```
"C:\Users\Reactor1967\vcns\work\91702A.BAS"
REM even numbers are even when divided by two and flips odd if not
REM odd numbers are odd when divided by two and flips even if not
REM rule of thumb divide by two even will be even when flips and odd
REM will be odd when flips.
REM flip refers to subtracting 104 from r#
REM v1# = v# + (v# - r1#) + 1
REM v2# = v# + (v# - r1#) + 3
REM v3# = v# + (v# - r1#) + 2
REM v4# = v# + (v# - r1#) + 4
REM add# 104 when over 100
v# = 1
r# = 1
RANDOMIZE TIMER
CLS
DO
z = INT(RND * 2) + 0
z = 1
REM z = 0
test1 = v# + (v# - r#) >= (r# + 100)
test2 = (z = 0) AND (test1 = 0)
test3 = (z = 0) AND (test1 = -1)
test4 = (z = 1) AND (test1 = 0)
test5 = (z = 1) AND (test1 = -1)
IF test2 = -1 THEN v# = v# + (v# - r#) + 1
IF test3 = -1 THEN v# = v# + (v# - r#) + 3
IF test4 = -1 THEN v# = v# + (v# - r#) + 4
IF test5 = -1 THEN v# = v# + (v# - r#) + 2
COLOR 2, 0
PRINT , , z; v#; r#; test1; v# / 2
REM PRINT , , z; v#; r#; v# / 2
IF test1 = -1 THEN r# = r# + 104
a$ = INKEY$
REM INPUT a$
IF a$ = "s" THEN STOP
LOOP
```

```
"C:\Users\Reactor1967\vcns\work\91302D.BAS"
DECLARE SUB d1 (v1#, v#, r#, god!, bit!)
DIM bank1(1000)
DIM bank2(1000)
gh = 1
bank1(gh) = v#
bank2(gh) = r#
v# = 1
r# = 1
RANDOMIZE TIMER
CLS
DO
test = v# + (v# - r# + 1) >= (r# + 100)
IF test = 0 THEN z = 0
IF test = -1 THEN z = 1
IF v# - r# = 1 THEN flag = 1 ELSE flag = 0
IF v# - r# = 1 THEN z = INT(RND * 2) + 0
IF z = 0 THEN v# = v# + (v# - r# + 1)
IF z = 1 THEN v# = v# + (v# - r# + 2)
COLOR 2, 0
IF flag = 0 THEN PRINT , , v#; r#
IF flag = 1 THEN PRINT , , v#; r#; z
gh = gh + 1
bank1(gh) = v#: bank2(gh) = r#
a$ = INKEY$
INPUT a$
IF a$ = "s" THEN STOP
decode = ((v# / 2) - INT(v# / 2) = 0) AND (a$ = "d")
IF decode = -1 THEN EXIT DO
test = (flag = 0) AND (z = 1)
IF test = -1 THEN r# = r# + 102
LOOP
DO
IF god = 0 THEN CALL d1(v1#, v#, r#, god, bit)
dist# = v# - r#
IF (dist# / 2) - INT(dist# / 2) = .5 THEN dist# = dist# - 1
dist# = dist# / 2
dist# = dist# + 1
v# = v# - dist#
test = (v# / 2) - INT(v# / 2) = .5 AND (god = 0) AND (r# > 1)
IF test = -1 THEN r# = r# - 102
IF v# - r# = 1 THEN god = 0
```

```
COLOR 2, 0
IF god = 0 THEN PRINT , , v#; r#
IF god = -1 THEN PRINT , , v#; r#; bit
gh = gh - 1
IF v# <> bank1(gh) THEN STOP: IF r# <> bank2(gh) THEN STOP
a$ = INKEY$
ex = (v# = 1) AND (r# = 1)
IF ex = -1 THEN EXIT DO
IF v# <= 0 THEN PRINT , , "Error"
IF v# <= 0 THEN STOP
INPUT a$
IF a$ = "s" THEN STOP
LOOP


SUB d1 (v1#, v#, r#, god, bit)
v1# = v#
test0 = (v# - r#) <= 10
dist# = v1# - r#
IF (dist# / 2) - INT(dist# / 2) = .5 THEN dist# = dist# - 1
dist# = dist# / 2
dist# = dist# + 1
v1# = v1# - dist#
bit = (v1# / 2) - INT(v1# / 2)
test1 = ((v1# - r#) = 3) OR ((v1# - r#) = 4)
dist# = v1# - r#
IF (dist# / 2) - INT(dist# / 2) = .5 THEN dist# = dist# - 1
dist# = dist# / 2
dist# = dist# + 1
v1# = v1# - dist#
test2 = ((v1# - r#) = 1)
dist# = v1# - r#
IF (dist# / 2) - INT(dist# / 2) = .5 THEN dist# = dist# - 1
dist# = dist# / 2
dist# = dist# + 1
v1# = v1# - dist#
test3 = v1# - r# = 0
god = (test1 = -1) AND (test2 = -1) AND (test3 = -1) AND (test0 = -1)
IF bit = .5 THEN bit = 1
END SUB




"C:\Users\Reactor1967\vcns\work\91302C.BAS"
v# = 1
r# = 1
CLS
```

```
RANDOMIZE TIMER
DO
test = v# + (v# - r#) >= (r# + 100)
IF test = 0 THEN z = 0
IF test = -1 THEN z = 1
IF v# - r# = 1 THEN z = INT(RND * 2) + 0
REM IF v# - r# = 1 THEN bit = z
REM IF v# - r# = 1 THEN flag = 1 ELSE flag = 0
dist# = v# - r#
IF z = 0 THEN v# = v# + (v# - r# + 1)
IF z = 1 THEN v# = v# + (v# - r# + 2)
COLOR 2, 0
PRINT , , z; v#; r#; v# - r#
REM IF v# - r# = 1 THEN STOP
INPUT a$
REM a$ = INKEY$
IF a$ = "s" THEN STOP
IF a$ = "d" THEN EXIT DO
REM ---------------------------------------
REM crossover
REM ---------------------------------------
IF test = -1 THEN r# = r# + 102
LOOP
DO
dist# = v# - r#
IF (dist# / 2) - INT(dist# / 2) = .5 THEN dist# = dist# - 1
dist# = dist# / 2
dist# = dist# + 1
v# = v# - dist#
COLOR 2, 0
PRINT , , v#; r#; v# - r#
test = ((v# / 2) - INT(v# / 2) = .5) AND (v# - r# <> 3) AND (v# - r# <> 4)
IF test = -1 THEN r# = r# - 102
ex = (v# = 1) AND (r# = 1)
IF ex = -1 THEN EXIT DO
IF v# <= 0 THEN EXIT DO
INPUT a$
REM a$ = INKEY$
IF a$ = "s" THEN STOP
LOOP


"C:\Users\Reactor1967\vcns\work\91102D.BAS"
v# = 1
r# = 1
```

```
dt = 0
CLS
RANDOMIZE TIMER
DO
z = INT(RND * 2) + 0
test = v# + (v# - r#) >= (r# + 100)
IF test = 0 THEN z = 0
IF test = -1 THEN z = 1
cdtest = ((v# + (v# - r#)) - r#) > 29 AND ((v# + (v# - r#)) - r#) <= 38 AND
bitstore = 0
IF cdtest = -1 THEN z = INT(RND * 2) + 0
IF cdtest = -1 THEN PRINT "---------------------------------------------"
IF z = 0 THEN v# = v# + (v# - r#) + 1
IF z = 1 THEN v# = v# + (v# - r#) + 2
COLOR 2, 0
PRINT , , z; v#; r#; v# - r#
IF cdtest = -1 THEN PRINT "---------------------------------------------"
INPUT a$
REM a$ = INKEY$
IF a$ = "s" THEN STOP
IF a$ = "d" THEN EXIT DO
IF test = -1 THEN r# = r# + 102
bitstore = z
LOOP
decode:
DO
h# = v# - r#
btest = (h# >= 30) AND (h# <= 40)
IF btest = -1 THEN bit = (v# / 2) - INT(v# / 2)
IF bit = .5 THEN bit = 1
dist# = v# - r#
test = (dist# / 2) - INT(dist# / 2)
IF test = .5 THEN dist# = dist# - 1
dist# = dist# / 2
dist# = dist# + 1
v# = v# - dist#
test1 = (v# / 2) - INT(v# / 2) = .5 AND (v# - r#) < 30
test2 = (v# / 2) - INT(v# / 2) = .5 AND (v# - r#) > 40
IF test1 = -1 THEN r# = r# - 102
IF test2 = -1 THEN r# = r# - 102
btest2 = (btest = -1) AND (v# / 2) - INT(v# / 2) = 0
IF btest2 = -1 THEN dt = 1 ELSE dt = 0
IF dt = 0 THEN PRINT v#; r#
IF dt = 1 THEN PRINT v#; r#; bit
IF v# = 1 THEN STOP
IF v# <= 0 THEN STOP
```

```
INPUT a$
REM a$ = INKEY$
IF a$ = "s" THEN STOP
LOOP




"C:\Users\Reactor1967\vcns\work\91102C.BAS"
v# = 23
r# = 9
CLS
DO
test = (v# + (v# - r#)) >= (r# + 100)
IF test = 0 THEN z = 0
IF test = -1 THEN z = 1
REM IF (v# + (v# - r#) + 2) - r# = 6 THEN PRINT , , "Here"
REM IF (v# + (v# - r#) + 2) - r# = 6 THEN z = INT(RND * 2) + 0
IF z = 0 THEN v# = v# + (v# - r#) + 2
IF z = 1 THEN v# = v# + (v# - r#) + 4
COLOR 2, 0
PRINT , , z; v#; r#; v# - r#
IF z = 1 THEN r# = r# + 102
INPUT a$
REM a$ = INKEY$
IF a$ = "s" THEN STOP
x = INT(RND * 2) + 0
LOOP




"C:\Users\Reactor1967\vcns\work\91102B.BAS"
REM Lesson Learned here. Inducing changes into the data
REM stream induces ripples into the system like a rock
REM being thrown into water. It seems the only way
REM to work out these ripples is to induce counter
REM changes by coding the reverse back into the system.
r# = 3
d# = 1
RANDOMIZE TIMER
CLS
count = 0
DO
IF d# - store# = 6 THEN z = 0
IF d# - store# = 12 THEN z = 1
rstore# = r#
IF z = 0 THEN r# = r# + ((r# - d#) * 2) + 2
```

```
IF z = 1 THEN r# = r# + ((r# - d#) * 2) + 4
IF z = 2 THEN r# = r# + ((r# - d#) * 2) + 6
COLOR 2, 0
PRINT , , z; r# - rstore#; r#; d#; d# - store#; r# / 3; r# - d#
store# = d#
DO
d# = d# + 6
LOOP UNTIL d# > r#
d# = d# - 6
INPUT a$
REM a$ = INKEY$
IF a$ = "s" THEN STOP
REM count = count + 1
LOOP




"C:\Users\Reactor1967\vcns\work\91102A.BAS"
r# = 3
d# = 1
RANDOMIZE TIMER
CLS
DO
z = INT(RND * 2) + 0
IF z = 0 THEN r# = r# + ((r# - d#) * 2) + 2
IF z = 1 THEN r# = r# + ((r# - d#) * 2) + 4
IF z = 2 THEN r# = r# + ((r# - d#) * 2) + 6
COLOR 2, 0
PRINT , , z; r#; d#; d# - store#
store# = d#
DO
d# = d# + 6
LOOP UNTIL d# > r#
d# = d# - 6
REM INPUT a$
a$ = INKEY$
IF a$ = "s" THEN STOP
LOOP




"C:\Users\Reactor1967\vcns\work\91002F.BAS"
r# = 1
v# = 1
RANDOMIZE TIMER
CLS
```

```
DO
z = INT(RND * 2) + 0
test = (v# + (v# - r#)) >= (r# + 4)
IF test = -1 THEN z = 2
IF z = 0 THEN v# = v# + ((v# - r#) * 1) + 1
IF z = 1 THEN v# = v# + ((v# - r#) * 1) + 2
IF z = 2 THEN v# = v# + ((v# - r#) * 2) + 3
COLOR 2, 0
PRINT , , z; v#; r#; v# / 3
IF v# - r# = 3 THEN STOP
REM a$ = INKEY$
IF a$ = "s" THEN STOP
IF test = -1 THEN r# = r# + 6
LOOP




"C:\Users\Reactor1967\vcns\work\91002E.BAS"
r# = 1
v# = 1

CLS
RANDOMIZE TIMER
DO
REM z = INT(RND * 2) + 0
test = v# + (v# - r#) >= (r# + 100)
IF test = 0 THEN z = 0 ELSE z = 1
dist# = v# - r#
IF z = 0 THEN v# = v# + (v# - r#) + 1
IF z = 1 THEN v# = v# + (v# - r#) + 2
test1 = (v2# / 2) - INT(v2# / 2) = (dist# / 2) - INT(dist# / 2)
IF test1 = -1 THEN dist# = dist# - 1
store# = r2#
r2# = v2# - dist#
z1 = INT(RND * 2) + 0
IF z1 = 0 THEN v2# = v2# + dist# + 1
IF z1 = 1 THEN v2# = v2# + dist# + 2
COLOR 2, 0
PRINT , z; v#; r#; v# - r#; z1; v2#; r2#; r2# - store#
IF test = -1 THEN r# = r# + 102
REM INPUT A$
a$ = INKEY$
IF a$ = "s" THEN STOP
LOOP
```

```
"C:\Users\Reactor1967\vcns\work\91002D.BAS"
r# = 1
v# = 1
CLS
RANDOMIZE TIMER
DO
z = INT(RND * 2) + 0
IF z = 0 THEN v# = v# + (v# - r#) + 1
IF z = 1 THEN v# = v# + (v# - r#) + 2
COLOR 2, 0
PRINT , , "1"; z; v#; r#; v# - r#
z = INT(RND * 4) + 0
z = 1
IF z = 0 THEN v# = v# + ((v# - r#) * 3) + 1
IF z = 1 THEN v# = v# + ((v# - r#) * 3) + 2
IF z = 2 THEN v# = v# + ((v# - r#) * 3) + 3
IF z = 3 THEN v# = v# + ((v# - r#) * 3) + 4
COLOR 2, 0
PRINT , , "2"; z; v#; r#; v# - r#
REM INPUT a$
IF a$ = "s" THEN STOP
test = (v# / 2) - INT(v# / 2)
IF test = 0 THEN r# = v# - 3
IF test = .5 THEN r# = v# - 2
LOOP



"C:\Users\Reactor1967\vcns\work\91002C.BAS"
v# = 3
r# = 1
RANDOMIZE TIMER
CLS
speed# = 10
dist# = 2
store# = r#
DO
FOR xd = 1 TO 2
z = INT(RND * 2) + 0
IF xd = 2 THEN z = 1
IF z = 0 THEN v# = v# + (v# - r#) + 1
IF z = 1 THEN v# = v# + (v# - r#) + 2
COLOR 2, 0
PRINT , , xd; v#; r#; v# - r#
```

```
NEXT xd
r# = v# - 2
PRINT v#; r#; r# - store#; (r# - store#) - accel#
accel# = r# - store#
INPUT a$
REM a$ = INKEY$
IF a$ = "s" THEN STOP
LOOP




"C:\Users\Reactor1967\vcns\work\91002B.BAS"
v# = 100
r# = 99
RANDOMIZE TIMER
CLS
DO
count = 0
DO
IF r# >= v# THEN EXIT DO
count = count + 1
a$ = INKEY$
IF a$ = "s" THEN STOP
r# = r# + 6
LOOP
IF r# > v# THEN r# = r# - 6
FOR xd = 1 TO 2
z = INT(RND * 3) + 0
IF z = 0 THEN v# = v# + ((v# - r#) * 2) + 0
IF z = 1 THEN v# = v# + ((v# - r#) * 2) + 2
IF z = 2 THEN v# = v# + ((v# - r#) * 2) + 6
COLOR 2, 0
PRINT , , z; v#; r#; count - 1
a$ = INKEY$
IF a$ = "s" THEN STOP
a$ = ""
NEXT xd
LOOP




"C:\Users\Reactor1967\vcns\work\91002A.BAS"
v# = 100
d1# = 1
d2# = 2
RANDOMIZE TIMER
```

```
CLS
DO
test = (v# / 2) - INT(v# / 2)
IF test = .5 THEN dist# = d2#
IF test = 0 THEN dist# = d1#
d1# = d1# + 2
d2# = d2# + 2
r# = v# - dist#
IF test = .5 THEN z = 1
IF test = 0 THEN z = 0
REM z = INT(RND * 2) + 0
IF z = 0 THEN v# = v# + (v# - r# + 1)
IF z = 1 THEN v# = v# + (v# - r# + 2)
COLOR 2, 0
PRINT , , v#; r#; dist#; d1#; d2#
a$ = INKEY$
IF a$ = "s" THEN STOP
LOOP



"C:\Users\Reactor1967\vcns\work\90702G.BAS"
v# = 1
r# = 1
v2# = 100
CLS
RANDOMIZE TIMER
DO
test = v# + (v# - r#) >= (r# + 50)
IF test = -1 THEN z = 1 ELSE z = 0
IF z = 1 THEN GOSUB code:
IF z = 0 THEN v# = v# + (v# - r# + 1)
IF z = 1 THEN v# = v# + (v# - r# + 2)
COLOR 2, 0
PRINT , , v#; r#; v2#
IF z = 1 THEN r# = r# + 52
REM INPUT a$
a$ = INKEY$
IF a$ = "s" THEN STOP
LOOP
code:
IF v2# - r# >= 200 THEN RETURN
x = INT(RND * 2) + 0
IF x = 0 THEN v2# = v2# + (v2# - r# + 1)
IF x = 1 THEN v2# = v2# + (v2# - r# + 2)
a$ = INKEY$
```

```basic
IF a$ = "s" THEN STOP
RETURN




"C:\Users\Reactor1967\vcns\work\90702F.BAS"
REM do not wast your time here
REM this is incomplete.
v# = 10
r# = 9
RANDOMIZE TIMER
CLS
DO
z = INT(RND * 2) + 0
IF z = 0 THEN v# = v# + (v# - r#) + 1
IF z = 1 THEN v# = v# + (v# - r#) + 2
COLOR 2, 0
PRINT , , z; v#; r1#; r#
dist# = dist# + 1

a$ = INKEY$
IF a$ = "s" THEN STOP
LOOP




"C:\Users\Reactor1967\vcns\work\90702E.BAS"
v# = 10
dist# = 0
RANDOMIZE TIMER
CLS
DO
z = INT(RND * 2) + 0
IF z = 0 THEN dist# = dist# + 1
IF z = 1 THEN dist# = dist# + 2
r# = v# - dist#
IF z = 0 THEN v# = v# + (v# - r#) + 1
IF z = 1 THEN v# = v# + (v# - r#) + 2
COLOR 2, 0
PRINT , , z; v#; r#; dist#
a$ = INKEY$
IF a$ = "s" THEN STOP
LOOP
```

```
"C:\Users\Reactor1967\vcns\work\90702D.BAS"
v# = 10
r# = 9
CLS
DO
z = INT(RND * 3)
r# = v#
IF (r# / 2) - INT(r# / 2) = 0 THEN r# = r# - 1
DO
IF z = 0 THEN v1# = v# + ((v# - r#) * 2) + 1
IF z = 1 THEN v1# = v# + ((v# - r#) * 2) + 2
IF z = 2 THEN v1# = v# + ((v# - r#) * 2) + 3
IF (v1# - r#) > dist# THEN EXIT DO
r# = r# - 2
IF r# <= 0 THEN STOP
LOOP
dist# = (v1# - r#)
v# = v1#
COLOR 2, 0
PRINT , , z; v#; r#; v# / 3
a$ = INKEY$
IF a$ = "s" THEN STOP
LOOP




"C:\Users\Reactor1967\vcns\work\90702C.BAS"
v# = 1
r# = 1
dist# = 0
RANDOMIZE TIMER
CLS
DO
test = (v# / 2) - INT(v# / 2)
z = INT(RND * 2) + 0
IF z = 0 THEN v# = v# + ((v# - r#) * 1) + 1
IF z = 1 THEN v# = v# + ((v# - r#) * 1) + 2
COLOR 2, 0
PRINT , , z; v#; r#; store#
a$ = INKEY$
IF a$ = "s" THEN STOP
IF a$ = "d" THEN EXIT DO
test1 = (test = 0) AND (z = 0)
test2 = (test = 0) AND (z = 1)
test3 = (test = .5) AND (z = 0)
test4 = (test = .5) AND (z = 1)
```

```
IF test1 = -1 THEN dist# = dist# + 2
IF test2 = -1 THEN dist# = dist# + 1
IF test3 = -1 THEN dist# = dist# + 1
IF test4 = -1 THEN dist# = dist# + 2
r# = v# - dist#
store# = dist#
LOOP
```

```
"C:\Users\Reactor1967\vcns\work\90702B.BAS"
v# = 1
r# = 1
RANDOMIZE TIMER
CLS
DO
dist# = v# - r#
z = INT(RND * 2) + 0
IF z = 0 THEN v# = v# + (v# - r#) + 1
IF z = 1 THEN v# = v# + (v# - r#) + 2
COLOR 2, 0
PRINT , , z; v#; r#; v# - r#; dist#
a$ = INKEY$
IF a$ = "s" THEN STOP
r# = v# - ((v# - r#) - dist#)
IF (r# / 2) - INT(r# / 2) = 0 THEN r# = r# - 1
LOOP
```

```
"C:\Users\Reactor1967\vcns\work\90702A.BAS"
v# = 10
r# = 9
RANDOMIZE TIMER
CLS
DO
z = INT(RND * 2) + 0
dist# = v# - r#
r# = v#
IF (r# / 2) - INT(r# / 2) = 0 THEN r# = r# - 1
DO
IF z = 0 THEN v1# = v# + (v# - r#) + 1
IF z = 1 THEN v1# = v# + (v# - r#) + 2
IF v1# - r# > dist# THEN EXIT DO
r# = r# - 2
IF r# <= 0 THEN STOP
```

```
LOOP
a# = v# - r#
IF z = 0 THEN v# = v# + (v# - r#) + 1
IF z = 1 THEN v# = v# + (v# - r#) + 2
COLOR 2, 0
PRINT , , v#; r#; v# - r#; a#
a$ = INKEY$
IF a$ = "s" THEN STOP
LOOP
```

```
"C:\Users\Reactor1967\vcns\work\90602C.BAS"
v# = 8
r# = 1
RANDOMIZE TIMER
CLS
DO
a# = v# - r#
test = (v# / 2) - INT(v# / 2)
IF test = 0 THEN b# = (a# - 1)
IF test = .5 THEN b# = (a# - 2)
b# = b# / 2
z = INT(RND * 2) + 0
IF z = 0 THEN v# = v# + (v# - r#) + 1
IF z = 1 THEN v# = v# + (v# - r#) + 2
COLOR 2, 0
PRINT , , v#; r#; a#; b#
r# = v# - (a# - b#)
a$ = INKEY$
IF a$ = "s" THEN STOP
LOOP
```

```
"C:\Users\Reactor1967\vcns\work\90602B.BAS"
REM Attempting to use the dist# between v# and R# to
REM control R#. Downside is how to do that while decoding.
v# = 100
r# = 1
CLS
RANDOMIZE TIMER
DO
z = INT(RND * 2) + 0
dist# = v# - r#
```

```
test = (v# / 2) - INT(v# / 2)
test1 = (test = 0) AND (z = 0)
test2 = (test = 0) AND (z = 1)
test3 = (test = .5) AND (z = 0)
test4 = (test = .5) AND (z = 1)
REM IF test1 = -1 THEN dist# = dist# + 4
REM IF test2 = -1 THEN dist# = dist# + 3
REM IF test3 = -1 THEN dist# = dist# + 3
REM IF test4 = -1 THEN dist# = dist# + 4
r# = v#
IF (r# / 2) - INT(r# / 2) = 0 THEN r# = r# - 1
DO
IF z = 0 THEN v1# = v# + (v# - r# + 1)
IF z = 1 THEN v1# = v# + (v# - r# + 2)
test = (v1# - r#) > dist#
REM test = (v1# - r#) = dist#
IF test = -1 THEN EXIT DO
r# = r# - 2
IF r# <= 0 THEN STOP
a$ = INKEY$
IF a$ = "s" THEN STOP
LOOP
dist# = v# - r#
IF z = 0 THEN v# = v# + (v# - r#) + 1
IF z = 1 THEN v# = v# + (v# - r#) + 2
COLOR 2, 0
PRINT , , z; v#; r#; v# - r#; dist#
INPUT a$
REM a$ = INKEY$
IF a$ = "s" THEN STOP
LOOP
```

"C:\Users\Reactor1967\vcns\work\90602A.BAS"
REM This program codes in two phases
REM 1. coding for data
REM 2. testing for running up r
REM When entering code just code data one bit only
REM Then enter testing phase. Code a binary 0 first
REM then if v - r does not meet
REM a specific condition code a binary one and run r up
REM Stay in this testing phase until v - r meets the
REM specific condition then goto code phase.
REM when decoding enter into testing and keep running r
REM down till test gives you a binary 0 then read your data.

```
REM then start testing again.
v# = 1
r# = 1
speed# = 1
CLS
RANDOMIZE TIMER
DO
REM coding phase.
z = INT(RND * 2) + 0
IF z = 0 THEN v# = v# + (v# - r# + 1)
IF z = 1 THEN v# = v# + (v# - r# + 2)
COLOR 2, 0
PRINT , , v#; r#; "Code"
REM testing phase
v# = v# + (v# - r# + 1)
COLOR 2, 0
PRINT , , v#; r#; "testing"
DO
test = v# + (v# - r#) >= (r# + 100)
IF test = 0 THEN EXIT DO
d# = (v# - r#)
v# = v# + (v# - r# + 2)
d# = (d# * 2) - 4
r# = r# + d#
COLOR 2, 0
PRINT , , v#; r#; "testing"
REM INPUT a$
a$ = INKEY$
IF a$ = "s" THEN STOP
LOOP
a$ = INKEY$
REM INPUT a$
IF a$ = "s" THEN STOP
LOOP


"C:\Users\Reactor1967\vcns\work\90602.BAS"
v# = 1
r# = 1
v2# = 100
cm# = r#
CLS
DO
test = ((v2# - r#) <= 100) AND (cm# <> r#)
REM test = (v# + (v# - r#) >= v2#)
```

```
DO
IF test = -1 THEN x = INT(RND * 2) + 0
test3 = (test = -1) AND (x = 0)
test4 = (test = -1) AND (x = 1)
IF test3 = -1 THEN v2# = v2# + (v2# - r#) + 1
IF test4 = -1 THEN v2# = v2# + (v2# - r#) + 2
xu = (test = -1) AND (v2# - r# > 100)
IF test = 0 THEN EXIT DO
a$ = INKEY$
IF a$ = "s" THEN STOP
LOOP UNTIL xu = -1
IF (v# + (v# - r#)) >= (r# + 50) THEN z = 1
IF (v# + (v# - r#)) < (r# + 50) THEN z = 0
IF z = 0 THEN v# = v# + ((v# - r#) * 1) + 1
IF z = 1 THEN v# = v# + ((v# - r#) * 1) + 2
COLOR 2, 0
PRINT , , z; v#; r#; v2#
a$ = INKEY$
REM INPUT a$
IF a$ = "s" THEN STOP
IF a$ = "d" THEN EXIT DO
cm# = r#
IF z = 1 THEN r# = r# + 52
count = count + 1
LOOP UNTIL v2# + (v2# - r#) >= 999999999999999#
GOTO sk:
DO
test = (v# / 2) - INT(v# / 2)
IF test = .5 THEN r# = r# - 52
sk:
dist# = v# - r#
IF (dist# / 2) - INT(dist# / 2) = .5 THEN dist# = dist# - 1
dist# = dist# / 2
dist# = dist# + 1
v# = v# - dist#
COLOR 2, 0
PRINT , , v#; r#; v2#
INPUT a$
REM a$ = INKEY$
IF a$ = "s" THEN STOP
IF v# = 1 THEN STOP
LOOP
decode:
DO
dist# = v2# - r#
IF (dist# / 2) - INT(dist# / 2) = .5 THEN dist# = dist# - 1
```

```
dist# = dist# / 2
dist# = dist# + 1
v2# = v2# - dist#
a$ = INKEY$
IF a$ = "s" THEN STOP
LOOP UNTIL (v2# - r# < 100)
RETURN




"C:\Users\Reactor1967\vcns\work\90502D.BAS"
v# = 1
r# = 1
RANDOMIZE TIMER
CLS
DO
z = INT(RND * 2) + 0
test = v# + (v# - r#) >= (r# + 100)
IF z = 0 THEN v# = v# + (v# - r#) + 1
IF z = 1 THEN v# = v# + (v# - r#) + 2
COLOR 2, 0
PRINT , , z; v#; r#
a$ = INKEY$
IF a$ = "s" THEN STOP
IF test = -1 THEN r# = r# + 102
LOOP




"C:\Users\Reactor1967\vcns\work\90502C.BAS"
REM if can conquare self coding and decoding in any base then there is
REM possibility of cycling a second v with r. This is an attempt to do that.
REM try keeping v and r at specific distances when coding and decoding.
REM see if that works.
REM use multiple conditions. Increment R + 100's so it can be use more
REM than once. Find other conditions if needed.
v# = 1
r# = 1
store# = r# + 6
RANDOMIZE TIMER
CLS
DO
count = 0
DO
test1 = (v# >= store#)
IF test1 = -1 THEN r# = store#
```

```
IF test1 = -1 THEN store# = store# + 6
IF test = -1 THEN count = count + 1
LOOP UNTIL v# < store#
z = INT(RND * 3) + 0
IF z = 0 THEN v# = v# + ((v# - r#) * 2) + 1
IF z = 1 THEN v# = v# + ((v# - r#) * 2) + 2
IF z = 2 THEN v# = v# + ((v# - r#) * 2) + 3
COLOR 2, 0
PRINT , , count; z; v#; r#; v# - r#
a$ = INKEY$
REM INPUT a$
IF a$ = "s" THEN STOP
LOOP


"C:\Users\Reactor1967\vcns\work\90502B.BAS"
v# = 1
r# = 1
v2# = 100
CLS
count = 1
RANDOMIZE TIMER
DO
test = (v2# - r#) <= 100
IF test = -1 THEN x = INT(RND * 2) + 0
test3 = (test = -1) AND (x = 0)
test4 = (test = -1) AND (x = 1)
IF test3 = -1 THEN v2# = v2# + (v2# - r#) + 1
IF test4 = -1 THEN v2# = v2# + (v2# - r#) + 2
IF (v# + (v# - r#)) >= (r# + 50) THEN z = 1
IF (v# + (v# - r#)) < (r# + 50) THEN z = 0
IF z = 0 THEN v# = v# + ((v# - r#) * 1) + 1
IF z = 1 THEN v# = v# + ((v# - r#) * 1) + 2
COLOR 2, 0
PRINT , , z; v#; r#; v2#
a$ = INKEY$
REM INPUT a$
IF a$ = "s" THEN STOP
IF a$ = "d" THEN EXIT DO
IF z = 1 THEN r# = r# + 52
LOOP UNTIL v2# + (v2# - r#) >= 999999999999999#
DO
test = (v2# - r#) > 50
IF test = -1 THEN GOSUB decode:
skip:
```

```
dist# = (v# - r#)
IF (dist# / 2) - INT(dist# / 2) = .5 THEN dist# = dist# - 1
dist# = dist# / 2
dist# = dist# + 1
v# = v# - dist#
test = (v# / 2) - INT(v# / 2)
IF test = .5 THEN r# = r# - 52
COLOR 2, 0
PRINT , , v#; r#; v2#
IF v2# = 100 THEN STOP
IF v# = 1 THEN STOP
INPUT a$
REM a$ = INKEY$
IF a$ = "s" THEN STOP
LOOP
STOP
decode:
dist# = (v2# - r#)
IF (dist# / 2) - INT(dist# / 2) = .5 THEN dist# = dist# - 1
dist# = dist# / 2
dist# = dist# + 1
v2# = v2# - dist#
RETURN




"C:\Users\Reactor1967\vcns\work\90502A.BAS"
v# = 1
r# = 1
CLS
DO
PRINT v#; v# + ((v# - r#) * 2) + 1; v# + ((v# - r#) * 2) + 2; v# + ((v# - r#) * 2)
+ 3
PRINT (v# + ((v# - r#) * 2) + 1) - v#
v# = v# + 1
INPUT a$
IF a$ = "s" THEN STOP
LOOP




"C:\Users\Reactor1967\vcns\work\90402E.BAS"
REM this is coding R only. You can find your previous R by just dividing it
REM by three. If your remainder is .3 then add 1 and cut off the fraction
REM If its even do nothing. If the remainder is .6 then just cut off the
```

```
REM fraction.
r# = 97
CLS
RANDOMIZE TIMER
redo:
DO
z = INT(RND * 3) + 0
IF z = 0 THEN r# = r# + ((r# - 2) * 2) + 2
IF z = 1 THEN r# = r# + ((r# - 2) * 2) + 4
IF z = 2 THEN r# = r# + ((r# - 2) * 2) + 6
COLOR 2, 0
PRINT , , z; r#
REM INPUT a$
a$ = INKEY$
IF a$ = "s" THEN STOP
LOOP UNTIL r# + ((r# - 2) * 2) >= 999999999999999#
```

```
"C:\Users\Reactor1967\vcns\work\90402D.BAS"
v# = 1
r# = 1
v2# = 102
RANDOMIZE TIMER
CLS
DO
test = v2# - r# < 100
IF test = -1 THEN z = INT(RND * 2) + 0
test1 = (test = -1) AND (z = 0)
test2 = (test = -1) AND (z = 1)
IF test1 = -1 THEN v2# = v2# + (v2# - r#) + 1
IF test2 = -1 THEN v2# = v2# + (v2# - r#) + 2
test = (v# + (v# - r#)) >= (r# + 50)
IF test = 0 THEN v# = v# + (v# - r# + 1)
IF test = -1 THEN v# = v# + (v# - r# + 2)
COLOR 2, 0
PRINT , , v#; r#; v2#
REM IF test = -1 THEN r# = r# + 52
IF test = -1 THEN r# = v#
a$ = INKEY$
IF a$ = "s" THEN STOP
LOOP
```

```
"C:\Users\Reactor1967\vcns\work\90402C.BAS"
REM this is a good ideal except it has no symetery so its no good.
v2# = 102
r# = 1
RANDOMIZE TIMER
CLS
DO
z = INT(RND * 2) + 0
vstore# = v2#
IF z = 0 THEN v2# = v2# + (v2# - r#) + 1
IF z = 1 THEN v2# = v2# + (v2# - r#) + 3
COLOR 2, 0
PRINT , , z; v2#; r#; v2# - vstore#
IF z = 0 THEN r# = r# + 102
IF z = 1 THEN r# = r# + 104
REM INPUT a$
a$ = INKEY$
IF a$ = "s" THEN STOP
LOOP
```

```
"C:\Users\Reactor1967\vcns\work\90402B.BAS"
v# = 1
r# = 1
count = 1
CLS
RANDOMIZE TIMER
DO
z = INT(RND * 4) + 0
vstore# = v#
IF z = 0 THEN v# = v# + ((v# - r#) * 3) + 1
IF z = 1 THEN v# = v# + ((v# - r#) * 3) + 2
IF z = 2 THEN v# = v# + ((v# - r#) * 3) + 3
IF z = 3 THEN v# = v# + ((v# - r#) * 3) + 4
COLOR 2, 0
PRINT , , z; v#; r#; r# - rstore#
rstore# = r#
r# = r# + (v# - vstore#) - 1
a$ = INKEY$
IF a$ = "s" THEN STOP
LOOP
```

"C:\Users\Reactor1967\vcns\work\90402.BAS"

```
v# = 1
r# = 1
speed# = -51
v2# = 100
CLS
count = 1
DO
af# = v2# - r#
test = (v2# - r#) <= 100
IF v2# - (r# + 52) <= 100 THEN store2# = (v2# - r#)
IF test = -1 THEN store# = v2# - r#
IF test = -1 THEN count = 1
IF test = -1 THEN x = INT(RND * 2) + 0
test3 = (test = -1) AND (x = 0)
test4 = (test = -1) AND (x = 1)
IF test3 = -1 THEN v2# = v2# + (v2# - r#) + 1
IF test4 = -1 THEN v2# = v2# + (v2# - r#) + 2
IF (v# + (v# - r#)) >= (r# + 50) THEN z = 1
IF (v# + (v# - r#)) < (r# + 50) THEN z = 0
IF z = 0 THEN v# = v# + ((v# - r#) * 1) + 1
IF z = 1 THEN v# = v# + ((v# - r#) * 1) + 2
COLOR 2, 0
PRINT , , z; v#; r#; v2#
a$ = INKEY$
REM INPUT a$
IF a$ = "s" THEN STOP
IF a$ = "d" THEN EXIT DO
speed# = r#
IF z = 1 THEN r# = r# + 52
count = count + 1
LOOP UNTIL v2# + (v2# - r#) >= 999999999999999#
GOTO sk:
DO
test = (v# / 2) - INT(v# / 2)
IF test = .5 THEN r# = r# - 52
sk:
REM IF v2# - r# >= 200 THEN GOSUB decode:
dist# = v# - r#
IF (dist# / 2) - INT(dist# / 2) = .5 THEN dist# = dist# - 1
dist# = dist# / 2
dist# = dist# + 1
v# = v# - dist#
COLOR 2, 0
PRINT , , v#; r#
REM INPUT a$
a$ = INKEY$
```

```
IF a$ = "s" THEN STOP
IF v# = 1 THEN STOP
LOOP
decode:
dist# = v2# - r#
IF (dist# / 2) - INT(dist# / 2) = .5 THEN dist# = dist# - 1
dist# = dist# / 2
dist# = dist# + 1
v2# = v2# - dist#
RETURN



"C:\Users\Reactor1967\vcns\work\90302D.BAS"
v# = 1
r# = 1
v2# = 100
CLS
DO
af# = v2# - r#
test = (v2# - r#) <= 100
IF test = -1 THEN x = INT(RND * 2) + 0
test3 = (test = -1) AND (x = 0)
test4 = (test = -1) AND (x = 1)
IF test3 = -1 THEN v2# = v2# + (v2# - r#) + 1
IF test4 = -1 THEN v2# = v2# + (v2# - r#) + 2
IF (v# + (v# - r#)) >= (r# + 50) THEN z = 1
IF (v# + (v# - r#)) < (r# + 50) THEN z = 0
IF z = 0 THEN v# = v# + ((v# - r#) * 1) + 1
IF z = 1 THEN v# = v# + ((v# - r#) * 1) + 2
COLOR 2, 0
PRINT , , z; v#; r#; v2#
a$ = INKEY$
REM INPUT a$
IF a$ = "s" THEN STOP
IF a$ = "d" THEN EXIT DO
IF z = 1 THEN r# = r# + 52
LOOP UNTIL v2# + (v2# - r#) >= 999999999999999#
GOTO sk:
DO
test = (v# / 2) - INT(v# / 2)
IF test = .5 THEN r# = r# - 52
sk:
REM IF v2# - r# >= 200 THEN GOSUB decode:
dist# = v# - r#
IF (dist# / 2) - INT(dist# / 2) = .5 THEN dist# = dist# - 1
```

```
dist# = dist# / 2
dist# = dist# + 1
v# = v# - dist#
COLOR 2, 0
PRINT , , v#; r#
REM INPUT a$
a$ = INKEY$
IF a$ = "s" THEN STOP
IF v# = 1 THEN STOP
LOOP
decode:
dist# = v2# - r#
IF (dist# / 2) - INT(dist# / 2) = .5 THEN dist# = dist# - 1
dist# = dist# / 2
dist# = dist# + 1
v2# = v2# - dist#
RETURN




"C:\Users\Reactor1967\vcns\work\90302C.BAS"
v# = 5
r# = 1
CLS
RANDOMIZE TIMER
DO
z = INT(RND * 2) + 0
vs# = v#
IF z = 0 THEN v# = v# + (v# - r#) + 1
IF z = 1 THEN v# = v# + (v# - r#) + 2
dist# = v# - vs#
IF (dist# / 2) - INT(dist# / 2) = .5 THEN dist# = dist# - 1
z1 = z#
IF dist# = 2 THEN z = 0
IF dist# = 4 THEN z = 1
IF dist# = 6 THEN z = 2
DO
r# = r# + 2
t# = r# / 3
t# = t# - INT(t#)
t# = t# * 10
t# = INT(t#)
test1 = (z = 0) AND (t# = 0)
test2 = (z = 1) AND (t# = 4)
test3 = (z = 2) AND (t# = 6)
IF test1 = -1 THEN EXIT DO
```

```
IF test2 = -1 THEN EXIT DO
IF test3 = -1 THEN EXIT DO
a$ = INKEY$
IF a$ = "s" THEN STOP
LOOP
PRINT z1; v#; r#; t#
LOOP




"C:\Users\Reactor1967\vcns\work\90302B.BAS"
store# = 1
r# = 5
RANDOMIZE TIMER
CLS
DO
z = INT(RND * 3) + 0
DO
r# = r# + 2
t# = r# / 3
t# = t# - INT(t#)
t# = t# * 10
t# = INT(t#)
test1 = (z = 0) AND (t# = 0)
test2 = (z = 1) AND (t# = 3)
test3 = (z = 2) AND (t# = 6)
IF test1 = -1 THEN EXIT DO
IF test2 = -1 THEN EXIT DO
IF test3 = -1 THEN EXIT DO
a$ = INKEY$
IF a$ = "s" THEN STOP
LOOP
COLOR 2, 0
PRINT , , z; r#; t#; r# - store#
store# = r#
LOOP




"C:\Users\Reactor1967\vcns\work\90302A.BAS"
v# = 1
r# = 1
v2# = 1
CLS
RANDOMIZE TIMER
```

```
DO
z = INT(RND * 2) + 0
vs2# = v2#
IF z = 0 THEN v2# = v2# + (v2# - v#) + 2
IF z = 1 THEN v2# = v2# + (v2# - v#) + 4
vs# = v#
IF z = 0 THEN v# = v# + ((v# - r#) * 1) + 2
IF z = 1 THEN v# = v# + ((v# - r#) * 1) + 4
COLOR 2, 0
PRINT , , v2#; v#; r#; v# - vs#; v2# - vs2#; r# - store#
a$ = INKEY$
IF a$ = "s" THEN STOP
dist# = v# - vs#
store# = r#
r# = r# + dist#
LOOP




"C:\Users\Reactor1967\vcns\work\83102D.BAS"
REM set R up so it can be coded just like you code v in the other
REM bases except that it is always even and the remainder you get
REM when you devide it by some number tells you how much you added
REM or subtracted from r to get your previous r. It is self coding
REM just like vcframe3.bas program is self coding and decoding. R to
REM must be self coding and decoding to be able to code data in v.
v# = 12345
CLS
DO
PRINT v#; v# / 5
v# = v# + 2
INPUT a$
IF a$ = "s" THEN STOP
LOOP




"C:\Users\Reactor1967\vcns\work\83102C.BAS"
v# = 1
r# = 0
CLS
RANDOMIZE TIMER
DO
z = INT(RND * 2) + 0
test = (v# / 2) - INT(v# / 2)
d# = v# - r#
```

```
IF z = 0 THEN v# = v# + r# + 1
IF z = 1 THEN v# = v# + r# + 2
COLOR 2, 0
PRINT , , z; v#; d#; r#
a$ = INKEY$
IF a$ = "s" THEN STOP
test1 = (z = 0) AND (test = 0)
test2 = (z = 0) AND (test = .5)
test3 = (z = 1) AND (test = 0)
test4 = (z = 1) AND (test = .5)
IF test1 = -1 THEN r# = r# + 2
IF test2 = -1 THEN r# = r# + 1
IF test3 = -1 THEN r# = r# + 1
IF test4 = -1 THEN r# = r# + 2
LOOP


"C:\Users\Reactor1967\vcns\work\83102B.BAS"
v# = 2
r# = 1
RANDOMIZE TIMER
CLS
add# = 1
DO
z = INT(RND * 2) + 0
store# = add#
test = (v# / 2) - INT(v# / 2)
test1 = (z = 0) AND (test = 0)
test2 = (z = 0) AND (test = .5)
test3 = (z = 1) AND (test = 0)
test4 = (z = 1) AND (test = .5)
IF test1 = -1 THEN add# = add# + 2
IF test2 = -1 THEN add# = add# + 1
IF test3 = -1 THEN add# = add# + 1
IF test4 = -1 THEN add# = add# + 2
IF z = 0 THEN v# = v# + ((v# - r#) * 1) + 1
IF z = 1 THEN v# = v# + ((v# - r#) * 1) + 2
COLOR 2, 0
PRINT , , v#; r#; store#
r# = v# - add#
REM INPUT a$
a$ = INKEY$
IF a$ = "s" THEN STOP
LOOP
```

```
"C:\Users\Reactor1967\vcns\work\83102A.BAS"
v# = 2
r# = 1
RANDOMIZE TIMER
CLS
DO
z = INT(RND * 2) + 0
vstore# = v#
IF z = 0 THEN v# = v# + ((v# - r#) * 1) + 1
IF z = 1 THEN v# = v# + ((v# - r#) * 1) + 2
PRINT v#; r#
d1# = v# - vstore#
d2# = v# - r#
dist# = d1# + d2#
IF (dist# / 2) - INT(dist# / 2) = .5 THEN dist# = dist# - 1
dist# = dist# / 2
dist# = dist# + 1
r# = v# - dist#
a$ = INKEY$
REM INPUT a$
IF a$ = "s" THEN STOP
LOOP UNTIL v# + (v# - r#) >= 99999999999999#
```

```
"C:\Users\Reactor1967\vcns\work\83002D.BAS"
v# = 1
r# = 1
CLS
RANDOMIZE TIMER
add# = 0
DO
z = INT(RND * 2) + 0
dist# = v# - r#
v1# = v#
IF z = 0 THEN v# = v# + (v# - r#) + 1
IF z = 1 THEN v# = v# + (v# - r#) + 2
COLOR 2, 0
PRINT , , v#; z; r#; v# - r#; add#
dist# = dist# - (v# - v1#)
r# = v# - dist#
r# = r# - add#
add# = add# + 2
a$ = INKEY$
```

```
IF a$ = "s" THEN STOP
LOOP




"C:\Users\Reactor1967\vcns\work\83002C.BAS"
v# = 1
speed# = 2
CLS
RANDOMIZE TIMER
DO
z = INT(RND * 2) + 0
test = (v# / 2) - INT(v# / 2)
store# = v# - speed#
a$ = INKEY$
IF a$ = "s" THEN STOP
r# = v# - speed#
IF z = 0 THEN v# = v# + speed# + 1
IF z = 1 THEN v# = v# + speed# + 2
COLOR 2, 0
PRINT , , v#; z; r#; speed#
test1 = (test = 0) AND (z = 0)
test2 = (test = 0) AND (z = 1)
test3 = (test = .5) AND (z = 0)
test4 = (test = .5) AND (z = 1)
IF test1 = -1 THEN speed# = speed# + 2
IF test2 = -1 THEN speed# = speed# + 1
IF test3 = -1 THEN speed# = speed# + 1
IF test4 = -1 THEN speed# = speed# + 2
LOOP




"C:\Users\Reactor1967\vcns\work\83002B.BAS"
REM Lesson learned here.
REM Could use speed and r on a chart.
REM use r to decode to previous v
REM then use v - r as a code to tell the
REM distance to the previous r.
REM repeat till decode is done.
REM try to create that kind of chart where
REM code from v2 to v1 then v1 - r2 = distance to r1
REM then r1 becomes r2 and the process repeats itself.
v# = 100
speed# = 0
CLS
```

```
RANDOMIZE TIMER
DO
z = INT(RND * 2) + 0
rstore# = r#
test1 = (v# / 2) - INT(v# / 2)
test2 = (speed# / 2) - INT(speed# / 2)
test3 = (test1 = 0) AND (test2 = 0)
test4 = (test1 = 0) AND (test2 = .5)
test5 = (test1 = .5) AND (test2 = 0)
test6 = (test1 = .5) AND (test2 = .5)
REM ----------------------------------------
IF test3 = -1 THEN speed# = speed# + 1
IF test4 = -1 THEN speed# = speed# + 2
IF test5 = -1 THEN speed# = speed# + 2
IF test6 = -1 THEN speed# = speed# + 1
REM ----------------------------------------
REM IF test3 = -1 THEN speed# = speed# - 1
REM IF test4 = -1 THEN speed# = speed# - 2
REM IF test5 = -1 THEN speed# = speed# - 2
REM IF test6 = -1 THEN speed# = speed# - 1
vstore# = v#
REM ----------------------------------------
IF z = 0 THEN v# = v# + (speed# + 1)
IF z = 1 THEN v# = v# + (speed# + 2)
REM ----------------------------------------
REM IF z = 0 THEN v# = v# - (speed# + 1)
REM IF z = 1 THEN v# = v# - (speed# + 2)
REM ----------------------------------------
rstore# = r#
r# = vstore# - speed#
REM ----------------------------------------
REM r# = vstore# + speed#
COLOR 2, 0
REM ----------------------------------------
IF z = 0 THEN sb# = 1 ELSE sb# = 2
REM PRINT , , v#; z; r#; v# - r#; ((r# - rstore#) + (v# - vstore#)) - sb#; speed#
PRINT , , v#; z; r#; speed#
REM ----------------------------------------
REM PRINT , , v#; z; r#; speed#; rstore# - r#
a$ = INKEY$
IF a$ = "s" THEN STOP
IF a$ = "d" THEN EXIT DO
LOOP
REM (r2 - r1) + (v1 - r2) = v1 - r1
REM (v2 - v1) + (v2 - r1) + 1 = v2 - r2
dist# = v# - r#
```

```
IF (dist# / 2) - INT(dist# / 2) = .5 THEN dist# = dist# - 1
dist# = dist# / 2
dist# = dist# + 1
v2# = v#
v# = v# - dist#
d1# = v2# - r#
d2# = v2# - v#
test# = (v# / 2) - INT(v# / 2)
IF test = 0 THEN s# = 1
IF test = .5 THEN s# = 2
r# = v# - (((d1# - d2#) * 2) - s#)
PRINT r#
```

```
"C:\Users\Reactor1967\vcns\work\83002A.BAS"
v# = 1
r# = 1
add# = 0
CLS
RANDOMIZE TIMER
DO
z = INT(RND * 2) + 0
v1# = v#
IF z = 0 THEN v# = v# + ((v# - r#) * 1) + 1
IF z = 1 THEN v# = v# + ((v# - r#) * 1) + 2
COLOR 2, 0
PRINT , , v#; r#
IF v# + (v# - r#) >= 999999999999999# THEN EXIT DO
REM INPUT a$
a$ = INKEY$
IF a$ = "s" THEN STOP
d1# = v# - v1#
d2# = v# - r#
dist# = d1# + d2#
IF (dist# / 2) - INT(dist# / 2) = .5 THEN dist# = dist# - 1
dist# = dist# / 2
dist# = dist# + 1
r# = v# - dist#
LOOP UNTIL v# + (v# - r#) >= 999999999999999#
dist# = v# - r#
IF (dist# / 2) - INT(dist# / 2) = .5 THEN dist# = dist# - 1
dist# = dist# / 2
dist# = dist# + 1
v1# = v#
v2# = v# - dist#
```

ccclx

```
d1# = v2# - r#
d2# = v2# - v1#
dist# = d1# - d2#
dist# = dist# * 2
dist# = dist# - 1
r# = v1# - dist#
PRINT v2#; r#




"C:\Users\Reactor1967\vcns\work\82902D.BAS"
REM Lesson learned here.
REM Could use speed and r on a chart.
REM use r to decode to previous v
REM then use v - r as a code to tell the
REM distance to the previous r.
REM repeat till decode is done.
REM try to create that kind of chart where
REM code from v2 to v1 then v1 - r2 = distance to r1
REM then r1 becomes r2 and the process repeats itself.
v# = 100
speed# = 0
CLS
RANDOMIZE TIMER
DO
z = INT(RND * 2) + 0
rstore# = r#
test1 = (v# / 2) - INT(v# / 2)
test2 = (speed# / 2) - INT(speed# / 2)
test3 = (test1 = 0) AND (test2 = 0)
test4 = (test1 = 0) AND (test2 = .5)
test5 = (test1 = .5) AND (test2 = 0)
test6 = (test1 = .5) AND (test2 = .5)
REM ---------------------------------------
IF test3 = -1 THEN speed# = speed# + 1
IF test4 = -1 THEN speed# = speed# + 2
IF test5 = -1 THEN speed# = speed# + 2
IF test6 = -1 THEN speed# = speed# + 1
REM ---------------------------------------
REM IF test3 = -1 THEN speed# = speed# - 1
REM IF test4 = -1 THEN speed# = speed# - 2
REM IF test5 = -1 THEN speed# = speed# - 2
REM IF test6 = -1 THEN speed# = speed# - 1
vstore# = v#
REM ---------------------------------------
```

```
IF z = 0 THEN v# = v# + (speed# + 1)
IF z = 1 THEN v# = v# + (speed# + 2)
REM ---------------------------------------
REM IF z = 0 THEN v# = v# - (speed# + 1)
REM IF z = 1 THEN v# = v# - (speed# + 2)
REM ---------------------------------------
rstore# = r#
r# = vstore# - speed#
REM ---------------------------------------
REM r# = vstore# + speed#
COLOR 2, 0
REM ---------------------------------------
IF z = 0 THEN sb# = 1 ELSE sb# = 2
PRINT , , v#; z; r#; v# - r#; ((r# - rstore#) + (v# - vstore#)) - sb#; speed#
IF v# - r# = 280 THEN STOP
REM 277
REM ---------------------------------------
REM PRINT , , v#; z; r#; speed#; rstore# - r#
REM INPUT a$
a$ = INKEY$
IF a$ = "s" THEN STOP
LOOP
REM (r2 - r1) + (v1 - r2) = v1 - r1
REM (v2 - v1) + (v2 - r1) + 1 = v2 - r2




"C:\Users\Reactor1967\vcns\work\82902C.BAS"
v# = 2
r# = 2
CLS
RANDOMIZE TIMER
DO
dist# = v# - r#
z = INT(RND * 2) + 0
r# = v#
REM r# = 1
IF (r# / 2) - INT(r# / 2) = 0 THEN r# = r# - 1
DO
IF z = 0 THEN v1# = v# + (v# - r# + 1)
IF z = 1 THEN v1# = v# + (v# - r# + 2)
IF v1# - r# > dist# THEN EXIT DO
REM v1# - r# < dist# then exit do
r# = r# - 2
REM r# = r# + 2
a$ = INKEY$
```

```
IF a$ = "s" THEN STOP
LOOP
v# = v1#
COLOR 2, 0
PRINT , , v#; r#; v# - r#
REM INPUT a$
a$ = INKEY$
IF a$ = "s" THEN STOP
LOOP
```

```
"C:\Users\Reactor1967\vcns\work\82902B.BAS"
CLS
RANDOMIZE TIMER
v# = 1
r# = 1
CLS
DO
z = INT(RND * 2) + 0
IF x = 1 THEN z = 2
IF z = 0 THEN v# = v# + ((v# - r#) * 2) + 1
IF z = 1 THEN v# = v# + ((v# - r#) * 2) + 2
IF z = 2 THEN v# = v# + ((v# - r#) * 2) + 3
x = 0
COLOR 2, 0
IF z = 2 THEN PRINT , , z; v#; r#; r# - rstore#; v# - r# ELSE PRINT , , z; v#;
r#; v# - r#
IF v# - r# > 100 THEN GOSUB catchup:
INPUT a$
IF a$ = "s" THEN STOP
LOOP
catchup:
rstore# = r#
DO
r# = r# + 2
test = v# - r# < 4
a$ = INKEY$
IF a$ = "s" THEN STOP
LOOP UNTIL test = -1
x = 1
RETURN
```

```
"C:\Users\Reactor1967\vcns\work\82902A.BAS"
v# = 1
r# = 1
DIM bank#(100)
g1# = 0
g2# = 0
CLS
DO
v1# = v#
DO
v1# = v1# + 2
test# = (v1# / 3) - INT(v1# / 3)
test# = test# * 10
test# = INT(test#)
IF test# = 3 THEN EXIT DO
a$ = INKEY$
IF a$ = "s" THEN STOP
LOOP
v2# = v1#
DO
v2# = v2# + 2
test# = (v2# / 3) - INT(v2# / 3)
test# = test# * 10
test# = INT(test#)
IF test# = 3 THEN EXIT DO
a$ = INKEY$
IF a$ = "s" THEN STOP
LOOP
COLOR 2, 0
PRINT , , v#; "="; v1#; v2#
a$ = INKEY$
REM INPUT a$
IF a$ = "s" THEN STOP
g1# = g1# + 1
bank#(g1#) = v1#
g1# = g1# + 1
bank#(g1#) = v2#
g2# = g2# + 1
v# = bank#(g2#)
LOOP UNTIL g1# + 2 > 100
```

"C:\Users\Reactor1967\vcns\work\82802E.BAS"

```
v# = 100
r# = 1
RANDOMIZE TIMER
CLS
DO
vstore# = v#
z = INT(RND * 2) + 0
IF z = 0 THEN v# = v# + (v# - INT(r#) + 1)
IF z = 1 THEN v# = v# + (v# - INT(r#) + 2)
COLOR 2, 0
PRINT , , z; v#; r#; r# - rstore#; v# - r#
a$ = INKEY$
IF a$ = "s" THEN STOP
IF a$ = "d" THEN EXIT DO
rstore# = r#
r# = v#
IF (r# / 2) - INT(r# / 2) = 0 THEN r# = r# - 1
DO
test# = (r# / 3) - INT(r# / 3)
test# = test# * 10
test# = INT(test#)
IF test# = 6 THEN EXIT DO
r# = r# - 2
IF r# < 0 THEN STOP
LOOP
LOOP


"C:\Users\Reactor1967\vcns\work\82802D.BAS"
v# = 10
r# = 1
RANDOMIZE TIMER
CLS
DO
z = INT(RND * 2) + 0
IF z = 0 THEN v# = v# + (v# - INT(r#) + 1)
IF z = 1 THEN v# = v# + (v# - INT(r#) + 2)
COLOR 2, 0
PRINT , , z; v#; r#; r# - rstore#
rstore# = r#
DO
r# = r# + 2
test = (r# / 3) - INT(r# / 3)
test = test * 10
test = INT(test)
```

```
REM PRINT test
IF test = 0 THEN EXIT DO
a$ = INKEY$
REM INPUT a$
IF a$ = "s" THEN STOP
LOOP
REM INPUT a$
a$ = INKEY$
IF a$ = "s" THEN STOP
LOOP UNTIL v# + (v# - r#) >= 999999999999999#
```

```
"C:\Users\Reactor1967\vcns\work\82802C.BAS"
v# = 10
r# = 1
RANDOMIZE TIMER
CLS
DO
z = INT(RND * 2) + 0
IF z = 0 THEN v# = v# + (v# - INT(r#) + 1)
IF z = 1 THEN v# = v# + (v# - INT(r#) + 2)
COLOR 2, 0
PRINT , , z; v#; r#; r# - rstore#
rstore# = r#
r# = v#
IF (r# / 2) - INT(r# / 2) = 0 THEN r# = r# - 1
DO
IF (v# / 3) - INT(v# / 3) = (r# / 3) - INT(r# / 3) THEN EXIT DO
a$ = INKEY$
IF a$ = "s" THEN STOP
IF r# > v# THEN STOP
r# = r# - 2
LOOP
REM INPUT a$
IF a$ = "s" THEN STOP
LOOP
```

```
"C:\Users\Reactor1967\vcns\work\82802B.BAS"
v# = 3
r# = 1
RANDOMIZE TIMER
CLS
DO
```

```
z = INT(RND * 2) + 0
r# = v#
IF (r# / 2) - INT(r# / 2) = 0 THEN r# = r# - 1
DO
IF z = 0 THEN v1# = v# + (v# - r#) + 1
IF z = 1 THEN v1# = v# + (v# - r#) + 2
REM PRINT v1#; v1# / 3; r# / 3
IF (v1# / 3) - INT(v1# / 3) = (r# / 3) - INT(r# / 3) THEN EXIT DO
REM INPUT a$
a$ = INKEY$
IF a$ = "s" THEN STOP
IF a$ = "d" THEN EXIT DO
r# = r# - 2
LOOP
IF z = 0 THEN v# = v# + (v# - INT(r#) + 1)
IF z = 1 THEN v# = v# + (v# - INT(r#) + 2)
COLOR 2, 0
PRINT , , z; v#; r#
REM INPUT a$
IF a$ = "s" THEN STOP
LOOP


"C:\Users\Reactor1967\vcns\work\82802A.BAS"
REM (v# - r# + N) / 2 = dist# from v2 to v1
v# = 11
r# = 1
RANDOMIZE TIMER
CLS
DO
z = INT(RND * 2) + 0
r1# = v#
IF (r1# / 2) - INT(r1# / 2) = 0 THEN r1# = r1# - 1
vstore# = v#
rstore# = r#
DO
IF z = 0 THEN v1# = v# + ((v# - r1#) * 1) + 1
IF z = 1 THEN v1# = v# + ((v# - r1#) * 1) + 2
t1 = (v1# - v#) - (r1# - r#) = 2 AND (z = 1)
t2 = (v1# - v#) - (r1# - r#) = 3 AND (z = 1)
t3 = (v1# - v#) - (r1# - r#) = 0 AND (z = 0)
t4 = (v1# - v#) - (r1# - r#) = 1 AND (z = 0)
IF (t1 = -1) OR (t2 = -1) OR (t3 = -1) OR (t4 = -1) THEN EXIT DO
a$ = INKEY$
REM INPUT a$
```

```
IF a$ = "s" THEN STOP
IF a$ = "d" THEN EXIT DO
r1# = r1# - 1
LOOP
r# = r1#
v# = v1#
COLOR 2, 0
PRINT "N="; z; "v="; v#; "r="; r#; "v2-v1="; v# - vstore#; "v-r="; v# - r#; "r2-
r1"; r# - rstore#; "v2-v1 - r2-r2"; (v# - vstore#) - (r# - rstore#)
a$ = INKEY$
IF a$ = "s" THEN STOP
LOOP




"C:\Users\Reactor1967\vcns\work\82702A.BAS"
v# = 1
r# = 1
CLS
RANDOMIZE TIMER
test# = 0
DO
count# = count# + 1
z = INT(RND * 2) + 0
store# = v#
IF z = 0 THEN v# = v# + ((v# - r#) * 1) + 1
IF z = 1 THEN v# = v# + ((v# - r#) * 1) + 2
COLOR 2, 0
speed# = v# - store#
REM IF (speed# / 2) - INT(speed# / 2) = .5 THEN speed# = speed# - 1
PRINT , v#; r#; z; speed#; speed# - sp#
sp# = speed#
r# = r# + speed#
REM INPUT a$
a$ = INKEY$
IF a$ = "s" THEN STOP
LOOP UNTIL v# + (v# - r#) >= 999999999999999#




"C:\Users\Reactor1967\vcns\work\82602A.BAS"
v# = 1
r# = 1
test# = 0
CLS
RANDOMIZE TIMER
```

```
DO
z = INT(RND * 2) + 0
store# = v#
ty# = v# - r#
IF z = 0 THEN v# = v# + ((v# - r#) * 1) + 1
IF z = 1 THEN v# = v# + ((v# - r#) * 1) + 2
COLOR 2, 0
PRINT , v#; r#; z; speed#
rt# = speed#
speed# = v# - store#
IF (speed# / 2) - INT(speed# / 2) = .5 THEN speed# = speed# - 1
r# = r# + speed#
REM INPUT a$
a$ = INKEY$
IF a$ = "s" THEN STOP
LOOP UNTIL v# + (v# - r#) >= 999999999999999#



"C:\Users\Reactor1967\vcns\work\82402C.BAS"
v# = 1
r# = 1
CLS
RANDOMIZE TIMER
add# = 0
DO
z = INT(RND * 2) + 0
store# = v#
r1# = v#
IF (r1# / 2) - INT(r1# / 2) = 0 THEN r1# = r1# - 1
IF z = 0 THEN v# = v# + ((v# - r#) * 1) + 1
IF z = 1 THEN v# = v# + ((v# - r#) * 1) + 2
COLOR 2, 0
REM PRINT , , v#; z; r#; speed#; store2# - speed#; v# - r#; (v# - r#) - store3#
PRINT , , v#; z; r#; speed#; speed# - store2#
store2# = speed#
store3# = v# - r#
speed# = v# - store#
IF (speed# / 2) - INT(speed# / 2) = .5 THEN speed# = speed# - 1
r# = r# + speed#
REM INPUT a$
a$ = INKEY$
IF a$ = "s" THEN STOP
LOOP
```

```
"C:\Users\Reactor1967\vcns\work\82402B.BAS"
v# = 1
r# = 1
rate# = 0
speed# = 0
CLS
RANDOMIZE TIMER
DO
z = INT(RND * 2) + 0
r# = v#
DO
r# = r# - 1
IF z = 0 THEN v1# = v# + (v# - r# + 1) + speed#
IF z = 1 THEN v1# = v# + (v# - r# + 2) + speed#
test = (v1# - r#) = speed#
IF test = -1 THEN EXIT DO
a$ = INKEY$
IF a$ = "s" THEN STOP
LOOP UNTIL r# = 0
IF r# = 0 THEN STOP
IF z = 0 THEN v# = v# + (v# - r# + 1) + speed#
IF z = 1 THEN v# = v# + (v# - r# + 2) + speed#
COLOR 2, 0
PRINT , , speed#; r#; v#; v# - r#
rate# = rate# + 2
speed# = speed# + rate#
a$ = INKEY$
IF a$ = "s" THEN STOP
LOOP UNTIL v# + (v# - r#) >= 999999999999999#


"C:\Users\Reactor1967\vcns\work\82402A.BAS"
RANDOMIZE TIMER
dist# = 0
speed# = 0
add# = 0
x = 1
DO
COLOR 2, 0
z = INT(RND * 2) + 0
test1 = (dist# / 2) - INT(dist# / 2)
test2 = (speed# / 2) - INT(speed# / 2)
testa = (z = 0) AND (test1 = 0) AND (test2 = 0)
```

```
testb = (z = 0) AND (test1 = 0) AND (test2 = .5)
testc = (z = 0) AND (test1 = .5) AND (test2 = 0)
testd = (z = 0) AND (test1 = .5) AND (test2 = .5)
teste = (z = 1) AND (test1 = 0) AND (test2 = 0)
testf = (z = 1) AND (test1 = 0) AND (test2 = .5)
testg = (z = 1) AND (test1 = .5) AND (test2 = 0)
testh = (z = 1) AND (test1 = .5) AND (test2 = .5)
IF testa = -1 THEN rate# = 2 + add#
IF testb = -1 THEN rate# = 1 + add#
IF testc = -1 THEN rate# = 1 + add#
IF testd = -1 THEN rate# = 2 + add#
IF testc = -1 THEN rate# = 1 + add#
IF testd = -1 THEN rate# = 2 + add#
IF teste = -1 THEN rate# = 1 + add#
IF testf = -1 THEN rate# = 2 + add#
IF testg = -1 THEN rate# = 2 + add#
IF testh = -1 THEN rate# = 1 + add#
speed# = speed# + rate#
dist# = dist# + speed#
COLOR 2, 0
PRINT , , rate#; speed#; dist#; z; dist# - ((speed# - 1) * 2); dist# - ((speed# - 1)
* 2) - ar#
ar# = dist# - ((speed# - 1) * 2)
REM IF add# >= 100 THEN x = 0
REM IF add# <= 2 THEN x = 1
IF x = 1 THEN add# = add# + 2
IF x = 0 THEN add# = add# - 2
a$ = INKEY$
IF a$ = "s" THEN STOP
LOOP




"C:\Users\Reactor1967\vcns\work\82302D.BAS"
RANDOMIZE TIMER
dist# = 0
speed# = 0
DO
COLOR 2, 0
z = INT(RND * 2) + 0
test1 = (dist# / 2) - INT(dist# / 2)
test2 = (speed# / 2) - INT(speed# / 2)
testa = (z = 0) AND (test1 = 0) AND (test2 = 0)
testb = (z = 0) AND (test1 = 0) AND (test2 = .5)
testc = (z = 0) AND (test1 = .5) AND (test2 = 0)
testd = (z = 0) AND (test1 = .5) AND (test2 = .5)
```

```
teste = (z = 1) AND (test1 = 0) AND (test2 = 0)
testf = (z = 1) AND (test1 = 0) AND (test2 = .5)
testg = (z = 1) AND (test1 = .5) AND (test2 = 0)
testh = (z = 1) AND (test1 = .5) AND (test2 = .5)
IF testa = -1 THEN rate# = 2
IF testb = -1 THEN rate# = 1
IF testc = -1 THEN rate# = 1
IF testd = -1 THEN rate# = 2
IF testc = -1 THEN rate# = 1
IF testd = -1 THEN rate# = 2
IF teste = -1 THEN rate# = 1
IF testf = -1 THEN rate# = 2
IF testg = -1 THEN rate# = 2
IF testh = -1 THEN rate# = 1
speed# = speed# + rate#
dist# = dist# + speed#
storer# = r#
r# = speed# - 1
r# = r# * 2
r# = dist# - r#
IF (r# / 2) - INT(r# / 2) = 0 THEN r# = r# - 1
PRINT , rate#; speed#; dist#; z; r#; r# - storer#; (r# - storer#) - accelr#; dist# -
r#; (dist# - r#) - dr#
a$ = INKEY$
IF a$ = "s" THEN STOP
dr# = (dist# - r#)
accelr# = (r# - storer#)
testz = (dist# / 2) - INT(dist# / 2)
testx = (testz = .5) AND (z = 0)
testy = (testz = 0) AND (z = 1)
IF testx = -1 THEN STOP
IF testy = -1 THEN STOP
LOOP


"C:\Users\Reactor1967\vcns\work\82302C.BAS"
s1# = 1
s2# = 2
CLS
RANDOMIZE TIMER
v# = 1
test = (v# / 2) - INT(v# / 2)
IF test = 0 THEN r# = v# - s1#
IF test = .5 THEN r# = v# - s2#
DO
```

```
z = INT(RND * 2) + 0
accelv# = v# - storev#
storev# = v#
IF z = 0 THEN v# = v# + (v# - r# + 1)
IF z = 1 THEN v# = v# + (v# - r# + 2)
COLOR 2, 0
PRINT , , r# - storer# - accelr#; r# - storer#; r#; v# - r#; v#; v# - storev# -
accelv#; v# - storev#
IF x = 0 THEN s1# = s1# + 2
IF x = 0 THEN s2# = s2# + 2
IF x = 1 THEN s1# = s1# - 2
IF x = 1 THEN s2# = s2# - 2
REM IF s1# >= 100 THEN x = 1
REM IF s1# <= 5 THEN x = 0
test = (v# / 2) - INT(v# / 2)
accelr# = r# - storer#
storer# = r#
IF test = 0 THEN r# = v# - s1#
IF test = .5 THEN r# = v# - s2#
a$ = INKEY$
IF a$ = "s" THEN STOP
LOOP




"C:\Users\Reactor1967\vcns\work\82302B.BAS"
v# = 100
r# = 99
CLS
RANDOMIZE TIMER
DO
z = INT(RND * 2) + 0
r1# = r#
storer# = r#
DO
r1# = r1# + 1
rs# = r1# - r#
IF z = 0 THEN v1# = v# + (v# - r1# + 1)
IF z = 1 THEN v1# = v# + (v# - r1# + 2)
vs# = v1# - v#
IF rs# = vs# THEN EXIT DO
REM IF vs# - rs# = 1 THEN EXIT DO
a$ = INKEY$
IF a$ = "s" THEN STOP
LOOP UNTIL r1# >= v#
IF r1# > v# THEN STOP
```

```
IF (r1# / 2) - INT(r1# / 2) = 0 THEN r1# = r1# - 1
r# = r1#
storev# = v#
IF z = 0 THEN v# = v# + (v# - r# + 1)
IF z = 1 THEN v# = v# + (v# - r# + 2)
COLOR 2, 0
PRINT , , r# - storer#; r#; z; v# - storev#; v#; v# - r#
a$ = INKEY$
IF a$ = "s" THEN STOP
LOOP
```

```
"C:\Users\Reactor1967\vcns\work\82302A.BAS"
v# = 1
r# = 1
CLS
RANDOMIZE TIMER
CLS
ave# = 0
DO
z = INT(RND * 2) + 0
accelv# = v# - storev#
storev# = v#
IF z = 0 THEN v# = v# + (v# - r#) + 1
IF z = 1 THEN v# = v# + (v# - r#) + 2
COLOR 2, 0
REM Table of contents for meaning of print statement below.
REM acceleration of R#        : Speed of r# : R : V : Aceleration of v#       :
Speed of v# :distancevr:Speed of r# + Speed of v# = distance between v# and
r# Plus or Minus 1. z = binary data being stored.
REM PRINT (r# - storer#) - accelr#; r# - storer#; r#; v#; (v# - storev#) -
accelv#; v# - storev#; v# - r#; (r# - storer#) + (v# - storev#); z
PRINT , , (r# - storer#) - accelr#; r# - storer#; z; (v# - storev#) - accelv#; v# -
storev#
REM INPUT a$
a$ = INKEY$
IF a$ = "s" THEN STOP
dist# = v# - storev#
IF (dist# / 2) - INT(dist# / 2) = .5 THEN dist# = dist# - 1
accelr# = r# - storer#
storer# = r#
r# = r# + dist#
LOOP
```

```
"C:\Users\Reactor1967\vcns\work\82202B.BAS"
REM Lesson learned here. R and V have to be measured in some way to
REM                 control them. Using speed, time, acceleration
REM                 and deacceleration seems the most proper since
REM                 a non-linear numerical system uses time to store
REM                 numerical values in a numerical base.
REM Key here is to Learn to control the acceleration and de-acceleration of
REM both v# and r#. By doing that you can control r# and v# and code and
REM decode binary data if using base two. If not then base 3 or what ever.
v# = 1
r# = 1
dist# = v# - r#
count = 0
CLS
RANDOMIZE TIMER
again:
DO
count = count + 1
m2# = r#
DO
IF (v# - r#) >= dist# THEN r# = r# + 2
REM a$ = INKEY$
REM IF a$ = "s" THEN STOP
LOOP UNTIL v# - r# < dist#
IF (v# - r#) > 500 THEN x = 0
IF (v# - r#) < 10 THEN x = 1
IF x = 1 THEN r# = r# - 4
dist# = v# - r#
z = INT(RND * 2) + 0
m1# = v#
IF z = 0 THEN v# = v# + (v# - r#) + 1
IF z = 1 THEN v# = v# + (v# - r#) + 2
COLOR 2, 0
PRINT (r# - m2#) - accelr#; "Spd +r"; r# - m2#; "+r"; r#; "N"; z; "+v"; v#; (v#
- m1#) - accelv#; "Spd +v"; v# - m1#; "Dist v,r"; v# - r#; "+Time"; count
accelr# = (r# - m2#)
accelv# = (v# - m1#)
a$ = INKEY$
IF a$ = "s" THEN INPUT b$
IF b$ = "s" THEN STOP
IF b$ = "o" THEN EXIT DO
IF v# + (v# - r#) > 999999 THEN EXIT DO
a$ = ""
b$ = ""
LOOP
```

```
a$ = ""
b$ = ""
switch# = v# - r#
r# = v# + switch#
DO
count = count + 1
m2# = r#
DO
IF (r# - v#) >= dist# THEN r# = r# - 2
REM a$ = INKEY$
REM IF a$ = "s" THEN STOP
LOOP UNTIL r# - v# < dist#
IF (r# - v#) > 500 THEN x = 0
IF (r# - v#) < 10 THEN x = 1
IF x = 1 THEN r# = r# + 4
dist# = r# - v#
z = INT(RND * 2) + 0
m1# = v#
IF z = 0 THEN v# = v# - (r# - v#) - 1
IF z = 1 THEN v# = v# - (r# - v#) - 2
COLOR 2, 0
PRINT (r# - m2#) - accelr#; "Spd -r"; m2# - r#; "-r"; r#; "N"; z; "-v"; v#; (v# -
m1#) - accelv#; "Spd -v"; m1# - v#; "Dist v,r"; r# - v#; "+Time"; count
a$ = INKEY$
IF a$ = "s" THEN INPUT b$
IF b$ = "s" THEN STOP
IF b$ = "o" THEN EXIT DO
IF count = 1 THEN EXIT DO: REM Even go using Negative V is possible this
program avoids it to keep things simple.
IF v# - (r# - v#) < 0 THEN EXIT DO
a$ = ""
b$ = ""
LOOP
a$ = ""
b$ = ""
switch# = r# - v#
r# = v# - switch#
GOTO again:




"C:\Users\Reactor1967\vcns\work\82202A.BAS"
REM v# = v# + ((speed) * (base - 1)) + N
v# = 1
r# = 1
CLS
```

```
RANDOMIZE TIMER
DO
z = INT(RND * 2) + 0
DO
IF (v# - r#) >= 100 THEN r# = r# + 100
a$ = INKEY$
IF a$ = "s" THEN STOP
LOOP UNTIL (v# - r#) < 100
store1# = v#
IF z = 0 THEN v# = v# + (v# - r#) + 1
IF z = 1 THEN v# = v# + (v# - r#) + 2
COLOR 2, 0
PRINT , , store1# - r#; r#; z; v#; v# - r#
a$ = INKEY$
IF a$ = "s" THEN STOP
LOOP




"C:\Users\Reactor1967\vcns\work\82102G.BAS"
v# = 1
CLS
RANDOMIZE TIMER
count = 1
DO
z = INT(RND * 2) + 0
IF z = 0 THEN v# = v# + ((count + 1) * 1) + 1
IF z = 1 THEN v# = v# + ((count + 1) * 1) + 2
REM v# = v# + ((speed) * (base - 1)) + N
COLOR 2, 0
PRINT , , count; v#; z; (v# / count)
test1 = (count / 2) - INT(count / 2)
test2 = (v# / 2) - INT(v# / 2)
test3 = (test1 = 0) AND (test2 = 0)
test4 = (test1 = 0) AND (test2 = .5)
test5 = (test1 = .5) AND (test2 = 0)
test6 = (test1 = .5) AND (test2 = .5)
IF test3 = -1 THEN count = count + 2
IF test4 = -1 THEN count = count + 1
IF test5 = -1 THEN count = count + 1
IF test6 = -1 THEN count = count + 2
a$ = INKEY$
IF a$ = "s" THEN STOP
LOOP
```

```
"C:\Users\Reactor1967\vcns\work\82102F.BAS"
f# = 2
DO
f# = f# * 2
LOOP UNTIL f# * 2 >= 2000
g# = f# / 2
CLS
count = 1
redo:
DO
PRINT f#; "="; g#; count
count = count + 1
f# = f# - 1
g# = g# - 1
a$ = INKEY$
IF a$ = "s" THEN STOP
LOOP UNTIL g# = 0
IF (f# / 2) - INT(f# / 2) = .5 THEN STOP
IF f# = 2 THEN STOP
f# = f# / 2
g# = f# / 2
a$ = INKEY$
REM INPUT a$
IF a$ = "s" THEN STOP
GOTO redo:
```

```
"C:\Users\Reactor1967\vcns\work\82102E.BAS"
v# = 1
r# = 1
CLS
redo:
DO
IF v# + (v# - r#) >= 500000000000000# THEN z = 1 ELSE z = 0
IF z = 0 THEN v# = v# + ((v# - r#) * 1) + 1
IF z = 1 THEN v# = v# + ((v# - r#) * 1) + 2
COLOR 2, 0
PRINT , , v#; "+"
REM INPUT a$
a$ = INKEY$
IF a$ = "s" THEN STOP
LOOP UNTIL z = 1
REM INPUT a$
IF a$ = "s" THEN STOP
```

```
r# = 999999999999999#
DO
IF v# - (r# - v#) <= 500000000000000# THEN z = 1 ELSE z = 0
IF z = 0 THEN v# = v# - ((r# - v#) * 1) + 1
IF z = 1 THEN v# = v# - ((r# - v#) * 1) + 2
COLOR 2, 0
PRINT , , v#; "-"
a$ = INKEY$
IF a$ = "s" THEN STOP
LOOP UNTIL z = 1
REM INPUT a$
IF a$ = "s" THEN STOP
r# = 1
GOTO redo:




"C:\Users\Reactor1967\vcns\work\82102D.BAS"
REM                     COPYRIGHT C 2002
REM                  Lloyd Dudley Burris
REM ---------------------Initilize R#-------------------------------------
CLS
DIM r#(255)
r#(1) = 1
REM PRINT r#(1)
pt1# = 5
rf# = 100
count = 2
a# = 0
redoa:
DO
a# = a# + pt1#
test = (a# / 2) - INT(a# / 2)
r#(count) = a#
IF (r#(count) / 2) - INT(r#(count) / 2) = 0 THEN r#(count) = r#(count) + 1
REM PRINT r#(count); count
count = count + 1
REM INPUT a$
REM a$ = INKEY$
IF a$ = "s" THEN STOP
IF count >= 254 THEN EXIT DO
LOOP UNTIL a# = rf#
pt1# = pt1# * 10
rf# = rf# * 10
IF count >= 254 THEN GOTO redob:
IF a# + pt1# > 999999999999999# THEN GOTO redob:
```

```
GOTO redoa:
redob:
r#(254) = 950000000000001#
r#(255) = 999999999999999#
REM ----------------------Code P# up.---------------------------------------
range = 237
REM p# = 1000
p# = 100000000000001#
RANDOMIZE TIMER
c = 0
REM file$ = "vcout.txt"
REM OPEN file$ FOR OUTPUT AS #1
v# = 1
repeat:
count = 0
DO
bt = (p# / 2) - INT(p# / 2)
IF bt >= .5 THEN bt = 1
COLOR 2, 0
PRINT , , p#; "+"; bt; r#(range); count
REM WRITE #1, p#, bt, r#(range)
a$ = INKEY$
REM INPUT a$
IF a$ = "s" THEN SYSTEM
IF (p# + (p# - r#(range) + 1)) > (r#(range + 1)) THEN c = 1 ELSE c = 0
IF c = 0 THEN p# = p# + (p# - (r#(range)) + 1)
IF c = 1 THEN p# = p# + (p# - (r#(range)) + 2)
IF a$ = "s" THEN STOP
IF p# > (r#(range + 1)) THEN range = range + 1
count = count + 1
IF range = 254 THEN EXIT DO
v# = v# + 1
p# = v#
LOOP UNTIL p# + (p# - (r#(range)) + 2) >= 999999999999999#
STOP
REM CLOSE #1
bt = (p# / 2) - INT(p# / 2)
IF bt >= .5 THEN bt = 1
PRINT , , p#; "+"; bt; r#(range); count
a$ = INKEY$
REM INPUT a$
IF a$ = "s" THEN SYSTEM
IF a$ = "d" THEN GOTO Decodedown:
REM ---------------------Code P# down.---------------------------------
range = range + 1
count = 0
```

```
DO
bt = (p# / 2) - INT(p# / 2)
IF bt >= .5 THEN bt = 1
COLOR 2, 0
PRINT , , p#; "-"; bt; r#(range); count
REM a$ = INKEY$
REM INPUT a$
REM IF a$ = "s" THEN STOP
REM c = INT(RND * 2) + 0
IF (p# - (r#(range) - p# + 1)) < (r#(range - 1)) THEN c = 1 ELSE c = 0
IF c = 0 THEN p# = p# - ((r#(range)) - p# + 1)
IF c = 1 THEN p# = p# - ((r#(range)) - p# + 2)
IF p# < (r#(range - 1)) THEN range = range - 1
count = count + 1
IF range <= 238 THEN EXIT DO
LOOP UNTIL p# < 150000000000000#
bt = (p# / 2) - INT(p# / 2)
IF bt >= .5 THEN bt = 1
PRINT , , p#; "-"; bt; r#(range); count
a$ = INKEY$
REM INPUT a$
IF a$ = "s" THEN SYSTEM
IF a$ = "d" THEN GOTO decode2:
range = range - 1
GOTO repeat:
REM ----------------------------DECODING AREA------------------------------
REM ---------------------------RESTRICTED AREA:PROGRAM AUTHOR
ONLY!!!------
decode1:
REM----------------------------[DECODING + TO -]---------------------------
Decodedown: REM from + to -
DO
bt = (p# / 2) - INT(p# / 2)
IF bt >= .5 THEN bt = 1
IF bt = 1 THEN range = range - 1
COLOR 2, 0
PRINT , , p#; "-"; bt; r#(range)
IF bt = 0 THEN p# = p# - ((p# - r#(range) + 1) / 2)
IF bt = 1 THEN p# = p# - ((p# - r#(range) + 2) / 2)
a$ = INKEY$
REM INPUT a$
IF a$ = "s" THEN STOP
bt = (p# / 2) - INT(p# / 2)
IF bt >= .5 THEN bt = 1
test = (bt = 1) AND range <= 237
IF p# = 100000000000001# THEN GOTO finish:
```

```
LOOP UNTIL test = -1
bt = (p# / 2) - INT(p# / 2)
IF bt >= .5 THEN bt = 1
COLOR 2, 0
PRINT , , p#; "-"; bt; r#(range)
REM ----------------------------[DECODING - TO +]---------------------------
range = range + 1
Decodeup: REM from - to +
decode2:
DO
bt = (p# / 2) - INT(p# / 2)
IF bt >= .5 THEN bt = 1
IF bt = 1 THEN range = range + 1
COLOR 2, 0
PRINT , , p#; "+"; bt; r#(range)
IF bt = 0 THEN p# = p# + ((r#(range) - p# + 1) / 2)
IF bt = 1 THEN p# = p# + ((r#(range) - p# + 2) / 2)
a$ = INKEY$
REM INPUT a$
IF a$ = "s" THEN STOP
bt = (p# / 2) - INT(p# / 2)
IF bt >= .5 THEN bt = 1
test = (bt = 1) AND range >= 255
LOOP UNTIL test = -1
bt = (p# / 2) - INT(p# / 2)
IF bt >= .5 THEN bt = 1
COLOR 2, 0
PRINT , , p#; "+"; bt; r#(range)
range = range - 1
GOTO decode1:
finish:
PRINT , , "DECODE FINISHED!!!!"
SYSTEM


"C:\Users\Reactor1967\vcns\work\82102C.BAS"
r# = 1
v# = 1
CLS
RANDOMIZE TIMER
DO
z = INT(RND * 2) + 0
IF z = 0 THEN v# = v# + (v# - r#) + 1
IF z = 1 THEN v# = v# + (v# - r#) + 2
COLOR 2, 0
```

```
PRINT , , r#; z; v#
a$ = INKEY$
IF a$ = "s" THEN STOP
LOOP UNTIL v# + (v# - r#) >= 999999999999999#
```

```
"C:\Users\Reactor1967\vcns\work\82102B.BAS"
DIM dist#(1000)
a# = 2
CLS
PRINT , , "initilizing vcns array...."
FOR count = 1 TO 1000
REM PRINT a#
dist#(count) = a#
a# = a# + 2
NEXT count
CLS
RANDOMIZE TIMER
count = 1
v# = 1
DO
z = INT(RND * 2) + 0
store# = v#
IF z = 0 THEN v# = v# + (dist#(count)) + 1
IF z = 1 THEN v# = v# + (dist#(count)) + 2
count = count + 1
COLOR 2, 0
PRINT , , z; v#; store#; dist#(count - 1); count
a$ = INKEY$
IF a$ = "s" THEN STOP
LOOP UNTIL count = 1001
```

```
"C:\Users\Reactor1967\vcns\work\82102A.BAS"
REM Lesson learned here:(Knock on wood).
REM Try to subliment all even numbers for
REM v# - r# so can reverse the equations and
REM code down in value with v# with out making
REM r# greater then v# so can keep r# stable.
REM anyway give it a try.
REM - tried it fault is you have to know the numerical value
REM   of your vector number to make it work.
v# = 1
r# = 1
```

```
CLS
RANDOMIZE TIMER
count = 1
redo:
DO
z = INT(RND * 2) + 0
dist# = v# - r#
IF (dist# / 2) - INT(dist# / 2) = .5 THEN dist# = dist# - 1
REM dist# = dist# / 2
IF z = 0 THEN v# = v# + (dist#) + 1
IF z = 1 THEN v# = v# + (dist#) + 2
COLOR 2, 0
PRINT , count; r#; z; v#; dist#
a$ = INKEY$
IF a$ = "s" THEN STOP
count = count + 1
LOOP UNTIL v# + ((v# - r#) * 1) >= 999999999999999#
COLOR 2, 0
PRINT , , "----------------------"
DO
z = INT(RND * 2) + 0
dist# = v# - r#
IF (dist# / 2) - INT(dist# / 2) = .5 THEN dist# = dist# - 1
store# = dist#
dist# = dist# / 2
IF z = 0 THEN v# = v# - (dist#) + 1
IF z = 1 THEN v# = v# - (dist#) + 2
COLOR 2, 0
PRINT , count; r#; z; v#; store#
a$ = INKEY$
IF a$ = "s" THEN STOP
count = count - 1
IF count = 0 THEN EXIT DO
LOOP UNTIL v# - INT((((v# - r#) * 1)) / 2) <= 1
COLOR 2, 0
PRINT , , "------------------------"
GOTO redo:




"C:\Users\Reactor1967\vcns\work\82002C.BAS"
v# = 1
r# = 1
RANDOMIZE TIMER
```

```
CLS
count = 0
DO
count = count + 1
z = INT(RND * 2) + 0
IF z = 0 THEN v# = v# + (v# - r# + 1)
IF z = 1 THEN v# = v# + (v# - r# + 2)
COLOR 2, 0
PRINT , , r1#; r2#; v#; v# - r#; count
a$ = INKEY$
IF a$ = "s" THEN STOP
r1# = v# - count
r2# = v# + 1 - count
IF (r1# / 2) - INT(r1# / 2) = .5 THEN r# = r1#
IF (r2# / 2) - INT(r2# / 2) = .5 THEN r# = r2#
LOOP




"C:\Users\Reactor1967\vcns\work\82002B.BAS"
REM Lesson Learned here. Code in such a way that when
REM decoding can devide the dist# between v# - r# into
REM 2 or some function of that and subtract it from R to
REM get your previous R#. Before actually coding test
REM this senero test different R's like did in earlier
REM programs then use the ones that actually codes and
REM fits the conditions then use its next v# with tested
REM r#'s to move on.
v# = 1
r# = 1
CLS
RANDOMIZE TIMER
count = 0
redo:
DO
z = INT(RND * 2) + 0
store# = v#
store2# = r#
IF z = 0 THEN v# = v# + (v# - r# + 1)
IF z = 1 THEN v# = v# + (v# - r# + 2)
COLOR 2, 0
PRINT , , r#; v#; count; r# - (INT((v# - r#) / 2))
REM a$ = INKEY$
INPUT a$
IF a$ = "s" THEN STOP
```

```
IF a$ = "d" THEN EXIT DO
count = count + 1
r# = v# - INT(v# / count)
LOOP UNTIL v# + (v# - r#) >= 999999999999999#




"C:\Users\Reactor1967\vcns\work\82002.BAS"
v# = 1
r# = 1
CLS
RANDOMIZE TIMER
count = 0
redo:
DO
z = INT(RND * 2) + 0
store# = v#
store2# = r#
IF z = 0 THEN v# = v# + (v# - r# + 1)
IF z = 1 THEN v# = v# + (v# - r# + 2)
COLOR 2, 0
PRINT , , r#; v#; count
a$ = INKEY$
REM INPUT a$
IF a$ = "s" THEN STOP
IF a$ = "d" THEN EXIT DO
test1 = (v# / 2) - INT(v# / 2)
test2 = (count / 2) - INT(count / 2)
test3 = (test1 = 0) AND (test2 = 0)
test4 = (test1 = 0) AND (test2 = .5)
test5 = (test1 = .5) AND (test2 = 0)
test6 = (test1 = .5) AND (test2 = .5)
IF test3 = -1 THEN count = (count + 1)
IF test4 = -1 THEN count = (count + 2)
IF test5 = -1 THEN count = (count + 2)
IF test6 = -1 THEN count = (count + 1)
r# = v# - count
LOOP UNTIL v# + (v# - r#) >= 999999999999999#




"C:\Users\Reactor1967\vcns\work\81702B.BAS"
v# = 100
r# = 100
```

```
CLS
RANDOMIZE TIMER
DO
z = INT(RND * 2) + 0
r1# = v#
IF (r1# / 2) - INT(r1# / 2) = 0 THEN r1# = r1# - 1
store2# = r#
dist# = v# - r#
DO
r1# = r1# - 2
IF z = 0 THEN v1# = v# + (v# - r1# + 1)
IF z = 1 THEN v1# = v# + (v# - r1# + 2)
a$ = INKEY$
IF a$ = "s" THEN STOP
LOOP UNTIL (v1# - r1#) >= dist#
r# = r1#
store1# = v#
IF z = 0 THEN v# = v# + (v# - r# + 1)
IF z = 1 THEN v# = v# + (v# - r# + 2)
COLOR 2, 0
PRINT , , r# - store2#; r#; z; v#; v# - store1#; v# - r#
a$ = INKEY$
IF a$ = "s" THEN STOP
LOOP




"C:\Users\Reactor1967\vcns\work\81702A.BAS"
REM Lesson learned here. That you can use average rate of
REM acceleration to increase R by. Get this down then use
REM it to let V# - R# get further apart and closer together
REM by controling the ave# rate of speed by which R increases
REM by.
v# = 1
r# = 1
CLS
RANDOMIZE TIMER
redo:
DO
z = INT(RND * 2) + 0
store1# = v#
IF z = 0 THEN dist# = (((v# - r#) * 1) + 2)
IF z = 1 THEN dist# = (((v# - r#) * 1) + 4)
store2# = r#
REM  IF (dist# / 2) - INT(dist# / 2) = .5 THEN dist# = dist# - 1
dist# = (dist# / 2)
```

```
r# = r# + dist# - 1
REM IF (r# / 2) - INT(r# / 2) = 0 THEN r# = r# - 1
IF z = 0 THEN v# = v# + (((v# - r#) * 1) + 2)
IF z = 1 THEN v# = v# + (((v# - r#) * 1) + 4)
COLOR 2, 0
PRINT , r# - store2#; r#; z; v#; v# - store1#
a$ = INKEY$
IF a$ = "s" THEN EXIT DO
LOOP UNTIL v# + ((v# - r#) * 2) >= 999999999999999#
decode:
```

```
"C:\Users\Reactor1967\vcns\work\81602C.BAS"
v# = 100
r# = 99
CLS
RANDOMIZE TIMER
x# = 100
redo:
COLOR 2, 0
PRINT , , r#; z; v#
DO
z = INT(RND * 2) + 0
IF v# + ((v# - r#) * 2) >= (r# + x#) THEN z = 2
IF z = 0 THEN v# = v# + (((v# - r#) * 2) + 1)
IF z = 1 THEN v# = v# + (((v# - r#) * 2) + 2)
IF z = 2 THEN v# = v# + (((v# - r#) * 2) + 3)
COLOR 2, 0
PRINT , , r#; z; v#
a$ = INKEY$
IF a$ = "s" THEN STOP
IF z = 2 THEN r# = r# + x#
IF z = 2 THEN x# = (x# * 10)
LOOP UNTIL v# + ((v# - r#) * 2) >= 999999999999999#
```

```
"C:\Users\Reactor1967\vcns\work\81602B.BAS"
v# = 1
r# = 1
CLS
RANDOMIZE TIMER
x# = 2
redo:
DO
```

```
z = INT(RND * 2) + 0
IF v# + (v# - r#) >= (r# + x#) THEN z = 1 ELSE z = 0
IF z = 0 THEN v# = v# + (v# - r# + 1)
IF z = 1 THEN v# = v# + (v# - r# + 2)
COLOR 2, 0
PRINT , , r#; z; v#
a$ = INKEY$
IF a$ = "s" THEN STOP
IF z = 1 THEN r# = r# + x#
IF v# >= (x# * 10) THEN x# = x# * 10
LOOP UNTIL v# + (v# - r#) >= 799999999999999#
dist# = v# - r#
r# = v# + dist#
DO
z = INT(RND * 2) + 0
IF v# - (r# - v#) <= (r# - x#) THEN z = 1 ELSE z = 0
IF z = 0 THEN v# = v# - (r# - v# + 1)
IF z = 1 THEN v# = v# - (r# - v# + 2)
COLOR 2, 0
PRINT , , r#; z; v#
a$ = INKEY$
IF a$ = "s" THEN STOP
IF z = 1 THEN r# = r# - x#
IF v# <= (x# / 10) THEN x# = x# / 10
LOOP UNTIL v# - (r# - v#) <= 200000000000000#
dist# = r# - v#
r# = v# - dist#
GOTO redo:


"C:\Users\Reactor1967\vcns\work\81602A.BAS"
v# = 1
r# = 1
CLS
RANDOMIZE TIMER
DO
z = INT(RND * 2) + 0
store2# = v#
IF z = 0 THEN v# = v# + (v# - r# + 1)
IF z = 1 THEN v# = v# + (v# - r# + 2)
COLOR 2, 0
PRINT , , r#; v#; r# - store1#; v# - store2#
a$ = INKEY$
REM INPUT a$
IF a$ = "s" THEN STOP
```

```
test = (v# / 2) - INT(v# / 2)
store1# = r#
IF test = 0 THEN r# = r# + ((v# - r#) - 1)
IF test = .5 THEN r# = r# + ((v# - r#) - 2)
LOOP
```

"C:\Users\Reactor1967\vcns\work\81502E.BAS"
```
v# = 100
r# = 1
CLS
RANDOMIZE TIMER
DO
z = INT(RND * 3) + 0
r1# = r#
low1# = 999999999999999#
low2# = 0
DO
r1# = r1# + 2
IF z = 0 THEN v1# = v# + (((v# - r1#) * 2) + 1)
IF z = 1 THEN v1# = v# + (((v# - r1#) * 2) + 2)
IF z = 2 THEN v1# = v# + (((v# - r1#) * 2) + 3)
dist# = v1# - v#
IF dist# < low1# THEN low2# = r1#
IF dist# < low1# THEN low1# = dist#
stop$ = INKEY$
IF stop$ = "s" THEN STOP
REM PRINT r1# - r#; v1# - v#
LOOP UNTIL r1# + 2 >= v#
store1# = r#
r# = low2#
store2# = v#
IF z = 0 THEN v# = v# + (((v# - r#) * 2) + 1)
IF z = 1 THEN v# = v# + (((v# - r#) * 2) + 2)
IF z = 2 THEN v# = v# + (((v# - r#) * 2) + 3)
COLOR 2, 0
PRINT , , r#; v#; v# - r#; r# - store1#; v# - store2#
REM INPUT a$
a$ = INKEY$
IF a$ = "s" THEN STOP
LOOP
```

"C:\Users\Reactor1967\vcns\work\81502D.BAS"

```
v# = 1
r# = 1
RANDOMIZE TIMER
CLS
DO
z = INT(RND * 2) + 0
store1# = v#
IF z = 0 THEN v# = v# + (v# - r# + 1)
IF z = 1 THEN v# = v# + (v# - r# + 2)
COLOR 2, 0
PRINT , , r#; v#; v# - r#
a$ = INKEY$
REM INPUT a$
IF a$ = "s" THEN STOP
dist# = v# - store1#
IF (v# - r#) < 100 THEN r# = r# + (dist# - 1)
IF (v# - r#) > 100 THEN r# = r# + (dist# + 1)
LOOP
```

```
"C:\Users\Reactor1967\vcns\work\81502C.BAS"
REM Lesson learned here. Can make math calculations
REM for V# - R# and V2# - v1# and r2# - r1# such that
REM as v# codes the dist# between v# and R go up or down
REM depending on how much we add to r# which is a function
REM of v2# - v1#. This looks very promising here.
v# = 1
r# = 1
CLS
RANDOMIZE TIMER
DO
z = INT(RND * 2) + 0
store1# = v#
IF z = 0 THEN v# = v# + (v# - r# + 1)
IF z = 1 THEN v# = v# + (v# - r# + 2)
COLOR 2, 0
PRINT , r#; z; v#; v# - store#; r# - store2#; v# - store1#
dist# = v# - store1#
store2# = r#
r# = r# + dist#
REM IF (r# / 2) - INT(r# / 2) = 0 THEN r# = r# + 5: REM to go up. This is a
control. Comment it or uncomment it to use it. Make sure to comment out the
control below if using this one.
IF (r# / 2) - INT(r# / 2) = 0 THEN r# = r# - 1: REM to go down. This is a
control. Comment it or uncomment it to use it. Make sure to comment out the
```

control above if using this one.
REM INPUT a$: rem speed control
a$ = INKEY$: REM speed control
IF a$ = "s" THEN STOP
LOOP

"C:\Users\Reactor1967\vcns\work\81502B.BAS"
REM Lesson learned here!
REM can do chart so that when v2# - v1 = x
REM then r# = r# + y. Can use this chart
REM for coding or decoding.
REM can make this chart also for 0 to a 0
REM or 0 to a 1
REM or 1 to a 0
REM or 1 to a 1.
REM Try to work on this chart.
v# = 100
r# = 1
RANDOMIZE TIMER
CLS
DO
z = INT(RND * 2) + 0
r1# = r#
store1# = v#
store2# = r#
DO
r1# = r1# + 2
IF z = 0 THEN v1# = v# + (v# - r1# + 1)
IF z = 1 THEN v1# = v# + (v# - r1# + 2)
test1 = (v1# - v#) - (r1# - r#) = 0
test2 = (v1# - v#) - (r1# - r#) = 1
test3 = (v1# - v#) - (r1# - r#) = 2
test4 = (v1# - v#) - (r1# - r#) = 3
IF test1 = -1 THEN EXIT DO
IF test2 = -1 THEN EXIT DO
IF test3 = -1 THEN EXIT DO
IF test4 = -1 THEN EXIT DO
a$ = INKEY$
IF a$ = "s" THEN STOP
LOOP
r# = r1#
v# = v1#
COLOR 2, 0
PRINT , , r#; z; v#; r# - store2#; v# - store1#

```
REM IF v# - store1# = 219 THEN STOP
a$ = INKEY$
REM INPUT a$
IF a$ = "s" THEN STOP
LOOP




"C:\Users\Reactor1967\vcns\work\81502.BAS"
v# = 1
r# = 1
CLS
RANDOMIZE TIMER
j# = 101
count = 1
DO
z = INT(RND * 2) + 0
store# = v#
store3# = v#
IF z = 0 THEN v# = v# + (((v# - r#) * 1) + 1)
IF z = 1 THEN v# = v# + (((v# - r#) * 1) + 2)
COLOR 2, 0
PRINT , , r#; v#; r# - store2#; v# - store3#
store2# = r#
r# = r# + (v# - store#)
IF (r# / 2) - INT(r# / 2) = 0 THEN r# = r# - 1
REM a$ = INKEY$
INPUT a$
IF a$ = "s" THEN STOP
LOOP




"C:\Users\Reactor1967\vcns\work\81402.BAS"
r# = 1
v# = 1
CLS
RANDOMIZE TIMER
DO
z = INT(RND * 2) + 0
test1 = (v# / 2) - INT(v# / 2)
test2# = v# + (v# - r# + 1)
test3# = v# + (v# - r# + 2)
test4 = (test2# / 2) - INT(test2# / 2)
test5 = (test3# / 2) - INT(test3# / 2)
test6 = (test1 = 0) AND (test4 = 0) AND (z = 0)
```

```
test7 = (test1 = 0) AND (test4 = .5) AND (z = 1)
test8 = (test1 = .5) AND (test4 = 0) AND (z = 0)
test9 = (test1 = .5) AND (test4 = .5) AND (z = 1)
test10 = (test1 = 0) AND (test5 = 0) AND (z = 0)
test11 = (test1 = 0) AND (test5 = .5) AND (z = 1)
REM test12 = (test1 = .5) AND (test5 = 0) AND (z = 0)
test13 = (test1 = .5) AND (test5 = .5) AND (z = 1)
COLOR 2, 0
store# = v#
IF (test6 = -1) OR (test7 = -1) OR (test8 = -1) OR (test9 = -1) THEN v# = v# +
(v# - r# + 1)
IF (test10 = -1) OR (test11 = -1) OR (test12 = -1) OR (test13 = -1) THEN v# =
v# + (v# - r# + 2)
PRINT , , r#; v#
REM IF (test6 = -1) OR (test7 = -1) OR (test8 = -1) OR (test9 = -1) THEN r# =
r# + (v# - r#) - 1
REM IF (test10 = -1) OR (test11 = -1) OR (test12 = -1) OR (test13 = -1) THEN
r# = r# + (v# - r#) - 2
IF (test6 = -1) OR (test7 = -1) OR (test8 = -1) OR (test9 = -1) THEN r# = (r# +
(v# - store#)) - 1
IF (test10 = -1) OR (test11 = -1) OR (test12 = -1) OR (test13 = -1) THEN r# =
(r# + (v# - store#)) - 2
a$ = INKEY$
REM INPUT a$
IF a$ = "s" THEN STOP
LOOP




"C:\Users\Reactor1967\vcns\work\81302.BAS"
REM can look at the numbers as they code and tell how much to add to d#
REM so when decoding if know to subtract 1 or a 2 could use d# to decode.
REM this is a test later will try to get r to increment in a controled
REM fashion by look at v# as it codes or decrease in a controled fashion
REM when decoding by looking at v#.
RANDOMIZE TIMER
CLS
v# = 100
d# = 1
r# = 0
DO
store# = r#
z = INT(RND * 2) + 0
r1# = v# - (d# + 1)
r2# = v# - (d# + 2)
v1# = v# + (v# - r1# + 1)
```

v2# = v# + (v# - r1# + 2)
v3# = v# + (v# - r2# + 1)
v4# = v# + (v# - r2# + 2)
test0 = (v# / 2) - INT(v# / 2)
test1 = (v1# / 2) - INT(v1# / 2)
test2 = (v2# / 2) - INT(v2# / 2)
test3 = (v3# / 2) - INT(v3# / 2)
test4 = (v4# / 2) - INT(v4# / 2)
test5 = (test0 = 0) AND (test1 = .5) AND (z = 1)
IF test5 = -1 THEN r# = (v# - (d# + 1))
IF test5 = -1 THEN v# = v1#
IF test5 = -1 THEN d# = d# + 1
IF test5 = -1 THEN GOTO skip:
test6 = (test0 = .5) AND (test1 = 0) AND (z = 0)
IF test6 = -1 THEN r# = (v# - (d# + 1))
IF test6 = -1 THEN v# = v1#
IF test6 = -1 THEN d# = d# + 1
IF test6 = -1 THEN GOTO skip:
test7 = (test0 = 0) AND (test2 = .5) AND (z = 1)
IF test7 = -1 THEN r# = (v# - (d# + 1))
IF test7 = -1 THEN v# = v2#
IF test7 = -1 THEN d# = d# + 1
IF test7 = -1 THEN GOTO skip:
test8 = (test0 = .5) AND (test2 = 0) AND (z = 0)
IF test8 = -1 THEN r# = (v# - (d# + 1))
IF test8 = -1 THEN v# = v2#
IF test8 = -1 THEN d# = d# + 1
IF test8 = -1 THEN GOTO skip:
test9 = (test0 = 0) AND (test3 = 0) AND (z = 0)
IF test9 = -1 THEN r# = (v# - (d# + 2))
IF test9 = -1 THEN v# = v3#
IF test9 = -1 THEN d# = d# + 2
IF test9 = -1 THEN GOTO skip:
test10 = (test0 = .5) AND (test3 = .5) AND (z = 1)
IF test10 = -1 THEN r# = (v# - (d# + 2))
IF test10 = -1 THEN v# = v3#
IF test10 = -1 THEN d# = d# + 2
IF test10 = -1 THEN GOTO skip:
test11 = (test0 = 0) AND (test4 = 0) AND (z = 0)
IF test11 = -1 THEN r# = (v# - (d# + 2))
IF test11 = -1 THEN v# = v4#
IF test11 = -1 THEN d# = d# + 2
IF test11 = -1 THEN GOTO skip:
test12 = (test0 = .5) AND (test4 = .5) AND (z = 1)
IF test12 = -1 THEN r# = (v# - (d# + 2))
IF test12 = -1 THEN v# = v4#

```
IF test12 = -1 THEN d# = d# + 2
IF test12 = -1 THEN GOTO skip:
skip:
COLOR 2, 0
PRINT , , r#; z; v#; d#; r# - store#
REM a$ = INKEY$
INPUT a$
IF a$ = "s" THEN STOP
REM IF d# > 100 THEN d# = 0
LOOP
```


```
"C:\Users\Reactor1967\vcns\work\81002.BAS"
d# = 1
v# = 100
r# = v# - d#
CLS
RANDOMIZE TIMER
DO
z = INT(RND * 2) + 0
IF z = 0 THEN v# = v# + (v# - r# + 1)
IF z = 1 THEN v# = v# + (v# - r# + 2)
COLOR 2, 0
PRINT , , r#; v#; d#
INPUT a$
IF a$ = "d" THEN EXIT DO
IF a$ = "s" THEN STOP
d# = d# + 1
r# = v# - d#
LOOP
decode:
PRINT "----------------------"
DO
dist# = v# - r#
IF (dist# / 2) - INT(dist# / 2) = .5 THEN dist# = dist# - 1
dist# = dist# / 2
dist# = dist# + 1
v# = v# - dist#
d# = d# - 1
r# = r# - d#
COLOR 2, 0
PRINT , , r#; v#; d#
INPUT a$
IF a$ = "s" THEN STOP
LOOP
```

```
"C:\Users\Reactor1967\vcns\work\12701.BAS"
p# = 1
CLS
RANDOMIZE TIMER
p# = 1
r# = 1
repeat:
r# = p#
count = 0
DO
c = INT(RND * 2) + 0
IF c = 0 THEN p# = p# + (p# - r# + 1)
IF c = 1 THEN p# = p# + (p# - r# + 2)
PRINT p#; p# - r#
count = count + 1
LOOP UNTIL p# + (p# - r# + 2) >= 999999999999999#
REM PRINT "----------------------------------------------"
REM PRINT "                              "; r#; count
a$ = INKEY$
REM INPUT a$
IF a$ = "s" THEN STOP
r# = p#
count = 0
DO
c = INT(RND * 2) + 0
IF c = 0 THEN p# = p# - (r# - p# + 1)
IF c = 1 THEN p# = p# - (r# - p# + 2)
PRINT p#; r# - p#
count = count + 1
REM INPUT a$
LOOP UNTIL p# - (r# - p# + 2) <= 0
REM PRINT "----------------------------------------------"
REM PRINT "                              "; r#; count
a$ = INKEY$
REM INPUT a$
IF a$ = "s" THEN STOP
GOTO repeat:


"C:\Users\Reactor1967\vcns\work\010703G.BAS"
v# = 6
```

```
dist# = 1
CLS
RANDOMIZE TIMER
DO
z = INT(RND * 2) + 0
test = (v# / 2) - INT(v# / 2)
test1 = (test = 0) AND (z = 0)
test2 = (test = 0) AND (z = 1)
test3 = (test = .5) AND (z = 0)
test4 = (test = .5) AND (z = 1)
IF test1 = -1 THEN dist# = dist# + 2
IF test2 = -1 THEN dist# = dist# + 1
IF test3 = -1 THEN dist# = dist# + 1
IF test4 = -1 THEN dist# = dist# + 2
IF z = 0 THEN v# = v# + (v# - r# + 1)
IF z = 1 THEN v# = v# + (v# - r# + 2)
x# = (v# - r#)
IF (v# / 2) - INT(v# / 2) = 0 THEN z2 = 0
IF (v# / 2) - INT(v# / 2) = .5 THEN z2 = 1
IF z2 = 0 THEN x# = x# - 1
IF z2 = 1 THEN x# = x# - 2
x# = x# / 2
PRINT , , z; v#; r#; x#; v# - r#
r# = v# - dist#
a$ = INKEY$
IF a$ = "s" THEN STOP
LOOP
```

"C:\Users\Reactor1967\vcns\work\010703F.BAS"
DECLARE SUB distgen (dist#, z!, v#)
REM Urekia, you have a distance bank with all your good
REM distances lined up in numerical order than you can
REM use. Just decide mathmatically how to use them.
REM --
REM No, what is needed here is how to count the distance
REM after coding so that if v# - r# after coding = 2501
REM which is a binary zero in this program and you want to
REM go to a binary one then the distance would be even so
REM you add 1 to 2501 = 2502 and test it for a viable distance.
REM if it is not add two and test again. when decoding be able to
REM do the same in revers. We are in essence counting up numerically
REM and counting down numerically in a more complex fashion. I know
REM that if I want to go from a binary zero to a binary zero we add
REM two to the distance then test it. If I want to go from a binary

```
REM 1 to a binary zero or vice versa we add one to the distance and
REM test it. When we decode we will know that we are going from a
REM binary zero to a binary one or vice versa and we will subtract
REM one from the distance and test it then if it does not meet
REM the critiea we subtract two then test it. That where I am.
REM ----------------------------------------------------------
REM                    How to count distance
REM rule 1: Odd distances represent even binary 0 numbers.
REM        even distances represent odd binary 1 numbers.
REM rule 2: A. even number to even number add 2 to distance
REM        b. even number to odd number add 1 to distance
REM        c. odd number to odd number add two to distance.
REM        d. odd number to even number add 1 to diantance.
REM rule 3: subtract N from distance and divide by two.
REM        result should be even if previous distance is even
REM        or odd if previous distance is odd.
REM rule 4: be able to do this in reverse when decoding.
REM ----------------------------------------------------------
v# = 6
dist# = 1
CLS
RANDOMIZE TIMER
DO
z = INT(RND * 2) + 0
dist# = dist# + 1
r# = (v# - dist#)
IF (r# / 2) - INT(r# / 2) = 0 THEN dist# = dist# + 1
r# = (v# - dist#)
IF z = 0 THEN v# = v# + (v# - r# + 1)
IF z = 1 THEN v# = v# + (v# - r# + 2)
PRINT , z; v#; r#; dist#; v# - r#
a$ = INKEY$
REM INPUT a$
IF a$ = "s" THEN STOP
LOOP




"C:\Users\Reactor1967\vcns\work\010703E.BAS"
DECLARE SUB distgen (dist#, z!, v#)
REM Urekia, you have a distance bank with all your good
REM distances lined up in numerical order than you can
REM use. Just decide mathmatically how to use them.
REM ----------------------------------------------------------
REM No, what is needed here is how to count the distance
REM after coding so that if v# - r# after coding = 2501
```

REM which is a binary zero in this program and you want to
REM go to a binary one then the distance would be even so
REM you add 1 to 2501 = 2502 and test it for a viable distance.
REM if it is not add two and test again. when decoding be able to
REM do the same in revers. We are in essence counting up numerically
REM and counting down numerically in a more complex fashion. I know
REM that if I want to go from a binary zero to a binary zero we add
REM two to the distance then test it. If I want to go from a binary
REM 1 to a binary zero or vice versa we add one to the distance and
REM test it. When we decode we will know that we are going from a
REM binary zero to a binary one or vice versa and we will subtract
REM one from the distance and test it then if it does not meet
REM the critiea we subtract two then test it. That where I am.
REM --
REM   How to count distance
REM rule 1: Odd distances represent even binary 0 numbers.
REM   even distances represent odd binary 1 numbers.
REM rule 2: A. even number to even number add 2 to distance
REM   b. even number to odd number add 1 to distance
REM   c. odd number to odd number add two to distance.
REM   d. odd number to even number add 1 to diantance.
REM rule 3: subtract N from distance and divide by two.
REM   result should be even if previous/current distance is even
REM   or odd if previous/current distance is odd.
REM rule 4: be able to do this in reverse when decoding.
REM --
v# = 6
r# = 1
dist# = 5
CLS
RANDOMIZE TIMER
DO
z = INT(RND * 2) + 0
test = (v# / 2) - INT(v# / 2)
IF test = 0 THEN z2 = 0
IF test = .5 THEN z2 = 1
DO
test1 = (z = 0) AND (z2 = 0)
test2 = (z = 0) AND (z2 = 1)
test3 = (z = 1) AND (z2 = 0)
test4 = (z = 1) AND (z2 = 1)
IF test1 = -1 THEN add# = 2
IF test2 = -1 THEN add# = 1
IF test3 = -1 THEN add# = 1
IF test4 = -1 THEN add# = 2
dist# = dist# + add#

```
redo:
x# = dist#
IF z = 0 THEN x# = x# - 1
IF z = 1 THEN x# = x# - 2
x# = x# / 2
IF (x# / 2) - INT(x# / 2) = 0 THEN z3 = 0
IF (x# / 2) - INT(x# / 2) = .5 THEN z3 = 1
IF z3 = z2 THEN EXIT DO
dist# = dist# + 2
a$ = INKEY$
IF a$ = "s" THEN STOP
GOTO redo:
a$ = INKEY$
IF a$ = "s" THEN STOP
LOOP
x# = dist#
IF z = 0 THEN x# = x# - 1
IF z = 1 THEN x# = x# - 2
x# = x# / 2
r# = v# - x#
IF z = 0 THEN v# = v# + (v# - r# + 1)
IF z = 1 THEN v# = v# + (v# - r# + 2)
PRINT , , z; v#; r#; dist#; v# - r#
a$ = INKEY$
IF a$ = "s" THEN STOP
LOOP

SUB distgen (dist#, z, v#)
test = (v# / 2) - INT(v# / 2)
test1 = (test = 0) AND (z = 0)
test2 = (test = 0) AND (z = 1)
test3 = (test = .5) AND (z = 0)
test4 = (test = .5) AND (z = 1)
IF test1 = -1 THEN dist# = dist# + 2
IF test2 = -1 THEN dist# = dist# + 1
IF test3 = -1 THEN dist# = dist# + 1
IF test4 = -1 THEN dist# = dist# + 2
END SUB




"C:\Users\Reactor1967\vcns\work\010703D.BAS"
v# = 1
r# = 1
CLS
DO
```

```
x1# = v# + (v# - r# + 1)
x2# = v# + (v# - r# + 2)
PRINT , , v#; x1#; x2#; x1# - r#; x2# - r#
a$ = INKEY$
INPUT a$
IF a$ = "s" THEN STOP
v# = v# + 1
LOOP
```

"C:\Users\Reactor1967\vcns\work\010703C.BAS"
DECLARE SUB distgen (dist#, z!, v#)
REM Urekia, you have a distance bank with all your good
REM distances lined up in numerical order than you can
REM use. Just decide mathmatically how to use them.
REM --
REM No, what is needed here is how to count the distance
REM after coding so that if v# - r# after coding = 2501
REM which is a binary zero in this program and you want to
REM go to a binary one then the distance would be even so
REM you add 1 to 2501 = 2502 and test it for a viable distance.
REM if it is not add two and test again. when decoding be able to
REM do the same in revers. We are in essence counting up numerically
REM and counting down numerically in a more complex fashion. I know
REM that if I want to go from a binary zero to a binary zero we add
REM two to the distance then test it. If I want to go from a binary
REM 1 to a binary zero or vice versa we add one to the distance and
REM test it. When we decode we will know that we are going from a
REM binary zero to a binary one or vice versa and we will subtract
REM one from the distance and test it then if it does not meet
REM the critiea we subtract two then test it. That where I am.
REM ---
REM   How to count distance
REM rule 1: Odd distances represent even binary 0 numbers.
REM   even distances represent odd binary 1 numbers.
REM rule 2: A. even number to even number add 2 to distance
REM   b. even number to odd number add 1 to distance
REM   c. odd number to odd number add two to distance.
REM   d. odd number to even number add 1 to diantance.
REM rule 3: subtract N from distance and divide by two.
REM   result should be even if previous distance is even
REM   or odd if previous distance is odd.
REM rule 4: be able to do this in reverse when decoding.
REM ---

```
v# = 2
r# = 1
dist# = 1
CLS
RANDOMIZE TIMER
DO
z = INT(RND * 2) + 0
CALL distgen(dist#, z, v#)
IF z = 0 THEN v# = v# + (v# - r# + 1)
IF z = 1 THEN v# = v# + (v# - r# + 2)
PRINT , z; v#; r#; dist#; v# - r#
r# = v# - dist#
a$ = INKEY$
REM INPUT a$
IF a$ = "s" THEN STOP
LOOP

SUB distgen (dist#, z, v#)
test = (v# / 2) - INT(v# / 2)
test1 = (test = 0) AND (z = 0)
test2 = (test = 0) AND (z = 1)
test3 = (test = .5) AND (z = 0)
test4 = (test = .5) AND (z = 1)
IF test1 = -1 THEN dist# = dist# + 2
IF test2 = -1 THEN dist# = dist# + 1
IF test3 = -1 THEN dist# = dist# + 1
IF test4 = -1 THEN dist# = dist# + 2
END SUB




"C:\Users\Reactor1967\vcns\work\010703B.BAS"


"C:\Users\Reactor1967\vcns\work\010703A.BAS"
v# = 10
r# = 1
dist# = 0
CLS
RANDOMIZE TIMER
DO
z = INT(RND * 2) + 0
dist# = dist# + 2
r# = v# - dist#
IF z = 0 THEN v# = v# + (v# - r# + 2)
IF z = 1 THEN v# = v# + (v# - r# + 4)
```

```
PRINT , , z; v#; r#; dist#; v# - r#; (v# - r#) - store#
store# = v# - r#
a$ = INKEY$
IF a$ = "s" THEN STOP
LOOP




"C:\Users\Reactor1967\vcns\work\010604A.BAS"
REM The program codes both data and system instructions and is decode-able.
REM So what I have learned is that my data has to go into my very first
REM r range because the vectors(v#) there will not repeat with different
REM r values. The rest of the vectors(v#) code a binary one when it flips
REM a r range and a binary 0 when it does not. Now, what I did to do is
REM have this program flip and code down in value either using a control
REM code only or a mirror image of this program except in reverse. What
REM is left to work out is my age old question(WHEN DO I FLIP THE
DAMM THING.)
REM Since I am using key files here maybe I can use base three and code
REM for data, system instructions, and one value for flip. My problem with
REM base3 was I can not flip my r's neatly to the next range but using
REM key files telling me what my coding r's are for my coding vectors
REM and telling me what my decoding r's are for my decode vectors fixes
REM all my problems. I can't go as fast or as far I have to stay within
REM the ranges of my key files but my problems are fixed. And I have to
REM remember what my goals are here and that is just to be able to code
REM and decode raw data with the systems I have developed with vcns.
REM I need to point out that decoding here did not need my key files
REM but when using higer bases a key file will be needed for decode.
REM  A key file is "The list of r# values that are used for coding or
REM             decoding specific v# values." This file can also contain
REM             0ther information such as N values or what ever else
REM             is needed.
REM Also, it is possible to use mathmathics, equations, alogrithms(misspelled)
REM to tell what r'values go with v values during coding and decoding but
REM im not einstein im just trying to accomplish a specific task.
v# = 1000
r# = 1000
CLS
a$ = "c:\key1.txt"
RANDOMIZE TIMER
DO
CLOSE #1
OPEN a$ FOR INPUT AS #1
store = r#
DO
```

```
INPUT #1, a#, b#
test1 = (a# = v#)
IF test1 = -1 THEN EXIT DO
LOOP UNTIL EOF(1)
CLOSE #1
IF test1 = 0 THEN STOP
r# = b#
test = (v# > 1099) AND (r# <> store)
IF test = -1 THEN z = 1 ELSE z = 0
IF v# <= 1099 THEN z = INT(RND * 2) + 0
IF z = 0 THEN v# = v# + (v# - r# + 1)
IF z = 1 THEN v# = v# + (v# - r# + 2)
m# = (v# / 4) - INT(v# / 4): m# = m# * 10: m# = INT(m#)
PRINT , , z; m#; v#; r#; v# - r#; test
z$ = INKEY$
REM INPUT z$
IF z$ = "s" THEN CLOSE #1
IF z$ = "s" THEN STOP
LOOP UNTIL v# >= 9950
PRINT "------------------------------------------------------"
INPUT "Ready for decode"; floyd$
DO
m# = (v# / 4) - INT(v# / 4): m# = m# * 10: m# = INT(m#)
IF m# = 0 THEN z = 1
IF m# = 5 THEN z = 1
IF m# = 2 THEN z = 0
IF m# = 7 THEN z = 0
PRINT , , z; m#; v#; r#; v# - r#; test
z$ = INKEY$
REM INPUT z$
IF z$ = "s" THEN STOP
test = ((m# = 0) OR (m# = 5))
dist# = (v# - r#)
IF (dist# / 2) - INT(dist# / 2) = .5 THEN dist# = dist# - 1
dist# = dist# / 2
dist# = dist# + 1
v# = v# - dist#
IF test = -1 THEN r# = r# - 100
IF r# < 1000 THEN r# = 1000
LOOP UNTIL v# <= 1000
m# = (v# / 4) - INT(v# / 4): m# = m# * 10: m# = INT(m#)
IF m# = 0 THEN z = 1
IF m# = 5 THEN z = 1
IF m# = 2 THEN z = 0
IF m# = 7 THEN z = 0
PRINT , , z; m#; v#; r#; v# - r#; test
```

cdv

```
"C:\Users\Reactor1967\vcns\work\010603J.BAS"
DECLARE SUB codevr (v#, r#, z!)
REM (THIS ALMOST WORKS EXCEPT i CAN,T FIND OUT WHAT THE
N IS BEFORE DECODING)
REM IF I COULD TELL N THIS THING HERE COULD AND WOULD
DECODE!!!.
REM N IS THE BINARY VALUE BEING DECODED.
REM Maybe try changing the equation.
REM problem here is R is going from even to odd and odd to even.
REM that causes stability problems. Its best to keep R just even or
REM just odd.
v# = 10
dist# = 2
CLS
RANDOMIZE TIMER
DO
z = INT(RND * 2) + 0
IF z = 0 THEN dist# = dist# + 2
IF z = 1 THEN dist# = dist# + 4
x# = dist#
IF z = 0 THEN x# = x# - 2
IF z = 1 THEN x# = x# - 4
x# = INT(x# / 2)
r# = v# - x#
REM This fixes the problem but creates another.......
REM IF (r# / 2) - INT(r# / 2) = .5 THEN r# = r# - 1
REM -------------------------------------------------
REM CALL codevr(v#, r#, z)
distb# = (v# - r#)
IF z = 0 THEN v# = v# + (v# - r# + 2)
IF z = 1 THEN v# = v# + (v# - r# + 4)
REM PRINT , (v# / 4); z; v#; r#; dist#; (r# + 2) / 4; (r# + 4) / 4
REM PRINT , z; v#; r#; distb#; dist# - distb#; dist#; v# - r#
PRINT , z; v#; r#; dist#; v# - r#
REM IF dist# <> (v# - r#) THEN STOP
a$ = INKEY$
REM INPUT a$
IF a$ = "s" THEN STOP
LOOP

SUB codevr (v#, r#, z)
x1# = r#
x2# = 0
```

```
x3# = 0
DO
x1# = x1# + 1
x2# = x1# + (x1# - r# + 2)
x3# = x1# + (x1# - r# + 4)
IF x1# = v# THEN EXIT DO
a$ = INKEY$
IF a$ = "s" THEN STOP
IF x1# > v# THEN PRINT "OUT OF RANGE ERROR!!!"
IF x1# > v# THEN STOP
LOOP
IF z = 0 THEN v# = x1#
IF z = 1 THEN v# = x2#
END SUB


"C:\Users\Reactor1967\vcns\work\012404E.BAS"
v# = 1

r# = 1

a$ = "c:\test.txt"

OPEN a$ FOR INPUT AS #1

INPUT #1, a#, b#, c#, d#, e#

CLS

RANDOMIZE TIMER

redo:

DO

z = INT(RND * 2) + 0

IF z = 0 THEN v# = v# + ((v# - r#) * 2) + 1

IF z = 1 THEN v# = v# + ((v# - r#) * 2) + 2

IF z = 2 THEN v# = v# + ((v# - r#) * 2) + 2

PRINT , , z; v#; r#; v# - r#

f$ = INKEY$

INPUT f$
```

```
IF f$ = "s" THEN STOP

LOOP UNTIL v# >= d#

CLOSE #1

OPEN a$ FOR INPUT AS #1

DO

INPUT #1, a#, b#, c#, d#, e#

test = (a# <= v#) AND (b# >= v#)

IF test = -1 THEN GOTO hello:

LOOP UNTIL EOF(1)

CLOSE #1

STOP

hello:

CLOSE #1

r# = e#

GOTO redo:
```

```
"C:\Users\Reactor1967\vcns\work\012404D.BAS"

REM encode decode key file creating program.

v# = 1

r# = 1

a$ = "c:\test.txt"

CLOSE

OPEN a$ FOR OUTPUT AS #1

CLOSE #1
```

```
CLS

redo:

a1# = v#

c1# = v# + ((v# - r#) * 2) + 3

FOR count = 1 TO 300

x# = v# + ((v# - r#) * 2) + 3

v# = v# + 1

NEXT count

b1# = (v# - 1)

d1# = x#

e1# = r#

OPEN a$ FOR APPEND AS #1

WRITE #1, a1#, b1#, c1#, d1#, e1#

CLOSE #1

v# = x# + 1

r# = v#

PRINT , , a1#; b1#; c1#; d1#; e1#

z$ = INKEY$

REM INPUT z$

IF z$ = "s" THEN CLOSE #1

IF z$ = "s" THEN STOP

GOTO redo:
```

"C:\Users\Reactor1967\vcns\work\012404C.BAS"

cdix

```
REM Your going to have to redo this:

REM a decoding file is fine but it needs to also be able to find r#

REM for each range of v# coding. A coding and decoding file may need

REM combined

REM When making the combined file each range of coding v# needs to be 1

REM higher than the last range of decoding v#.

v# = 1

r# = 1

RANDOMIZE TIMER

CLS

a$ = "c:\test.txt"

OPEN a$ FOR INPUT AS #1

FOR count = 1 TO 2

INPUT #1, a#, b#, c#

NEXT count

redo:

DO

z = INT(RND * 3) + 0

IF z = 0 THEN v# = v# + ((v# - r#) * 2) + 1

IF z = 1 THEN v# = v# + ((v# - r#) * 2) + 2

IF z = 2 THEN v# = v# + ((v# - r#) * 2) + 3

m# = (v# / 3) - INT(v# / 3)

m# = m# * 10

m# = INT(m#)

PRINT , , z; v#; r#; v# - r#

z$ = INKEY$
```

```
INPUT z$

IF z$ = "s" THEN STOP

LOOP UNTIL v# >= c#

r# = c#

INPUT #1, a#, b#, c#

IF v# > b# THEN STOP

GOTO redo:
```

```
"C:\Users\Reactor1967\vcns\work\012404B.BAS"

REM This program attempts to create a range coding & decoding program

REM spoke up in the prevous basic programs.

v# = 1

r# = 1

CLS

a$ = "c:\test.txt"

OPEN a$ FOR OUTPUT AS #1

CLOSE #1

low# = 1

high# = 300

redo:

CLOSE
```

```
low# = v# + ((v# - r#) * 2) + 1

DO

x# = v# + ((v# - r#) * 2) + 3

v# = v# + 1

PRINT v#; r#; x#

z$ = INKEY$

IF z$ = "s" THEN STOP

LOOP UNTIL v# >= high#

high# = x#

PRINT , , low#, high#, r#

CLOSE #1

OPEN a$ FOR APPEND AS #1

WRITE #1, low#, high#, r#

CLOSE #1

low# = high#

high# = low# + 300

v# = low#

r# = low#

INPUT "Ready y/n"; z$

IF z$ = "n" THEN z$ = "s"

IF z$ = "s" THEN STOP

GOTO redo:
```

"C:\Users\Reactor1967\vcns\work\012404A.BAS"

```
REM This is a range program. The ideal here is that each range of v#

REM as only 1 r# to code 1 and after coding each range of v# has only

REM 1 r# to decode with. So Code what you can before going out of rang#

REM then add to v# a specific amount that will put v# into the next

REM coding range. Now using base 3 to decode when ever we see a n=3 then we

REM know to subtract from v# look at its range to find its r# then start

REM decoding again.

REM when jumping from 1 coding range to the next v# = v# + x#

REM when jumping from 1 decoding range to the next v# = v# - x#

REM r# is all ways specific for each range.

REM v# - Coding range r1

REM v# - decoding range r1

REM v# - coding range r2

REM v# - decoding range r2

REM v# - coding range r3

REM v# - decoding range r3

REM each v# has only 1 coding range

REM each v# has only 1 decoding range

REM v# and a number you add to it to take it from its decoding range

REM to its next coding range when coding.

REM When decoding subtract this number v# in its coding range to

REM take it to its next coding range.

CLS

v# = 1

r# = 1
```

```
y# = 101

low# = 9999999

high# = 0

redo:

low# = v#

DO

x# = v# + ((v# - r#) * 2) + 1

IF x# > high# THEN high# = x#

v# = v# + 1

LOOP UNTIL v# >= y#

PRINT , , low#; high# + 2; r#; "Decoding Ranges"

z$ = INKEY$

REM INPUT z$

IF z$ = "s" THEN STOP

v# = high# + 3

r# = high# + 3

y# = v# + 300

GOTO redo:
```

```
"C:\Users\Reactor1967\vcns\work\012304B.BAS"

DECLARE SUB test (v#, r#)

v# = 1

r# = 1
```

```
a# = 1

b# = 1

CLS

RANDOMIZE TIMER

DO

z = INT(RND * 3) + 0

REM IF t = 1 THEN z = 2

sv# = v#

REM z = 2

IF z = 0 THEN v# = v# + ((v# - r#) * 2) + 1

IF z = 1 THEN v# = v# + ((v# - r#) * 2) + 2

IF z = 2 THEN v# = v# + ((v# - r#) * 2) + 3

m# = (v# / 3) - INT(v# / 3)

m# = m# * 10

m# = INT(m#)

x# = x# + ((v# - sv#) - (r# - sr#))

PRINT , , z; m#; v#; r#; count

a$ = INKEY$

IF a$ = "s" THEN STOP

IF a$ = "d" THEN GOTO here:

IF v# = a# THEN GOTO skip:

t = 0

sr# = r#

count = 0

DO

count = count + 1
```

```
a# = a# + 1

IF a# - b# >= 99 THEN b# = b# + 99

REM a$ = INKEY$

REM IF a$ = "s" THEN STOP

LOOP UNTIL v# = a#

skip:

r# = b#

LOOP

here:

PRINT "-------------------------------------------"

INPUT z$

IF z$ = "s" THEN STOP

DO

m# = (v# / 3) - INT(v# / 3)

m# = m# * 10

m# = INT(m#)

IF m# = 6 THEN z = 0

IF m# = 0 THEN z = 1

IF m# = 3 THEN z = 2

PRINT , , z; m#; v#; r#; v# - r#

a$ = INKEY$

INPUT a$

IF a$ = "s" THEN STOP

dist# = v# - r#

IF z = 0 THEN dist# = dist# - 1

IF z = 1 THEN dist# = dist# - 2
```

```
IF z = 2 THEN dist# = dist# - 3

dist# = dist# / 3

dist# = dist# * 2

IF z = 0 THEN dist# = dist# + 1

IF z = 1 THEN dist# = dist# + 2

IF z = 2 THEN dist# = dist# + 3

v# = v# - dist#

LOOP
```

"C:\Users\Reactor1967\vcns\work\012304A.BAS"

```
REM 1 = 2 3 4

REM 2 = 4 5 6

REM 3 = 7 8 9

REM 4 = 10 11 12

REM 5 = 6 7 8

REM 6 = 7 8 9

REM 7 = 10 11 12

REM 8 = 13 14 15

REM 9 = 10 11 12

REM the ideal here is that for each v# and r# v# has a n# value but

REM you need both v# & r# to tell what N is. Here a simple division trick

REM by v# or division by v# & r# does not work as with my previous
equations.

REM something more complicated needs to happen here. For each v# has a
```

REM specific N value depending on its r#. So v# can mean one N value with

REM one r# and the same v# could mean a different N value with a differnt

REM r#.

REM So that makes it possible for this to work here as long as the

REM process from switching r and v can be solved.

redo:

v# = 1

r# = 1

t = 0

CLS

RANDOMIZE TIMER

DO

z = INT(RND * 2) + 0

IF t = 1 THEN z = 2

v# = v# + (v# - r#) + z + 1

m1# = ((v# - r#) / 3) - INT((v# - r#) / 3): m1# = m1# * 10: m1# = INT(m1#)

m2# = (v# / 3) - INT(v# / 3): m2# = m2# * 10: m2# = INT(m2#)

PRINT , , z; m1#; m2#; v#; r#; v# - r#

a$ = INKEY$

REM INPUT a$

IF a$ = "s" THEN STOP

IF v# > (r# + 96) THEN t = 1 ELSE t = 0

IF v# > (r# + 96) THEN r# = r# + 99

LOOP

"C:\Users\Reactor1967\vcns\work\012204B.BAS"

REM before coding take the data and find the x value which is

REM x = x + ((v - sv) - (r -sr)) in reverse our how it would look decoding.

REM then use that to plot the speed of r while coding the data. X needs a

REM specific coding pattern that can be duplicated while decoding so

REM planning it in reverse before coding seems logical and like it might

REM work.

v# = 1

r# = 1

CLS

RANDOMIZE TIMER

DO

z = INT(RND * 3) + 0

sv# = v#

v# = v# + ((v# - r#) * 2) + z + 1

m# = (v# / 3) - INT(v# / 3)

m# = m# * 10

m# = INT(m#)

x# = x# + ((v# - sv#) - (r# - sr#))

PRINT , , z; m#; v#; r#; v# - r#; v# - sv#; r# - sr#; x#

a$ = INKEY$

IF a$ = "s" THEN STOP

sr# = r#

r# = r# + ((INT((v# - r#) / 3)) * 3)

LOOP

"C:\Users\Reactor1967\vcns\work\012204A.BAS"

REM Here are my list of options on this day as I can remember them.

REM 1. control x# = x# + ((v# - sv#) - (r# - sr#)) while coding in a

REM   specific pattern and reverse the pattern when decoding.

REM 2. Assign each v# a specific speed of r# but this is complicated.

REM 3. Have a control coding with intermediate coding of data.

REM 4. Develope more equations/systems/algorithms.

REM 5. Key files for coding and decoding.

REM 6. Saving partial information to disk and splitting it up as with

REM   encryption but the full advantages are lost here but its good

REM   encryption.

REM Write a book and use the proceeds to file for patent again this is not

REM in full working order I would be relying on others to finish my work

REM and without a working system the book may not sell but it would be of

REM interest to a lot of people.

REM --

REM x# = x# + ((v# - sv#) - (r# - sr#)) is a good tool variable to use

REM to monitor your coding. With the right equation/alogrityme/system

REM it good be used to control vcns.

v# = 1

r# = 1

x# = 0

CLS

```
RANDOMIZE TIMER

DO

z = INT(RND * 2) + 0

sv# = v#

IF z = 0 THEN v# = v# + ((v# - r#) * 2) + 1

IF z = 1 THEN v# = v# + ((v# - r#) * 2) + 2

IF z = 2 THEN v# = v# + ((v# - r#) * 2) + 3

x# = x# + ((v# - sv#) - (r# - sr#))

REM LOCATE 12, 30

REM PRINT z; v#; r#; v# - r#

REM LOCATE 13, 5

PRINT "|Speed of v#|"; v# - sv#; "|Speed of r#|"; r# - sr#; "|x# = x# + sv - sr|";
x#; "|sv - sr|"; (v# - sv#) - (r# - sr#); "|"

sr# = r#

a$ = INKEY$

REM INPUT a$

IF a$ = "s" THEN STOP

r# = r# + ((INT((v# - r#) / 3)) * 3)

LOOP




"C:\Users\Reactor1967\vcns\work\012104F.BAS"

DIM sr(9999)

CLS

RANDOMIZE TIMER
```

```
GOTO outcome:

redo:

v# = 1000

r# = 1000

DO

z = INT(RND * 2) + 0

IF lb = 1 THEN z = 2

IF z = 0 THEN v# = v# + ((v# - r#) * 2) + 1

IF z = 1 THEN v# = v# + ((v# - r#) * 2) + 2

IF z = 2 THEN v# = v# + ((v# - r#) * 2) + 3

m# = (v# / 3) - INT(v# / 3)

m# = m# * 10

m# = INT(m#)

PRINT , , z; m#; v#; r#; v# - r#

a$ = INKEY$

IF a$ = "s" THEN STOP

IF a$ = "d" THEN GOTO outcome:

dist# = v# - r#

dist# = INT(dist# / 3)

dist# = dist# * 3

IF v# > 9999 THEN EXIT DO

sr(v#) = dist#

lb = 0

test = (v# - r#) > 99

IF test = -1 THEN lb = 1

IF lb = 1 THEN r# = r# + dist#
```

```
LOOP UNTIL v# > 9999

GOTO redo:

s$ = "c:\test.txt"

OPEN s$ FOR OUTPUT AS #1

FOR count = 1 TO 10000

IF count > 9999 THEN EXIT FOR

WRITE #1, count, sr(count)

NEXT count

CLOSE #1

outcome:

v# = 1000

r# = 1000

s$ = "c:\test.txt"

OPEN s$ FOR INPUT AS #1

count = 0

DO

count = count + 1

INPUT #1, a#, b#

sr(count) = b#

LOOP UNTIL EOF(1)

DO

z = INT(RND * 2) + 0

test = ((v# + ((v# - r#) * 2) + z + 1) - r#) > 99

IF v# > 9999 THEN EXIT DO

IF test = -1 THEN r# = r# + sr(v#)

IF test = -1 THEN z = 2
```

```
IF z = 0 THEN v# = v# + ((v# - r#) * 2) + 1

IF z = 1 THEN v# = v# + ((v# - r#) * 2) + 2

IF z = 2 THEN v# = v# + ((v# - r#) * 2) + 3

m# = (v# / 3) - INT(v# / 3)

m# = m# * 10

m# = INT(m#)

PRINT , , z; m#; v#; r#; v# - r#

a$ = INKEY$

IF a$ = "s" THEN STOP

LOOP
```

"C:\Users\Reactor1967\vcns\work\012104E.BAS"

```
REM !!!!!!!!!!!!!!!!!!!!!!!!!! TEST SUCCESSFUL !!!!!!!!!!!!!!!!!!!!!!!!

REM Each v# as a assigned speed of r# but it may or may not be used.

REM when that speed of r# is used its coded into the next v# a system

REM instruction. So when decoding the program sees this instruction as

REM a base 3 number. Then it decodes to the previous v#. Gets the speed

REM of r# for that v# and decrements r# by that speed and keeps decoding.

REM ------------------------------------------------------------------------

DIM sr(9999)

REM v# is the common variable that is always known in this system so

REM within v# is the answer to the speed of r#. So with each v# has to
```

REM be assigned some system/number to tell what the speed of r# should be if

REM you wish to jump r# at that v#. So, something needs to be assigned

REM to each v# that works in a system to increment r# so when decoding

REM one look at v# tells you what your something was you used to inrement

REM v# and you use that something to derement v# and keep decoding.

REM LET THIS PROGRAM RUN A WHILE FOR GOING TO THE NEXT TEST!!!!

RANDOMIZE TIMER

CLS

redo:

V# = 1000

R# = 1000

lb = 0

LOW# = 999999999999999#

high# = 0

DO

z = INT(RND * 2) + 0

IF lb = 1 THEN z = 2

IF z = 0 THEN V# = V# + ((V# - R#) * 2) + 1

IF z = 1 THEN V# = V# + ((V# - R#) * 2) + 2

IF z = 2 THEN V# = V# + ((V# - R#) * 2) + 3

m# = (V# / 3) - INT(V# / 3)

m# = m# * 10

m# = INT(m#)

PRINT , , z; m#; V#; R#; V# - R#; lb; LOW#; high#

IF V# - R# < LOW# THEN LOW# = V# - R#

```
IF V# - R# > high# THEN high# = V# - R#

a$ = INKEY$

REM INPUT a$

IF a$ = "s" THEN STOP

IF a$ = "d" THEN GOTO outcome:

lb = 0

store# = R#

IF V# - R# > 99 THEN lb = 1

IF lb = 1 THEN dist# = INT((V# - R#) / 3)

IF lb = 1 THEN dist# = dist# * 3

IF lb = 1 THEN R# = R# + dist#

IF V# > 9999 THEN EXIT DO

sr(V#) = R# - store#

REM IF sr(v#) <= 0 THEN sr(v#) = r# - store#

LOOP UNTIL V# > 9999

GOTO redo:

GOTO outcome:

REM outcome:

CLS

FOR count = 1 TO 8999

PRINT , , count, sr(count)

INPUT a$

IF a$ = "s" THEN STOP

IF a$ = "d" THEN EXIT FOR

NEXT count

outcome:
```

```basic
PRINT "-------------------------------------------------"
CLS
V# = 1000
R# = 1000
DO
z = INT(RND * 2) + 0
IF lb = 1 THEN z = 2
IF z = 0 THEN V# = V# + ((V# - R#) * 2) + 1
IF z = 1 THEN V# = V# + ((V# - R#) * 2) + 2
IF z = 2 THEN V# = V# + ((V# - R#) * 2) + 3
m# = (V# / 3) - INT(V# / 3)
m# = m# * 10
m# = INT(m#)
PRINT , , z; m#; V#; R#; V# - R#; lb
IF V# - R# > high# THEN PRINT "OUCH HIGH"
IF V# - R# < LOW# THEN PRINT "OUCH LOW"
a$ = INKEY$
REM INPUT a$
IF a$ = "s" THEN STOP
lb = 0
IF V# > 9999 THEN EXIT DO
test = (sr(V#) > 0) AND (R# + sr(V#) <= V#)
IF test = -1 THEN lb = 1
IF lb = 1 THEN R# = R# + sr(V#)
LOOP UNTIL V# > 9999
```

"C:\Users\Reactor1967\vcns\work\012104D.BAS"

REM !!!!!!!!!!!!!!!!!!!!!!!!!! TEST SUCCESSFUL !!!!!!!!!!!!!!!!!!!!!!!!!

REM Each v# as a assigned speed of r# but it may or may not be used.

REM when that speed of r# is used its coded into the next v# a system

REM instruction. So when decoding the program sees this instruction as

REM a base 3 number. Then it decodes to the previous v#. Gets the speed

REM of r# for that v# and decrements r# by that speed and keeps decoding.

REM ---

DIM sr(2000)

REM v# is the common variable that is always known in this system so

REM within v# is the answer to the speed of r#. So with each v# has to

REM be assigned some system/number to tell what the speed of r# should be if

REM you wish to jump r# at that v#. So, something needs to be assigned

REM to each v# that works in a system to increment r# so when decoding

REM one look at v# tells you what your something was you used to inrement

REM v# and you use that something to derement v# and keep decoding.

REM LET THIS PROGRAM RUN A WHILE FOR GOING TO THE NEXT TEST!!!!

RANDOMIZE TIMER

CLS

redo:

v# = 1

r# = 1

lb = 0

```basic
DO

z = INT(RND * 2) + 0

IF lb = 1 THEN z = 2

IF z = 0 THEN v# = v# + ((v# - r#) * 2) + 1

IF z = 1 THEN v# = v# + ((v# - r#) * 2) + 2

IF z = 2 THEN v# = v# + ((v# - r#) * 2) + 3

m# = (v# / 3) - INT(v# / 3)

m# = m# * 10

m# = INT(m#)

PRINT , , z; m#; v#; r#; v# - r#; lb

a$ = INKEY$

REM INPUT a$

IF a$ = "s" THEN STOP

IF a$ = "d" THEN GOTO outcome:

lb = 0

store# = r#

IF v# - r# > 99 THEN lb = 1

IF lb = 1 THEN dist# = INT((v# - r#) / 3)

IF lb = 1 THEN dist# = dist# * 3

IF lb = 1 THEN r# = r# + dist#

IF sr(v#) <= 0 THEN sr(v#) = r# - store#

LOOP UNTIL v# > 1000

GOTO redo:

outcome:

CLS

FOR count = 1 TO 1000
```

```
PRINT , , count, sr(count)

INPUT a$

IF a$ = "s" THEN STOP

IF a$ = "d" THEN EXIT FOR

NEXT count

v# = 1

r# = 1

DO

z = INT(RND * 2) + 0

IF lb = 1 THEN z = 2

IF z = 0 THEN v# = v# + ((v# - r#) * 2) + 1

IF z = 1 THEN v# = v# + ((v# - r#) * 2) + 2

IF z = 2 THEN v# = v# + ((v# - r#) * 2) + 3

m# = (v# / 3) - INT(v# / 3)

m# = m# * 10

m# = INT(m#)

PRINT , , z; m#; v#; r#; v# - r#; lb

a$ = INKEY$

REM INPUT a$

IF a$ = "s" THEN STOP

lb = 0

test = sr(v#) > 0 AND (r# + sr(v#) <= v#)

IF test = -1 THEN lb = 1

IF lb = 1 THEN r# = r# + sr(v#)

LOOP UNTIL v# > 1000
```

"C:\Users\Reactor1967\vcns\work\012104C.BAS"

REM question? When decoding will v1#,v2#,and r2# always have the same r1#?

v# = 1

r# = 1

lb = 0

RANDOMIZE TIMER

CLS

DO

z = INT(RND * 2) + 0

IF lb = 1 THEN z = 2

IF z = 0 THEN v# = v# + ((v# - r#) * 2) + 1

IF z = 1 THEN v# = v# + ((v# - r#) * 2) + 2

IF z = 2 THEN v# = v# + ((v# - r#) * 2) + 3

m# = (v# / 3) - INT(v# / 3)

m# = m# * 10

m# = INT(m#)

PRINT , , z; m#; v#; r#; v# - r#; lb

a$ = INKEY$

INPUT a$

IF a$ = "s" THEN STOP

lb = 0

IF v# - r# <= 99 THEN lb = 0 ELSE lb = 1

IF lb = 1 THEN dist# = v# - r#

```
IF lb = 1 THEN dist# = INT((v# - r#) / 3)

IF lb = 1 THEN dist# = dist# * 3

IF lb = 1 THEN r# = r# + dist#

LOOP
```

```
"C:\Users\Reactor1967\vcns\work\012104B.BAS"

v# = 1

r# = 1

a$ = "c:\test.txt"

DO

OPEN a$ FOR APPEND AS #1

WRITE #1, v#, r#

PRINT , , v#; r#

CLOSE #1

v# = v# + 1

IF v# - r# <= 99 THEN GOTO skip:

DO

IF r# + 3 > v# THEN EXIT DO

r# = r# + v#

LOOP

skip:

LOOP UNTIL v# > 1000
```

```
"C:\Users\Reactor1967\vcns\work\012104A.BAS"

REM when v# - r# is out of range then r# = r# + v# + 1

REM so if every v# is to have only 1 r# to code with then we need something

REM like this r# = r# + a# + b# where a# can be anything concerving v#

REM and b# can be any calculated off set to balance everything.

v# = 1

r# = 1

x# = 1

low = 999999999999999#

high = 0

lb = 0

CLS

DO

z = INT(RND * 2) + 0

IF lb = 1 THEN z = 2

IF z = 0 THEN v# = v# + ((v# - r#) * 2) + 1

IF z = 1 THEN v# = v# + ((v# - r#) * 2) + 2

IF z = 2 THEN v# = v# + ((v# - r#) * 2) + 3

m# = (v# / 3) - INT(v# / 3)

m# = m# * 10

m# = INT(m#)

PRINT , , z; m#; v#; r#; v# - r#; low; high

IF v# - r# < low THEN low = v# - r#

IF v# - r# > high THEN high = v# - r#
```

```
REM IF v# - r# = 5 THEN INPUT f$

REM IF f$ = "s" THEN STOP

a$ = INKEY$

REM INPUT a$

IF a$ = "s" THEN STOP

lb = 0

IF v# - r# <= 99 THEN GOTO skip:

DO

IF x# + 3 > v# THEN EXIT DO

x# = x# + 3

LOOP

r# = x#: lb = 1

skip:

LOOP
```

```
"C:\Users\Reactor1967\vcns\work\012103B.BAS"

v# = 1

r# = 1

lb = 0

CLS

DO

z = INT(RND * 2) + 0

IF lb = 1 THEN z = 2

IF z = 0 THEN v# = v# + ((v# - r#) * 2) + 1
```

```
IF z = 1 THEN v# = v# + ((v# - r#) * 2) + 2
IF z = 2 THEN v# = v# + ((v# - r#) * 2) + 3
m# = (v# / 3) - INT(v# / 3)
m# = m# * 10
m# = INT(m#)
PRINT , , z; m#; v#; r#; v# - r#; low; high
a$ = INKEY$
REM INPUT a$
IF a$ = "s" THEN STOP
lb = 0
IF v# - r# <= 99 THEN lb = 0 ELSE lb = 1
IF lb = 1 THEN dist# = v# - r#
IF lb = 1 THEN dist# = INT((v# - r#) / 3)
IF lb = 1 THEN dist# = dist# * 3
IF lb = 1 THEN r# = r# + dist#
LOOP




"C:\Users\Reactor1967\vcns\work\011804A.BAS"
REM In the previous programs why not tune to the last variable as r# instead
REM of the fist since it is the most steady. That would make sense.
v# = 1
r# = 1
z = 1
CLS
```

```
RANDOMIZE TIMER

DO

REM z = INT(RND * 2) + 0

REM IF z = 0 THEN v# = v# + ((v# - r#) * 2) + 1

REM IF z = 1 THEN v# = v# + ((v# - r#) * 2) + 2

v# = v# + ((v# - r#) * 2) + 3

m# = (v# / 3) - INT(v# / 3)

m# = m# * 10

m# = INT(m#)

PRINT , , z; m#; v#; r#; v# - r#; count

a$ = INKEY$

INPUT a$

IF a$ = "s" THEN STOP

count = 0

DO

count = count + 1

IF (r# + 3) <= v# THEN r# = r# + 3

a$ = INKEY$

IF a$ = "s" THEN STOP

LOOP UNTIL (r# + 3) > v#

IF count = 1 THEN z = 0

IF count = 2 THEN z = 1

LOOP
```

```
REM sub-routine(Go with this.)

REM This is used as a utility for coding. read previous programs for more info.

REM create a specific number of v#'s in a specific order. See if this program

REM can decode with each v#. Also, coding still not right.

v# = 100

r1# = 1

r2# = 1

count = 1

z = 2

CLS

RANDOMIZE TIMER

redo:

DO

REM z = INT(RND * 2) + 0

x# = r1# + ((r1# - r2#) * 2) + 3

test = v# - x# < 10

IF test = -1 THEN z = 2

IF z = 0 THEN r1# = r1# + ((r1# - r2#) * 2) + 1

IF z = 1 THEN r1# = r1# + ((r1# - r2#) * 2) + 2

IF z = 2 THEN r1# = r1# + ((r1# - r2#) * 2) + 3

m# = ((r1# - 0) / 3) - INT((r1# - 0) / 3): m# = m# * 10: m# = INT(m#)

PRINT z; m#; r1#; r2#; r1# - r2#; count; r2# - sr#, v#

sr# = r2#

a$ = INKEY$

INPUT a$
```

```
IF a$ = "s" THEN STOP

count = 0

DO

count = count + 1

IF r2# + 3 <= r1# THEN r2# = r2# + 3

a$ = INKEY$

IF a$ = "s" THEN STOP

IF r2# > r1# THEN STOP

LOOP UNTIL r2# + 3 >= r1#

IF count = 1 THEN z = 0

IF count = 2 THEN z = 1

IF count = 3 THEN z = 2

LOOP UNTIL test = -1

REM STOP

v# = v# + INT(RND * 200) + 100

GOTO redo:
```

```
"C:\Users\Reactor1967\vcns\work\011704B.BAS"

v# = 1

r# = 1

CLS

count = 0

RANDOMIZE TIMER

redo:
```

```
z = 3

DO

REM z = INT(RND * 2) + 0

IF a$ = "d" THEN z = 3

IF z = 0 THEN v# = v# + ((v# - r#) * 2) + 1

IF z = 1 THEN v# = v# + ((v# - r#) * 2) + 2

IF z = 2 THEN v# = v# + ((v# - r#) * 2) + 3

m# = ((v# - 0) / 3) - INT((v# - 0) / 3): m# = m# * 10: m# = INT(m#)

PRINT , , z; m#; v#; r#; v# - r#; count

a$ = INKEY$

INPUT a$

IF a$ = "s" THEN STOP

count = 0

DO

count = count + 1

IF (r# + 3) <= v# THEN r# = r# + 3

REM a$ = INKEY$

REM INPUT a$

REM IF a$ = "s" THEN STOP

LOOP UNTIL r# + 3 > v#

IF count = 1 THEN z = 0

IF count = 2 THEN z = 1

IF count = 3 THEN z = 2

LOOP UNTIL a$ = "d"

m# = (v# / 3) - INT(v# / 3): m# = m# * 10: m# = INT(m#)

PRINT , , z; m#; v#; r#; v# - r#; count
```

```
INPUT z$

IF z$ = "s" THEN STOP

GOTO redo:
```

"C:\Users\Reactor1967\vcns\work\011704A.BAS"

```
REM Well This seems like its getting somewhere but it needs cleaned up

REM and tested. The ideal is to code in base3 with the left two most

REM numbers. When v# - r1# is out of range the right two most numbers

REM are used in base 3 to code v# - r1# back into range. the left two

REM numbers are for coding data and 1 system instruction and the right

REM two numbers are used for coding system instructions only.

DECLARE SUB adjust (v#, r1#, r2#, count!)

v# = 1

r1# = 1

r2# = 1

RANDOMIZE TIMER

CLS

DO

z = INT(RND * 2) + 0

IF ((v# + ((v# - r1#) * 2)) - r1#) >= 99 THEN z = 2

IF z = 0 THEN v# = v# + ((v# - r1#) * 2) + 1

IF z = 1 THEN v# = v# + ((v# - r1#) * 2) + 2

IF z = 2 THEN v# = v# + ((v# - r1#) * 2) + 3

m# = (v# / 3) - INT(v# / 3)
```

```
m# = m# * 10

m# = INT(m#)

PRINT , , z; m#; v#; r1#; r2#; v# - r1#; count

a$ = INKEY$

REM INPUT a$

IF a$ = "s" THEN STOP

IF v# - r1# >= 99 THEN CALL adjust(v#, r1#, r2#, count)

LOOP


SUB adjust (v#, r1#, r2#, count)

count = 1

z = 2

REM CLS

RANDOMIZE TIMER

DO

REM z = INT(RND * 2) + 0

IF z = 0 THEN r1# = r1# + ((r1# - r2#) * 2) + 1

IF z = 1 THEN r1# = r1# + ((r1# - r2#) * 2) + 2

IF z = 2 THEN r1# = r1# + ((r1# - r2#) * 2) + 3

m# = ((r1# - 0) / 3) - INT((r1# - 0) / 3): m# = m# * 10: m# = INT(m#)

REM PRINT z; m#; r1#; r2#; r1# - r2#; count; r2# - sr#, v#

sr# = r2#

a$ = INKEY$

REM INPUT a$

IF a$ = "s" THEN STOP

count = 0
```

```
DO

count = count + 1

IF r2# + 3 <= r1# THEN r2# = r2# + 3

a$ = INKEY$

IF a$ = "s" THEN STOP

IF r2# > r1# THEN STOP

LOOP UNTIL r2# + 3 >= r1#

IF count = 1 THEN z = 0

IF count = 2 THEN z = 1

IF count = 3 THEN z = 2

LOOP UNTIL v# - r1# < 10

END SUB
```

```
"C:\Users\Reactor1967\vcns\work\011604B.BAS"

REM This is a sub-routine or helper program I plan to use in my real world

REM coding program. You can use any base to code with but if depending

REM on the last value in the base as the exit instruction then that

REM instruction can only be used at the start and end of the program.

REM the rest of the values tell how much to increment or decrement r

REM by. R# goes through a count and N equals how many times r#

REM incrmented when coding and decreases when decoding. N tells you

REM how to code and decode.

v# = 1

r# = 1
```

```
count = 1

z = 2

CLS

RANDOMIZE TIMER

z = 2

DO

REM z = INT(RND * 2) + 0

IF z = 0 THEN v# = v# + ((v# - r#) * 2) + 1

IF z = 1 THEN v# = v# + ((v# - r#) * 2) + 2

IF z = 2 THEN v# = v# + ((v# - r#) * 2) + 3

m# = ((v# - 0) / 3) - INT((v# - 0) / 3): m# = m# * 10: m# = INT(m#)

PRINT , , z; m#; v#; r#; v# - r#; count; r# - sr#

sr# = r#

a$ = INKEY$

REM INPUT a$

IF a$ = "s" THEN STOP

IF a$ = "d" THEN GOTO ou:

count = 0

DO

count = count + 1

IF r# + 3 <= v# THEN r# = r# + 3

a$ = INKEY$

IF a$ = "s" THEN STOP

LOOP UNTIL r# + 3 >= v#

IF count = 1 THEN z = 0

IF count = 2 THEN z = 1
```

```
IF count = 3 THEN z = 2

LOOP

ou:

z = 2

IF z = 0 THEN v# = v# + ((v# - r#) * 2) + 1

IF z = 1 THEN v# = v# + ((v# - r#) * 2) + 2

IF z = 2 THEN v# = v# + ((v# - r#) * 2) + 3

m# = ((v# - 0) / 3) - INT((v# - 0) / 3): m# = m# * 10: m# = INT(m#)

PRINT , , z; m#; v#; r#; v# - r#; count; r# - sr#
```

```
"C:\Users\Reactor1967\vcns\work\011604A.BAS"

DECLARE SUB codeup (v#, r1#, r2#, count!)

REM I,ve done this before a while back. If codeup could work in base3 this

REM thing would be a cinch because codeup would code 0 & 1 when r2#
changed

REM and code 2 when it was time to exit back to the main code. Doing this

REM a decode would be possible.

v# = 1

r1# = 1

r2# = 1

RANDOMIZE TIMER

CLS

DO

z = INT(RND * 2) + 0
```

```
IF z = 0 THEN v# = v# + (v# - r1#) + 1

IF z = 1 THEN v# = v# + (v# - r1#) + 2

PRINT , , z; v#; r1#; r2#; v# - r1#; count

a$ = INKEY$

REM INPUT a$

IF a$ = "s" THEN STOP

IF v# - r1# > 4 THEN CALL codeup(v#, r1#, r2#, count)

LOOP


SUB codeup (v#, r1#, r2#, count)

count = 0

DO

count = count + 1

test2 = v# - r1# <= 4

IF test2 = -1 THEN EXIT DO

test = r1# >= (r2# + 3)

IF test = -1 THEN r2# = r2# + 4

IF test = 0 THEN r1# = r1# + (r1# - r2#) + 1

IF test = -1 THEN r1# = r1# + (r1# - r2#) + 2

a$ = INKEY$

IF a$ = "s" THEN STOP

LOOP

END SUB
```

```
v# = 1

r# = 1

RANDOMIZE TIMER

CLS

x# = 4

DO

z = INT(RND * 2) + 0

IF z = 0 THEN v# = v# + (v# - r#) + 1

 IF z = 1 THEN v# = v# + (v# - r#) + 2

m# = (v# / 4) - INT(v# / 4): m# = m# * 10: m# = INT(m#)

PRINT , z; m#; v#; r#; x#; v# - r#

a$ = INKEY$

REM INPUT a$

 IF a$ = "s" THEN STOP

IF a$ = "d" THEN EXIT DO

sr# = r#

test = v# >= (r# + (x# - 1))

IF test = -1 THEN x# = x# + 4

IF v# >= (r# + (x# - 1)) THEN r# = r# + x#

IF (r# - 1) > v# THEN STOP

LOOP
```

REM Well I have established the existance of the x value. The x value is the

REM range r looks for when it increases. Using the x value it can now be

REM possible to work like I have in other bases. With the x value a system

REM or equation(s) need to be establish to code the x value and decode the

REM x value as v# codes and decodes. Also, the x value can be used to code

REM data to store and system data by keep r open long enough to do both then

REM increase r#. Anyway its a step. I was working on another agenda but I

REM came across. Im working on how to read ahead on the data and create

REM a path sort of speak in coding that can be seen we decoding using the

REM known variables on a decode. The decode path. Its a path thats created

REM when coding thats readable when decoding. This path contains information

REM such as the speed of r and when to decrement and increment r. The known

REM variables are v2#, v2# - r2#, v2# - v1#, v1#, and misc variables which

REM may be added such as x#.

v# = 1

r# = 1

RANDOMIZE TIMER

CLS

x# = 4

DO

z = INT(RND * 2) + 0

IF z = 0 THEN v# = v# + (v# - r#) + 1

IF z = 1 THEN v# = v# + (v# - r#) + 2

m# = (v# / 4) - INT(v# / 4): m# = m# * 10: m# = INT(m#)

```
PRINT , , z; m#; v#; r#; x#

sr# = r#

a$ = INKEY$

REM INPUT a$

IF a$ = "s" THEN STOP

test = v# >= (r# + (x# - 1))

IF v# >= (r# + (x# - 1)) THEN r# = r# + x#

IF test = -1 THEN x# = x# + 4

IF (r# - 1) > v# THEN STOP

LOOP
```

"C:\Users\Reactor1967\vcns\work\011104A.BAS"

REM Starting with v# a equation, method or program needs to be written

REM that will bring r# to within range for coding to the next v# and

REM when ready to decode given v# the equation, medthod or program can

REM bring r# to within range for decode to the previous v#.

REM Maybe a key file can be made showing both the coding and decoding r#'s

REM So the ((v# - r#) after coding) = (((v# - r#) before coding) * base) + N)

REM And   ((v# - r#) before coding) = (((v# - r#) after coding) - N) / base)

REM v1# = v2# - ((v# - r#) after coding - (v# - r#) before coding)

REM --

REM There seems to be no relation from v2# - r1#.

REM Maybe creating some relationship so that r# or (v1# - r1#)

REM before coding has

```
REM some relationship with v2# - r1# after coding. This is a good approach
REM to try.
REM ------------------------------------------------------------
REM The only answer I can come up with right now to the above question
REM is a chart that takes the v1# - r1# after coding and gives the
REM v2# - r1# question. So the distance your going to code to has to
REM have some relationship to what v2# - r1# will be. You have to
REM know in advance what your going to do and how your going to do it'
REM which always seems to pop up as the only way to make this work.
REM ------------------------------------------------------------
v# = 1
r# = 1
RANDOMIZE TIMER
CLS
count# = 1
DO
DO
excellent3# = r#
IF count# + 12 > v# THEN EXIT DO
count# = count# + 12
a$ = INKEY$
IF a$ = "s" THEN STOP
LOOP
r# = count#
z = INT(RND * 3) + 0
excellent# = v# - r#
```

excellent2# = v#

IF z = 0 THEN v# = v# + ((v# - r#) * 2) + 1

IF z = 1 THEN v# = v# + ((v# - r#) * 2) + 2

IF z = 2 THEN v# = v# + ((v# - r#) * 2) + 3

m# = (v# / 3) - INT(v# / 3): m# = m# * 10: m# = INT(m#)

PRINT , , z; m#; v#; r#; v# - r#; excellent#; v# - excellent2#; v# - excellent3#

a$ = INKEY$

IF a$ = "s" THEN STOP

LOOP

"C:\Users\Reactor1967\vcns\work\010904A.BAS"

REM to use the methods here the distance between v# & r# have to change by

REM the multiple of the base being used.

v# = 100

r# = 1

RANDOMIZE TIMER

CLS

DO

z = INT(RND * 3) + 0

DO

x = INT(RND * 100) + 3

a$ = INKEY$

IF a$ = "s" THEN STOP

```basic
LOOP UNTIL ((x / 3) - INT(x / 3)) = 0
r# = v# - x
IF z = 0 THEN v# = v# + (v# - r#) + 1
IF z = 1 THEN v# = v# + (v# - r#) + 2
IF z = 2 THEN v# = v# + (v# - r#) + 3
m# = ((v# - r#) / 3) - INT((v# - r#) / 3): m# = m# * 100: m# = INT(m#)
PRINT , , z; v#; r#; m#
a$ = INKEY$
IF a$ = "s" THEN STOP
LOOP
```

"C:\Users\Reactor1967\vcns\work\010804A.BAS"

REM This program is a model for my ideal. To have two charts. On the

REM first chart data is coded at the very beginning in one range of r#

REM only. The rest of the chart is used for system instructions that

REM code a binary 1 every time r changes ranges but codes a zero the

REM rest of the time. When the chart flips it keeps v - r as small as

REM possible and codes only system instructions again 1 for when r

REM changes range and 0 the rest of the time. It codes from a high number

REM to a low number as close as possible to zero. The the chart flips back

REM to the first chart coding data first then instructions. Now when decoding

REM on the first chart when v# is less than or equal some range it knows

REM to flip to the second chart decoding from a low v# to a high v# when

```
REM again when v# is above or at a certain range it flips to the first

REM chart coding down. Problem here is the range at the top of the chart

REM on the second chart is not clear yet. I really need to use base 3 or

REM above for this.

DECLARE SUB up (v#, r#)

DECLARE SUB down (v#, r#)

v# = 1

r# = 1

CLS

RANDOMIZE TIMER

redo:

DO

store# = r#

CALL up(v#, r#)

IF r# = store# THEN z = 0

IF r# <> store# THEN z = 1

IF z = 0 THEN v# = v# + (v# - r#) + 1

IF z = 1 THEN v# = v# + (v# - r#) + 2

PRINT , , z; v#; r#; v# - r#

a$ = INKEY$

IF a$ = "s" THEN STOP

LOOP UNTIL v# >= 950

DO

store# = r#

CALL down(v#, r#)

IF r# = store# THEN z = 0
```

```
IF r# <> store# THEN z = 1

IF z = 0 THEN v# = v# - (r# - v#) + 1

IF z = 1 THEN v# = v# - (r# - v#) + 2

PRINT , , z; v#; r#; v# - r#

a$ = INKEY$

IF a$ = "s" THEN STOP

LOOP UNTIL v# <= 10

GOTO redo:


SUB down (v#, r#)

r# = 1000

count = 1000

DO

IF count <= (r# - 9) THEN r# = r# - 10

a$ = INKEY$: IF a$ = "s" THEN STOP

count = count - 1

LOOP UNTIL count <= v#

END SUB


SUB up (v#, r#)

r# = 1

FOR count = 1 TO 1000

IF count >= (r# + 99) THEN r# = r# + 100

IF count = v# THEN EXIT FOR

NEXT count

END SUB
```

"C:\Users\Reactor1967\vcns\work\010604A.BAS"

REM The program codes both data and system instructions and is decode-able.

REM So what I have learned is that my data has to go into my very first

REM r range because the vectors(v#) there will not repeat with different

REM r values. The rest of the vectors(v#) code a binary one when it flips

REM a r range and a binary 0 when it does not. Now, what I did to do is

REM have this program flip and code down in value either using a control

REM code only or a mirror image of this program except in reverse. What

REM is left to work out is my age old question(WHEN DO I FLIP THE
DAMM THING.)

REM Since I am using key files here maybe I can use base three and code

REM for data, system instructions, and one value for flip. My problem with

REM base3 was I can not flip my r's neatly to the next range but using

REM key files telling me what my coding r's are for my coding vectors

REM and telling me what my decoding r's are for my decode vectors fixes

REM all my problems. I can't go as fast or as far I have to stay within

REM the ranges of my key files but my problems are fixed. And I have to

REM remember what my goals are here and that is just to be able to code

REM and decode raw data with the systems I have developed with vcns.

REM I need to point out that decoding here did not need my key files

REM but when using higer bases a key file will be needed for decode.

REM A key file is "The list of r# values that are used for coding or

REM   decoding specific v# values." This file can also contain

```
REM            Other information such as N values or what ever else
REM            is needed.
REM Also, it is possible to use mathmathics, equations, alogrithms(misspelled)
REM to tell what r'values go with v values during coding and decoding but
REM im not einstein im just trying to accomplish a specific task.
v# = 1000
r# = 1000
CLS
a$ = "e:\qbx\bundle1\key1.txt"
RANDOMIZE TIMER
DO
CLOSE #1
OPEN a$ FOR INPUT AS #1
store = r#
DO
INPUT #1, a#, b#
test1 = (a# = v#)
IF test1 = -1 THEN EXIT DO
LOOP UNTIL EOF(1)
CLOSE #1
IF test1 = 0 THEN STOP
r# = b#
test = (v# > 1099) AND (r# <> store)
IF test = -1 THEN z = 1 ELSE z = 0
IF v# <= 1099 THEN z = INT(RND * 2) + 0
IF z = 0 THEN v# = v# + (v# - r# + 1)
```

cdlv

```
IF z = 1 THEN v# = v# + (v# - r# + 2)

m# = (v# / 4) - INT(v# / 4): m# = m# * 10: m# = INT(m#)

PRINT , , z; m#; v#; r#; v# - r#; test

z$ = INKEY$

REM INPUT z$

IF z$ = "s" THEN CLOSE #1

IF z$ = "s" THEN STOP

LOOP UNTIL v# >= 9950

PRINT "----------------------------------------------------"

INPUT "Ready for decode"; floyd$

DO

m# = (v# / 4) - INT(v# / 4): m# = m# * 10: m# = INT(m#)

IF m# = 0 THEN z = 1

IF m# = 5 THEN z = 1

IF m# = 2 THEN z = 0

IF m# = 7 THEN z = 0

PRINT , , z; m#; v#; r#; v# - r#; test

z$ = INKEY$

REM INPUT z$

IF z$ = "s" THEN STOP

test = ((m# = 0) OR (m# = 5))

dist# = (v# - r#)

IF (dist# / 2) - INT(dist# / 2) = .5 THEN dist# = dist# - 1

dist# = dist# / 2

dist# = dist# + 1

v# = v# - dist#
```

```
IF test = -1 THEN r# = r# - 100

IF r# < 1000 THEN r# = 1000

LOOP UNTIL v# <= 1000

m# = (v# / 4) - INT(v# / 4): m# = m# * 10: m# = INT(m#)

IF m# = 0 THEN z = 1

IF m# = 5 THEN z = 1

IF m# = 2 THEN z = 0

IF m# = 7 THEN z = 0

PRINT , , z; m#; v#; r#; v# - r#; test
```

```
"C:\Users\Reactor1967\vcns\work\010304A.BAS"

REM purpose of this program was to create a list where by the v's listed

REM have the r's listed besides them used for decode. This was a starting

REM process for learning how to create a chart for vectors showing their

REM coding r's(reference points.) and their decoding r's(reference points.).

CLS

a$ = "c:\list1.txt"

OPEN a$ FOR INPUT AS #1

CLOSE #1

v# = 1000

r# = v# - 499

LOCATE 13, 38

PRINT "Working..."

DO
```

```
dist# = v# - r#

IF (dist# / 2) - INT(dist# / 2) = .5 THEN dist# = dist# - 1

dist# = dist# / 2

dist# = dist# + 1

v2# = v# - dist#

OPEN a$ FOR APPEND AS #1

WRITE #1, v2#, r#

CLOSE #1

v# = v# - 1

IF v# < r# THEN r# = r# - 500

LOOP UNTIL v# = 0

CLS

LOCATE 13, 38

PRINT "Task Completed"

INPUT a$

SYSTEM




"C:\Users\Reactor1967\vcns\work\010304.BAS"

v# = 1

r# = 1

RANDOMIZE TIMER

CLS

DO

z = INT(RND * 2) + 0
```

```
IF z = 0 THEN v# = v# + (v# - r# + 1)

IF z = 1 THEN v# = v# + (v# - r# + 2)

m# = (v# / 4) - INT(v# / 4): m# = m# * 10: m# = INT(m#)

PRINT , , z; m#; v#; r#; v# - r#; r# - sr#; (r# - sr#) - dsr#

dsr# = (r# - sr#)

sr# = r#

r# = r# + INT((v# - r#) / 2) - z - 1

REM r# = r# + (v# - r#) - z - 1

a$ = INKEY$

REM INPUT a$

IF a$ = "s" THEN STOP

LOOP




"C:\Users\Reactor1967\vcns\work\010204A.BAS"

REM trying to find a progression for decoding vcns. well I learned something

REM it seems. That since im using 4 as a division to find n that 4 has to

REM be the highest distance from v to r then progress downward for n to

REM make v# solely divisible to find n.

REM -------------------------------------------------------------------

REM from looking at the last few programs it seems a chart with v# and

REM what the n value of v# should be plus what the distance from v# to

REM r# to decode the next v# is. Also what the distance from v# is to

REM code the next v#. When the distance from v# to r# is held steady

REM before coding all that is needed to decode is knowing the n value
```

REM which tells what the distance from v# to r# is for the next decode.

REM That seems better than working with the speed of r#.

```
v# = 100

z = 0

CLS

RANDOMIZE TIMER

timing = 0

DO

REM timing = timing + 1

IF z = 0 THEN r# = v# - 3

IF z = 1 THEN r# = v# - 4

REM IF timing = 5 THEN z = INT(RND * 2) + 0 ELSE z = 0

REM IF z = 0 THEN z = 1 ELSE z = 0

z = INT(RND * 2) + 0

IF z = 0 THEN N = 1

IF z = 1 THEN N = 2

v# = v# + (v# - r#) + N

m# = (v# / 4) - INT(v# / 4): m# = m# * 10: m# = INT(m#)

REM PRINT , , z; m#; v#; r#; v# - r#; r# - sr#; timing

PRINT , , z; m#; v#; r#; v# - r#; r# - sr#

REM IF timing > 10 THEN timing = 0

sr# = r#

a$ = INKEY$

IF a$ = "s" THEN STOP

IF a$ = "j" THEN EXIT DO

LOOP
```

cdlx

```
DO

REM timing = timing + 1

IF z = 0 THEN r# = v# - 3

IF z = 1 THEN r# = v# - 4

REM IF timing = 5 THEN z = INT(RND * 2) + 0 ELSE z = 0

REM IF z = 0 THEN z = 1 ELSE z = 0

z = 1

REM z = 0

REM z = INT(RND * 2) + 0

IF z = 0 THEN N = 1

IF z = 1 THEN N = 2

v# = v# + (v# - r#) + N

m# = (v# / 4) - INT(v# / 4): m# = m# * 10: m# = INT(m#)

REM PRINT , , z; m#; v#; r#; v# - r#; r# - sr#; timing

PRINT , , z; m#; v#; r#; v# - r#; r# - sr#

REM IF timing > 10 THEN timing = 0

sr# = r#

a$ = INKEY$

INPUT a$

IF a$ = "s" THEN STOP

LOOP
```

```
REM BINGO!!!!!!!!!!!! YEHAAAAAAAAAAAAAA!!!!!!!!!!!!!!!!!!!!

REM THE SPEED OF D# CHANGES EVERYTHING HERE JESUS
CHRIST BASE 3 WITH

REM BASE 2 EQUATIONS. DAMM IM GOOD!!!!!!!!!!!!!!!!


v# = 1

r# = 1

CLS

RANDOMIZE TIMER

DO

z = INT(RND * 10) + 0

d# = v# - r#

IF z = 0 THEN v# = v# + (v# - r#) + 1

IF z = 1 THEN v# = v# + (v# - r#) + 2

IF z = 2 THEN v# = v# + (v# - r#) + 3

IF z = 3 THEN v# = v# + (v# - r#) + 4

IF z = 4 THEN v# = v# + (v# - r#) + 5

IF z = 5 THEN v# = v# + (v# - r#) + 6

IF z = 6 THEN v# = v# + (v# - r#) + 7

IF z = 7 THEN v# = v# + (v# - r#) + 8

IF z = 8 THEN v# = v# + (v# - r#) + 9

IF z = 9 THEN v# = v# + (v# - r#) + 10

m# = ((v# - r#) / 10) - INT((v# - r#) / 10): m# = m# * 10: m# = INT(m#)

PRINT , z; m#; v#; r#; v# - r#; d#; d# - (r# - sr#); r# - sr#

sr# = r#

a$ = INKEY$
```

```
IF a$ = "s" THEN STOP

DO

d# = d# + 1

IF (d# / 10) - INT(d# / 10) = 0 THEN EXIT DO

LOOP

r# = v# - d#

LOOP




"C:\Users\Reactor1967\vcns\work\VRDSSUB.BAS"

CLS

INPUT "Enter v# - r#"; a

INPUT "enter desired d"; b

d = 0

DO

s = a - d

IF d - s = b THEN EXIT DO

a$ = INKEY$

IF a$ = "s" THEN STOP

d = d + 1

LOOP UNTIL d >= a

IF d >= a THEN PRINT "No result"

IF d >= a THEN SYSTEM

PRINT , , a; d; d - s; s; " = "; b

INPUT a$
```

```
"C:\Users\Reactor1967\vcns\work\TEMP.BAS"

v# = 1

r# = 1

N = 1

CLS

DO

d# = v# - r#

IF (d# / 4) - INT(d# / 4) = 0 THEN bs = 4 ELSE bs = 2

IF (d# / 4) - INT(d# / 4) <> 0 THEN z = INT(RND * 2) + 0

IF (d# / 4) - INT(d# / 4) = 0 THEN z = INT(RND * 4) + 0

IF z = 0 THEN v# = v# + (v# - r#) + N

IF z = 1 THEN v# = v# + (v# - r#) + N + 1

IF z = 2 THEN v# = v# + (v# - r#) + N + 2

IF z = 3 THEN v# = v# + (v# - r#) + N + 3

PRINT , , z; v#; r#; v# - r#; bs; d#; r# - sr#; N

sr# = r#

INPUT a$

IF a$ = "s" THEN STOP

N = N + 2

IF v# >= (r# + 99) THEN r# = r# + 100

LOOP
```

```
"C:\Users\Reactor1967\vcns\work\STUDY1.BAS"

CLS

REM cord# stands for coordinate on code decode chart something like that

REM would have to be used for this to be useful.

cord# = 0

vr# = 27

DO

x# = vr#

z = 0

redo:

DO

x# = x# + 1

test1 = ((x# / 2) - INT(x# / 2) = 0) AND (z = 1)

test2 = ((x# / 2) - INT(x# / 2) = .5) AND (z = 0)

IF test1 = -1 THEN EXIT DO

IF test2 = -1 THEN EXIT DO

a$ = INKEY$

IF a$ = "s" THEN STOP

LOOP

dist# = x#

IF (dist# / 2) - INT(dist# / 2) = .5 THEN dist# = dist# - 1

dist# = dist# / 2

dist# = dist# + 1
```

```
IF z = 0 THEN dist# = dist# - 1

IF z = 1 THEN dist# = dist# - 2

s# = vr# - dist#

test = (s# / 4) - INT(s# / 4) = 0

a$ = INKEY$: IF a$ = "s" THEN STOP

IF test = 0 THEN GOTO redo:

IF ((dist# * 2) + z + 1) - vr# = 1 THEN cord# = cord# + 1

PRINT , , vr#; ((dist# * 2) + z + 1) - vr#; (dist# * 2) + z + 1;

z = 1

x# = vr#

redo2:

DO

x# = x# + 1

test1 = ((x# / 2) - INT(x# / 2) = 0) AND (z = 1)

test2 = ((x# / 2) - INT(x# / 2) = .5) AND (z = 0)

IF test1 = -1 THEN EXIT DO

IF test2 = -1 THEN EXIT DO

a$ = INKEY$

IF a$ = "s" THEN STOP

LOOP

dist# = x#

IF (dist# / 2) - INT(dist# / 2) = .5 THEN dist# = dist# - 1

dist# = dist# / 2

dist# = dist# + 1

IF z = 0 THEN dist# = dist# - 1

IF z = 1 THEN dist# = dist# - 2
```

```
s# = vr# - dist#

test = (s# / 4) - INT(s# / 4) = 0

a$ = INKEY$: IF a$ = "s" THEN STOP

IF test = 0 THEN GOTO redo2:

PRINT (dist# * 2) + z + 1; cord#

x# = x# + 1

a$ = INKEY$

INPUT a$

IF a$ = "s" THEN STOP

vr# = vr# + 1

LOOP




REM BINGO!!!!!!!!!!!! YEHAAAAAAAAAAAAA!!!!!!!!!!!!!!!!!!!

REM THE SPEED OF D# CHANGES EVERYTHING HERE JESUS
CHRIST BASE 3 WITH

REM BASE 2 EQUATIONS. DAMM IM GOOD!!!!!!!!!!!!!!!!!


v# = 1

r# = 1

CLS

RANDOMIZE TIMER

DO

z = INT(RND * 10) + 0

d# = v# - r#
```

```
IF z = 0 THEN v# = v# + (v# - r#) + 1

IF z = 1 THEN v# = v# + (v# - r#) + 2

IF z = 2 THEN v# = v# + (v# - r#) + 3

IF z = 3 THEN v# = v# + (v# - r#) + 4

IF z = 4 THEN v# = v# + (v# - r#) + 5

IF z = 5 THEN v# = v# + (v# - r#) + 6

IF z = 6 THEN v# = v# + (v# - r#) + 7

IF z = 7 THEN v# = v# + (v# - r#) + 8

IF z = 8 THEN v# = v# + (v# - r#) + 9

IF z = 9 THEN v# = v# + (v# - r#) + 10

m# = ((v# - r#) / 10) - INT((v# - r#) / 10): m# = m# * 10: m# = INT(m#)

PRINT , z; m#; v#; r#; v# - r#; d#; d# - (r# - sr#); r# - sr#

sr# = r#

a$ = INKEY$

IF a$ = "s" THEN STOP

DO

d# = d# + 1

IF (d# / 10) - INT(d# / 10) = 0 THEN EXIT DO

LOOP

r# = v# - d#

LOOP


#


"C:\Users\Reactor1967\vcns\work\RFINDER.BAS"

v1# = 1
```

```
r1# = 1

DO

r1# = v1#

PRINT v1#; v1# + 199; r1#

test = (v# >= (v1#)) AND (v# <= (v1# + 199))

IF test = -1 THEN EXIT DO

a$ = INKEY$

INPUT a$

IF a$ = "s" THEN STOP

v1# = v1# + 200

LOOP

r# = r1#


SUB rfinder (v#, r#)

v1# = 1

r1# = 1

CLS

DO

r1# = v1#

REM PRINT v1#; v1# + 199; r1#

test = (v# >= (v1#)) AND (v# <= (v1# + 199))

IF test = -1 THEN EXIT DO

REM INPUT a$

IF a$ = "s" THEN STOP

v1# = v1# + 200

LOOP
```

```
r# = r1#

END SUB




"C:\Users\Reactor1967\vcns\work\PLAN.BAS"

v# = 1

r# = 1

range = 201

CLS

RANDOMIZE TIMER

high = 0

DO

z = INT(RND * 2) + 0

a1# = v# + (((v# - r#) * 2) + 2)

a2# = a1# + (((a1# - r#) * 2) + 2)

test = a2# > range

IF test = -1 THEN z = 3

IF z = 0 THEN v# = v# + ((v# - r#) * 2) + 1

IF z = 1 THEN v# = v# + ((v# - r#) * 2) + 2

IF z = 2 THEN v# = v# + ((v# - r#) * 2) + 3

IF v# - r# > high THEN high = v# - r#

PRINT , , z; v#; r#; v# - r#; high

a$ = INKEY$

REM INPUT a$

IF a$ = "s" THEN STOP
```

```
dist# = v# - r#

dist# = INT(dist# / 4)

IF z = 3 THEN range = range + 201

IF z = 3 THEN r# = r# + 201

IF z = 3 THEN v# = r# + dist#

IF z = 3 THEN PRINT , , z; v#; r#; v# - r#

LOOP




"C:\Users\Reactor1967\vcns\work\MM1026.BAS"

REM v1# = (((v2# - r2#) / base) * (base - 1)) + (N - 1):coding up

REM NEW DISCOVERY!!! IT IS POSSIBLE FOR V - R TO STAY
POSITIVE AS LONG

REM AS D STAYS RIGID TOO.

v# = 2

r# = 1

CLS

RANDOMIZE TIMER

DO

z = INT(RND * 2) + 0

d# = v# - r#

IF z = 0 THEN v# = v# + (v# - r#) + 2

IF z = 1 THEN v# = v# + (v# - r#) + 4

m# = ((v# - r#) / 4) - INT((v# - r#) / 4): m# = m# * 10: m# = INT(m#)

PRINT z; m#; v#; r#; v# - r#; d#; d# - (r# - sr#); r# - sr#
```

```
y# = (r# - sr#)

sr# = r#

a$ = INKEY$

REM INPUT a$

IF a$ = "s" THEN STOP

x# = INT((v# - r#) / 2) - 20

IF (x# / 2) - INT(x# / 2) = 0 THEN x# = x# + 1

DO

s# = (v# - r#) - x#

o# = 0: REM This controls if v# - r# gets larger or smaller.

test = (((x# + o#) - s#) > (d# - y#)) AND ((x# / 2) - INT(x# / 2) = .5)

REM test = (x# - s#) = ((d# - y#) + 2)

IF test = -1 THEN EXIT DO

a$ = INKEY$

IF a$ = "s" THEN STOP

x# = x# + 2

LOOP

r# = v# - x#

LOOP
```

```
REM        VECTOR COORDINATE NUMERICAL SYSTEM KEY
MAKING PROGRAM

REM ----------------------------------------------------------------------
```

REM (Changes to make. Add line numbers to the keys so that when decoding

REM (each vector tells what line number it came from. Coordinates sort of

REM (speaking. Since the r's repeat. Also, don,t use the traditional

REM (equations of v# = v# + ((v# - r#) * (base - 1)) + N use

REM v# = v# + (v# - r#) + N with N being used according to the base.

REM the reason for this is the traditional equations speed things up

REM too fast. And, code your data but leave one spot left in the base

REM for telling when to flip coding or decoding direction. Be able to

REM code up or down and be able to decode up or down. Several key files

REM may be needed for this. Rem to make all this work this program may

REM be needed to be altered considerably.

REM ---

CLS

INPUT "ENTER A PLUS FOR CODING UP(+) OR NEGATIVE FOR CODING DOWN(-)"; a1$

INPUT "ENTER YOUR BEGINNING NUMBER"; b1

INPUT "ENTER YOUR ENDING NUMBER"; b2

INPUT "ENTER YOUR R INCREMENT"; INCR

later# = b1

CLOSE #1

A2$ = "C:\KEY1.TXT"

OPEN A2$ FOR OUTPUT AS #1

r = b1

REM WRITING V AND CODING R'S

IF a1$ = "+" THEN X = r + INCR

IF a1$ = "-" THEN X = r - INCR

```
DO

IF a1$ = "+" THEN X = r + INCR

IF a1$ = "-" THEN X = r - INCR

WRITE #1, b1, r

IF a1$ = "+" THEN b1 = b1 + 1

IF a1$ = "-" THEN b1 = b1 - 1

TEST1 = (b1 >= X) AND (a1$ = "+")

TEST2 = (b1 <= X) AND (a1$ = "-")

IF TEST1 = -1 THEN r = r + INCR

IF TEST2 = -1 THEN r = r - INCR

TEST1 = (b1 <= b2) AND a1$ = "-"

TEST2 = (b1 >= b2) AND a1$ = "+"

IF TEST1 = -1 THEN EXIT DO

IF TEST2 = -1 THEN EXIT DO

LOOP

PRINT "You made it here."

PRINT "READY TO CREATE FILE 2"

INPUT a$

CLOSE #1

STOP

REM Now create key2. It will use keyd1. It will take the vectors and the

REM coding r's and code 2 vectors from each one vector. Each vector will

REM contain the r used to decode it and its n value.

CLOSE

REM -----------------------------

REM        Creating the file
```

```
A3$ = "C:\KEY2.TXT"

OPEN A3$ FOR OUTPUT AS #1

REM -----------------------------

REM -----------------------------

REM      Opening key1 File for input

A2$ = "C:\KEY1.TXT"

OPEN A2$ FOR INPUT AS #2

REM -----------------------------

DO

INPUT #2, v#, r#

PRINT , , v#; " WORKING... "

REM -----------------------------

REM This section has to be programed to what you want.

v1# = v# - ((r# - v#) * 2) + 1

v2# = v# - ((r# - v#) * 2) + 2

v3# = v# - ((r# - v#) * 2) + 3

WRITE #1, v1#, r#, " 0 "

WRITE #1, v2#, r#, " 1 "

WRITE #1, v3#, r#, "Flip"

REM -----------------------------

LOOP UNTIL EOF(2)

CLOSE #1

CLOSE #2

REM Now go throw and create a third file for all the numbers that repeat.

REM record how many times they repeat.

REM This will be the long one to run.
```

REM step one Count how many numbers in the file

REM Just know what your beginning number is and what your ending number is.

REM Then use a variable to represent one number and increment it or decrement

REM it and count how many of those numbers are in the file.

CLOSE

CLS

a# = later#

REM IF A1$ = "+" THEN b1 = b1 + 1: REM FOR REFERENCE

REM IF A1$ = "-" THEN b1 = b1 - 1: REM FOR REFERENCE

a4$ = "c:\key3.txt"

OPEN a4$ FOR OUTPUT AS #2

CLS

DO

PRINT , , a#; b2

A3$ = "C:\KEY2.TXT"

OPEN A3$ FOR INPUT AS #1

count = 0

DO

INPUT #1, v#, r#, n$

IF a# = v# THEN count = count + 1

IF a# = v# THEN n2$ = n$

LOOP UNTIL EOF(1)

CLOSE #1

WRITE #2, a#, count, n2$

```basic
IF a1$ = "+" THEN a# = a# + 1: REM FOR REFERENCE

IF a1$ = "-" THEN a# = a# - 1: REM FOR REFERENCE

count = 0

nasa1 = (a1$ = "+") AND (a# > b2)

nasa2 = (a1$ = "-") AND (a# < b2)

IF nasa1 = -1 THEN EXIT DO

IF nasa2 = -1 THEN EXIT DO

LOOP

CLOSE #1

CLOSE #2

CLOSE
```

```basic
"C:\Users\Reactor1967\vcns\work\GOWITH.BAS"

v1# = 1

v2# = 1

r# = 1

CLS

RANDOMIZE TIMER

redo:

REM z = 2

REM GOTO iraq:

REM When decoding here v2# = v1# = r# is the out condition.

DO

z = INT(RND * 2) + 0
```

```
test = ((v1# + ((v1# - r#) * 2) + 3) - r#) >= 99

IF test = -1 THEN z = 2

iraq:

IF z = 0 THEN v1# = v1# + ((v1# - r#) * 2) + 1

IF z = 1 THEN v1# = v1# + ((v1# - r#) * 2) + 2

IF z = 2 THEN v1# = v1# + ((v1# - r#) * 2) + 3

m# = (v1# / 3) - INT(v1# / 3)

m# = m# * 10

m# = INT(m#)

PRINT , , "+"; "|"; z; "|"; m#; "|"; v1#; "|"; v2#; "|"; r#; "|"

a$ = INKEY$

REM INPUT a$

IF a$ = "s" THEN STOP

LOOP UNTIL z = 2

REM PRINT "--------------------------------------------------"

REM When decoding it seems conditions like v1# n = 2 and v1# = r# then exit
decode.

REM might work.

DO

z = 0

test = (r# + 2) <= v2#

IF test = -1 THEN r# = r# + 3

IF test = -1 THEN z = 1 ELSE z = 0

IF z = 2 THEN sv# = v#

IF z = 0 THEN v2# = v2# + ((v2# - r#) * 1) + 1

IF z = 1 THEN v2# = v2# + ((v2# - r#) * 1) + 2
```

```basic
m# = (v1# / 4) - INT(v1# / 4)

m# = m# * 10

m# = INT(m#)

PRINT , , "-"; "|"; z; "|"; m#; "|"; v1#; "|"; v2#; "|"; r#; "|"

a$ = INKEY$

REM INPUT a$

IF a$ = "s" THEN STOP

LOOP UNTIL v2# >= v1#

REM PRINT "-----------------------------------------------"


GOTO redo:
```

```basic
"C:\Users\Reactor1967\vcns\work\FFINDER.BAS"

REM this program is intended to find out what to add to v# to throw it to

REM the next r# range and still have the same N.

REM It got deleted. Anyway I use the chart in r finder and test every

REM v# until it goes out of range this I look at the previous v# and find

REM out how much to add to throw it into the next range. I got 135.

v1# = 1

r1# = 1

DO

r1# = v1#

PRINT v1#; v1# + 199; r1#
```

```
test = (v# >= (v1#)) AND (v# <= (v1# + 199))

IF test = -1 THEN EXIT DO

a$ = INKEY$

INPUT a$

IF a$ = "s" THEN STOP

v1# = v1# + 200

LOOP

r# = r1#


SUB rfinder (v#, r#)

v1# = 1

r1# = 1

CLS

DO

r1# = v1#

REM PRINT v1#; v1# + 199; r1#

test = (v# >= (v1#)) AND (v# <= (v1# + 199))

IF test = -1 THEN EXIT DO

REM INPUT a$

IF a$ = "s" THEN STOP

v1# = v1# + 200

LOOP

r# = r1#

END SUB
```

```
"C:\Users\Reactor1967\vcns\work\12282256.BAS"

sr# = 0

CLS

RANDOMIZE TIMER

v# = 1

r# = 1

DO

z = INT(RND * 2) + 0

dale# = (v# - r#)

IF z = 0 THEN N = 1

IF z = 1 THEN N = 2

v# = v# + (v# - r#) + N

m# = (v# / 4) - INT(v# / 4): m# = m# * 10: m# = INT(m#)

PRINT z; m#; v#; r#; v# - r#; sr#; INT((sr# / (v# - sv#)) * 100); INT(((sr#) / (v# - r#)) * 100)

sv# = v#

a$ = INKEY$

REM INPUT a$

IF a$ = "s" THEN STOP

REM -------------------------------------------------- Push r back

test = r# + sr# + 4 <= v# - INT((v# - r#) / 2)

IF test = -1 THEN sr# = sr# + 4

REM --------------------------------------------------

r# = r# + sr#

LOOP
```

```
"C:\Users\Reactor1967\vcns\work\12282205.BAS"

CLS

RANDOMIZE TIMER

v# = 1

r# = 1

DO

z = INT(RND * 2) + 0

IF z = 0 THEN N = 1

IF z = 1 THEN N = 2

sv# = v#

v# = v# + (v# - r#) + N

m# = (v# / 4) - INT(v# / 4): m# = m# * 10: m# = INT(m#)

PRINT z; m#; v#; r#; v# - r#; v# - sv#; r# - sr#; (r# - sr#) - sr2#

sr2# = (r# - sr#)

sr# = r#

a$ = INKEY$

IF a$ = "s" THEN STOP

dist# = INT((v# - r#) / 2)

DO

test = (v# - (r# + 4)) < dist#

IF test = -1 THEN EXIT DO

r# = r# + 4

a$ = INKEY$
```

```
IF a$ = "s" THEN STOP

LOOP

LOOP




"C:\Users\Reactor1967\vcns\work\12281952.BAS"

dist# = 99

r# = 1

DO

a# = INT((dist#) / 2)

DO

test = (a# / 4) - INT(a# / 4) = 0

IF test = -1 THEN EXIT DO

a# = a# - 1

a$ = INKEY$

IF a$ = "s" THEN STOP

LOOP

sr# = r#

r# = r# + a#

PRINT , , dist#; r#; r# - sr#

a$ = INKEY$

INPUT a$

IF a$ = "s" THEN STOP

dist# = dist# + 1

LOOP
```

```
"C:\Users\Reactor1967\vcns\work\12281951.BAS"

v# = 48

r# = 1

CLS

RANDOMIZE TIMER

DO

z = INT(RND * 2) + 0

IF z = 0 THEN N = 1

IF z = 1 THEN N = 2

dist# = v# - r#

v# = v# + dist# + N

m# = (v# / 4) - INT(v# / 4): m# = m# * 10: m# = INT(m#)

PRINT z; m#; v#; r#; (v# - r#); v# - sv#; r# - sr#

REM IF (v# - r#) >= 7292 THEN STOP

sr# = r#

sv# = v#

a$ = INKEY$

REM INPUT a$

IF a$ = "s" THEN STOP

a# = INT((v# - r#) / 2)

DO

test = (a# / 4) - INT(a# / 4) = 0

IF test = -1 THEN EXIT DO
```

```
a# = a# - 1

a$ = INKEY$

IF a$ = "s" THEN STOP

LOOP

r# = r# + a#

LOOP
```

"C:\Users\Reactor1967\vcns\work\12272230.BAS"

```
REM NEW DISCOVERY ON M# ((V# - R#) / 4) - INT((V# - R#) /8)

REM This part is new I,ve never used this before but it works.

REM This is for when N1 and N2 are both even and when the speed of r is

REM even doing a ((v# - r#) / 4) - int((v# - r#) / 4) yeilds the same

REM result for both 0 and 1. The above example helps elaminate that.

REM But I still have the same probem when (v# / 4) - int(v# / 4) is

REM the same for both 0 and 1.

REM --------------------------------------------------------

REM This program is an example of how the speed of v# affects how

REM r# can keep up and and the speed of r# affects how we find our

REM N for v#. Both can be controlled with a mathmatical progression

REM the program is not the answer to that but it helps understand

REM how to get started finding that progression. What I need is to

REM find a progression so that I can divide v# by some number to find

REM N. And in turn use N to control the speed of r#.

REM --------------------------------------------------------
```

REM Here to decode we have to know the previous N value for v#.

REM since I can,t do that here I store that N value in memory so

REM that I can decode and prove my method of progression works and

REM will decode. But I need to find a new progression for r# when

REM coding v#

REM --

DIM bstore(10000)

bin = 0

v# = 4945

r# = (v# - 4)

X# = r#

RANDOMIZE TIMER

CLS

redo:

DO

REM --

r# = v# - 4

IF bin >= 10000 THEN EXIT DO

REM --

z = INT(RND * 2) + 0

bin = bin + 1

bstore(bin) = z

IF z = 0 THEN N = 1

IF z = 1 THEN N = 2

v# = v# + (v# - r#) + N

PRINT , , z; v#; r#; v# - r#; r# - sr#

```
sr# = r#
a$ = INKEY$
REM INPUT a$
IF a$ = "s" THEN STOP
LOOP
decode:
PRINT "-------------------------------------"
PRINT "BEGIN DECODE"
SLEEP 5
DO
z = bstore(bin)
bin = bin - 1
IF z = 0 THEN r# = r# - 5
IF z = 1 THEN r# = r# - 6
dist# = v# - r#
IF (dist# / 2) - INT(dist# / 2) = .5 THEN dist# = dist# - 1
dist# = dist# / 2
dist# = dist# + 1
PRINT , , z; v#; r#; v# - r#
v# = v# - dist#
IF v# = 4945 THEN EXIT DO
a$ = INKEY$
REM INPUT a$
IF a$ = "s" THEN STOP
LOOP
IF r# = X# THEN PRINT "Decode Complete"
```

IF r# <> X# THEN PRINT "Decode FAILED"

IF r# <> X# THEN STOP

SLEEP 5

PRINT "BEGIN Coding"

SLEEP 5

GOTO redo:

"C:\Users\Reactor1967\vcns\work\12262127.BAS"

REM If possible I need N to stand for (both the numeric value and the speed

REM of r# at the same time. If that can be set up that way.)

REM which would work if the speed of r# is always divisible by a specific

REM number namely the base or higher. Then when decoding and we see v#

REM and find n# then we would know the speed of r#.

REM so speed of r# = (v# - r#) + N which has to be divisible by the base or

REM a multiple of the base.

v# = 100

r# = 1

CLS

RANDOMIZE TIMER

DO

z = INT(RND * 8) + 0

N = z

v# = v# + (v# - r#) + N

```
m# = ((v# - r#) / 8) - INT((v# - r#) / 8)

m# = m# * 10

m# = INT(m#)

PRINT , , z; m#; v#; r#; v# - r#; r# - sr#

sr# = r#

a$ = INKEY$

IF a$ = "s" THEN STOP

r# = v# - 99

LOOP
```

```
"C:\Users\Reactor1967\vcns\work\11281615.BAS"

CLS

dist# = 99

r# = 1

DO

a# = INT((dist#) / 2)

DO

test = (a# / 4) - INT(a# / 4) = 0

IF test = -1 THEN EXIT DO

a# = a# - 1

a$ = INKEY$

IF a$ = "s" THEN STOP

LOOP

sr# = r#
```

```
r# = r# + a#

PRINT , , dist#; r#; r# - sr#

a$ = INKEY$

INPUT a$

IF a$ = "s" THEN STOP

dist# = dist# + 1

LOOP




"C:\Users\Reactor1967\vcns\work\11281614.BAS"

v# = 48

r# = 1

CLS

RANDOMIZE TIMER

DO

z = INT(RND * 2) + 0

IF z = 0 THEN N = 1

IF z = 1 THEN N = 2

dist# = v# - r#

v# = v# + dist# + N

m# = (v# / 4) - INT(v# / 4): m# = m# * 10: m# = INT(m#)

PRINT z; m#; v#; r#; (v# - r#); v# - sv#; r# - sr#

REM IF (v# - r#) >= 7292 THEN STOP

sr# = r#

sv# = v#
```

```
a$ = INKEY$

REM INPUT a$

IF a$ = "s" THEN STOP

a# = INT((v# - r#) / 2)

DO

test = (a# / 4) - INT(a# / 4) = 0

IF test = -1 THEN EXIT DO

a# = a# - 1

a$ = INKEY$

IF a$ = "s" THEN STOP

LOOP

r# = r# + a#

LOOP
```

"C:\Users\Reactor1967\vcns\work\11260913.BAS"

```
REM by dividing v# by several numbers not just one I can see if its

REM possible to find N. If thats possible then its possible to decode this.

v# = 1

r# = 1

y = 1

CLS

DO

z = INT(RND * 2) + 0

d# = v#
```

```
IF z = 0 THEN v# = v# + (v# - r#) + 1

IF z = 1 THEN v# = v# + (v# - r#) + 2

m# = (v# / 4) - INT(v# / 4): m# = m# * 10: m# = INT(m#)

PRINT , , z; m#; v#; r#; v# - r#; v# - d#

a$ = INKEY$

REM INPUT a$

IF a$ = "s" THEN STOP

IF v# >= (r# + 3) THEN r# = r# + 4

LOOP




"C:\Users\Reactor1967\vcns\work\11211118.BAS"

v# = 1

r# = 1

CLS

RANDOMIZE TIMER

DO

z = INT(RND * 2) + 0

IF z = 0 THEN v# = v# + (v# - r# + 4)

IF z = 1 THEN v# = v# + (v# - r# + 5)

REM m# = ((v# - r#) / 4) - INT((v# - r#) / 4): m# = m# * 10: m# = INT(m#)

m# = (v# / 4) - INT(v# / 4): m# = m# * 10: m# = INT(m#)

PRINT , , z; m#; v#; r#; v# - r#; v# - sv#; r# - sr#

sv# = v#

sr# = r#
```

```
a$ = INKEY$

REM INPUT a$

IF a$ = "s" THEN STOP

IF a$ = "d" THEN EXIT DO

IF v# - r# < 8 THEN r# = r# + 4

IF v# - r# >= 8 THEN r# = r# + 8

LOOP
```

```
"C:\Users\Reactor1967\vcns\work\11090441.BAS"

REM using probability it might be possibility to determine when r# will

REM flip. This is just calculating probability nothing else. Anyway

REM using probabibility the the variables would be the binary pattern

REM differences between v# and r#, d# ect...

CLS

RANDOMIZE TIMER

d# = 0

count = 0

DO

z = INT(RND * 2) + 0

count# = count# + 1

IF z = 0 THEN d# = d# + 1

x# = (d# / count#) - INT(d# / count#)

PRINT , , count#; d#; x#

a$ = INKEY$
```

IF a$ = "s" THEN STOP

LOOP

"C:\Users\Reactor1967\vcns\work\11040849.BAS"

REM Im considering here whether is possible to determine a relationship

REM between v# and r# so that r# is the integer of a division before

REM coding and r# is also the interger of a division before decoding.

REM hummm. Is that possible?

REM After studying this program it is possible.

REM To decode do this

REM before decoding v# - r# = (v# - ((int(v# / 100)) * 100)) + 100 Sometimes you do not add 100

REM This follow the decode equation which is v1# = v2# - ((((v# - r#) - N) / base) * (base * 2)) + N

v# = 5

r# = 1

CLS

RANDOMIZE TIMER

DO

z = INT(RND * 3) + 0

d2# = v# - r#

IF z = 0 THEN v# = v# + ((v# - r#) * 2) + 1

IF z = 1 THEN v# = v# + ((v# - r#) * 2) + 2

IF z = 2 THEN v# = v# + ((v# - r#) * 2) + 3

```
M# = ((v# - r#) / 4) - INT((v# - r#) / 4): M# = M# * 10: M# = INT(M#)

PRINT , , z; M#; v#; r#; d2#; v# - r#

s# = r#

r# = (INT(v# / 100)) * 100

a$ = INKEY$

REM INPUT a$

IF a$ = "s" THEN STOP

LOOP
```

"C:\Users\Reactor1967\vcns\work\11030114.BAS"

```
REM rotate N and that makes everything work out ok

d# = 0

CLS

N = 1

DO

IF (d# / 4) - INT(d# / 4) = 0 THEN PRINT d#; (d# * 2) + 1; (d# * 2) + 4;
"base4", N

IF (d# / 4) - INT(d# / 4) <> 0 THEN PRINT d#; (d# * 2) + 1; (d# * 2) + N + 1;
"base2", N

INPUT a$

IF a$ = "s" THEN STOP

d# = d# + 1

N = N + 2

IF N >= 16 THEN N = 1
```

LOOP

"C:\Users\Reactor1967\vcns\work\11022051.BAS"

REM In order for me to get my previous programs to work I have to be able to

REM move v# up from one r# range to another. I need to be able to do this

REM in such a way that v# can code at least one bit or find a way to code

REM v# up but as of now I don,t know how to do that in higher bases than two

REM because r# trys to skip a range sometimes.

max# = 200

CLS

RANDOMIZE TIMER

low# = 999999

redo:

v# = 1

r# = 1

DO

z = INT(RND * 2) + 0

test1 = (v# + ((v# - r#) * 2) + z + 1) > 201

test2 = (test1 = -1) AND (v# < low#)

IF test2 = -1 THEN low# = v#

IF test1 = -1 THEN EXIT DO

IF z = 0 THEN v# = v# + ((v# - r#) * 2) + 1

IF z = 1 THEN v# = v# + ((v# - r#) * 2) + 2

PRINT , , z; low#; v#

```
a$ = INKEY$

IF a$ = "s" THEN STOP

LOOP

GOTO redo:
```

"C:\Users\Reactor1967\vcns\work\11012001.BAS"

```
REM Inventor Notes: This program can work and work very well but it needs

REM some error checking procedures and a way to make sure it codes at least

REM one bit when jumping to another range. This can be done with a value

REM that is variable rather than constant or the ranges themselves can be

REM variable. Anyway this needs some thought.

REM Well I learned a good lesson. All these years I have been trying to

REM increase r#. Well, get a range of what r# should be for a range of v#.

REM Increase v# when it can no longer code in the r# range with out getting

REM off the chart. Leave r# along and increase v# to the next range of r#.

REM that works good. Now, what im going to try to do is take the top half of

REM the chart when v# increases and have it go to a preplaned v# on the bottom

REM half of the next v# to r# range.

DECLARE SUB rfinder (v#, r#)

DIM vbuffer(3)

DIM rbuffer(3)

count = 1

v# = 1
```

```
r# = 1

CLS

RANDOMIZE TIMER

redo:

DO

z = INT(RND * 2) + 0

d# = v# - r#

IF r# > v# THEN STOP

IF v# < r# THEN STOP

IF v# <= 0 THEN STOP

IF z = 0 THEN v# = v# + ((v# - r#) * 2) + 1

IF z = 1 THEN v# = v# + ((v# - r#) * 2) + 2

IF z = 2 THEN v# = v# + ((v# - r#) * 2) + 3

m# = ((v# - r#) / 3) - INT((v# - r#) / 3): m# = m# * 10: m# = INT(m#)

PRINT , z; m#; v#; r#; v# - r#

rbuffer(count) = r#

vbuffer(count) = v#: count = count + 1: IF count = 4 THEN count = 1

sr# = r#

a$ = INKEY$

REM INPUT a$

IF a$ = "s" THEN STOP

str# = r#

CALL rfinder(v#, r#)

IF str# <> r# THEN EXIT DO

LOOP

FOR l = 1 TO 2
```

```
count = count - 1
IF count = 0 THEN count = 3
NEXT l
v# = vbuffer(count)
r# = rbuffer(count)
REM PRINT v#; r#
IF r# > v# THEN STOP
IF v# < r# THEN STOP
IF v# <= 0 THEN STOP
d# = v# - r#
z = 2
v# = v# + ((v# - r#) * 2) + 3
m# = ((v# - r#) / 3) - INT((v# - r#) / 3): m# = m# * 10: m# = INT(m#)
PRINT , z; m#; v#; r#; v# - r#
sr# = r#
v# = v# + 135
CALL rfinder(v#, r#)
m# = ((v# - r#) / 3) - INT((v# - r#) / 3): m# = m# * 10: m# = INT(m#)
PRINT , z; m#; v#; r#; v# - r#
GOTO redo:

SUB rfinder (v#, r#)
v1# = 1
DO
test = v1# >= v#
IF test = -1 THEN EXIT DO
```

```
v1# = v1# + 200

REM a$ = INKEY$

REM IF a$ = "s" THEN STOP

LOOP

DO

IF v1# > v# THEN v1# = v1# - 200

a$ = INKEY$: IF a$ = "s" THEN STOP

LOOP UNTIL v1# <= v#

r1# = v1#

REM PRINT v1#; v1# + 199; r1#

test = (v# >= (v1#)) AND (v# <= (v1# + 199))

IF test = 0 THEN STOP

r# = r1#

IF r1# > v# THEN STOP

END SUB




"C:\Users\Reactor1967\vcns\work\11011526.BAS"

DECLARE SUB rfinder (v#, r#)

v# = 1

r# = 1

CLS

RANDOMIZE TIMER

DO

CALL rfinder(v#, r#)
```

d

```
DO

z = INT(RND * 2) + 0

IF flag = 1 THEN z = 2

d# = v# - r#

sv# = v#

sr# = r#

v# = v# + ((v# - r#) * 2) + z + 1

CALL rfinder(v#, r#)

IF sr# <> r# THEN EXIT DO

v# = sv#

r# = sr#

IF z = 0 THEN v# = v# + ((v# - r#) * 2) + 1

IF z = 1 THEN v# = v# + ((v# - r#) * 2) + 2

IF z = 2 THEN v# = v# + ((v# - r#) * 2) + 3

PRINT , , z; v#; r#; v# - r#

a$ = INKEY$

REM INPUT a$

IF a$ = "s" THEN STOP

flag = 0

LOOP

v# = sv#

r# = sr#

v# = v# + 20

flag = 1

LOOP
```

```
SUB rfinder (v#, r#)

v1# = 1

r1# = 1

DO

r1# = v1#

REM PRINT v1#; v1# + 199; r1#

test = (v# >= (v1#)) AND (v# <= (v1# + 199))

IF test = -1 THEN EXIT DO

a$ = INKEY$

REM INPUT a$

IF a$ = "s" THEN STOP

v1# = v1# + 200

LOOP

r# = r1#

END SUB
```

"C:\Users\Reactor1967\vcns\work\11011426.bas"

REM Inventor Notes:

REM make a chart that shows what each r# is for each range of v#. When

REM v# gets ready to leave that range code a base3 2. Then add 100 to v# or

REM so to throw it in another chart range. Code in that range again. When

REM decoding when the program reads that base3 2 it knows to subtract 100

REM or so from v#. Then look at what the r# range is for that v# then

REM continue decoding. L.B.

```
REM Ok, code your data with your v# but before you leave your v# range
REM code a N that represents leaving that range. Now what you add to your
REM v# make sure that your v# still represents the same N value so that when
REM you decode in your next v# range and see the N that means jump to the
REM next range you know to subtract from v# to throw v# into its next range.
REM so in base3 0 & 1 is the data and 2 means jump.
v# = 1
r# = 1
CLS
RANDOMIZE TIMER
DO
z = INT(RND * 2) + 0
IF flag = 1 THEN z = 2
d# = v# - r#
test = (v# - r#) >= 100
IF test = -1 THEN range# = INT(v# / 3)
IF test = -1 THEN range# = range# * 3
IF test = -1 THEN r# = range#
IF test = -1 THEN z = 2
IF z = 0 THEN v# = v# + ((v# - r#) * 2) + 1
IF z = 1 THEN v# = v# + ((v# - r#) * 2) + 2
IF z = 2 THEN v# = v# + ((v# - r#) * 2) + 3
PRINT , , z; v#; r#; v# - r#; d#; r# - sr#
INPUT a$
IF a$ = "s" THEN STOP
LOOP
```

```
"C:\Users\Reactor1967\vcns\work\11011421.bas"

v# = 1

r# = 1

CLS

RANDOMIZE TIMER

DO

range# = INT(v# / 3)

range# = range# * 3

r# = range#

flag = 1

DO

z = INT(RND * 2) + 0

IF flag = 1 THEN z = 2

d# = v# - r#

IF z = 0 THEN v# = v# + ((v# - r#) * 2) + 1

IF z = 1 THEN v# = v# + ((v# - r#) * 2) + 2

IF z = 2 THEN v# = v# + ((v# - r#) * 2) + 3

PRINT , , z; v#; r#; v# - r#; d#; r# - sr#

INPUT a$

IF a$ = "s" THEN STOP

flag = 0

LOOP UNTIL v# - r# >= 100
```

div

LOOP

```
"C:\Users\Reactor1967\vcns\work\11010332.BAS"

v# = 1

r# = 1

REM base 4

CLS

RANDOMIZE TIMER

DO

range# = (INT(v# / 100)) * 100: IF range# = 0 THEN range = 1#

r# = range#

DO

z = INT(RND * 4) + 0

IF z = 0 THEN v# = v# + ((v# - r#) * 3) + 1

IF z = 1 THEN v# = v# + ((v# - r#) * 3) + 2

IF z = 2 THEN v# = v# + ((v# - r#) * 3) + 3

IF z = 3 THEN v# = v# + ((v# - r#) * 3) + 4

m# = (v# / 4) - INT(v# / 4): m# = m# * 10: m# = INT(m#)

PRINT , , z; m#; v#; r#; v# - r#

a$ = INKEY$

REM INPUT a$

IF a$ = "s" THEN STOP

IF a$ = "d" THEN EXIT DO
```

```
test = (INT(v# / 100)) * 100 <> range

flag = 0

IF test = -1 THEN EXIT DO

LOOP

v# = v# + 100

a$ = INKEY$

REM INPUT a$

IF a$ = "s" THEN STOP

LOOP
```

"C:\Users\Reactor1967\vcns\work\11010144.bas"

REM Inventor Notes:

REM Here im coding base 4 and base two at the same time but another thing

REM that would help is to when ever r# changes go ahead and code a binary

REM 1 to change r# but instead of incrementing r# goto a chart that shows

REM what r should be for each range of v#. That way all this would not be

REM necessary. It would be a good ideal to know before you change r# so that

REM you could get r# for what your next intended range of v# should be so

REM that when you decode you decode to your range of v# that your next r is

REM in and when you see that binary 1 you know hey I got my v# go to my chart

REM and see what r# should be for it. Now doing that code base 4 or 5 or

REM something else so that you can put your system instructions in the lower

REM half of the base and your data instructions in the upper half of

```
REM the base. L.B.

v# = 1

r# = 1

N = 1

CLS

DO

d# = v# - r#

IF (d# / 4) - INT(d# / 4) = 0 THEN bs = 4 ELSE bs = 2

IF (d# / 4) - INT(d# / 4) <> 0 THEN z = INT(RND * 2) + 0

IF (d# / 4) - INT(d# / 4) = 0 THEN z = INT(RND * 4) + 0

IF z = 0 THEN v# = v# + (v# - r#) + N

IF z = 1 THEN v# = v# + (v# - r#) + N + 1

IF z = 2 THEN v# = v# + (v# - r#) + N + 2

IF z = 3 THEN v# = v# + (v# - r#) + N + 3

m# = (v# / 4) - INT(v# / 4): m# = m# * 10: m# = INT(m#)

PRINT z; m#; v#; r#; v# - r#; bs; d#; d# - (r# - sr#); r# - sr#; N

sr# = r#

a$ = INKEY$

REM INPUT a$

IF a$ = "s" THEN STOP

IF v# >= (r# + 99) THEN r# = r# + 100

REM N = N + 2

REM IF N >= 12 THEN N = 1


LOOP
```

```
"C:\Users\Reactor1967\vcns\work\11010118.bas"

REM Inventor notes. Its possible to rotate N which seems like that might

REM be helpful for program stability in some cases.

v# = 1

r# = 1

N = 1

CLS

DO

z = INT(RND * 2) + 0

d# = v# - r#

IF z = 0 THEN v# = v# + (v# - r#) + N

IF z = 1 THEN v# = v# + (v# - r#) + N + 1

N = N + 4

m# = (v# / 4) - INT(v# / 4): m# = m# * 10: m# = INT(m#)

PRINT z; m#; v#; r#; v# - r#; N

a$ = INKEY$

REM INPUT a$

IF a$ = "s" THEN STOP

IF (d# / 2) - INT(d# / 2) = .5 THEN d# = d# - 1

r# = r# + 2 + d#

LOOP
```

```
"C:\Users\Reactor1967\vcns\work\11010114.BAS"

d# = 0

CLS

DO

IF (d# / 4) - INT(d# / 4) = 0 THEN PRINT d#; (d# * 2) + 1; (d# * 2) + 4;
"base4"

IF (d# / 4) - INT(d# / 4) <> 0 THEN PRINT d#; (d# * 2) + 0; (d# * 2) + 1;
"base2"

INPUT a$

IF a$ = "s" THEN STOP

d# = d# + 1

LOOP
```

```
"C:\Users\Reactor1967\vcns\work\10311047.BAS"

REM Draw a straight Line. On that line list the number ranges for

REM distances to code to in specific bases. Below that line draw another

REM line. On that line list the specific d's ranges that code for each

REM base. Both lines must corrospond with each other.

d# = 1

DIM can1(100)

DIM can2(100)

DIM can3(100)

CLS

DO
```

```
PRINT d#;

can1(d#) = d#

FOR count = 1 TO 5

test = (d# / count) - INT(d# / count) = 0

IF test = -1 THEN PRINT count;

IF test = -1 THEN can3(d#) = (d# * count) + x#

IF test = -1 THEN can2(d#) = (d# * count) + 1

IF test = -1 THEN x# = count

NEXT count

PRINT (d# * count) + 1; (d# * count) + x#;

PRINT " "

d# = d# + 1

a$ = INKEY$

INPUT a$

IF a$ = "s" THEN STOP

LOOP UNTIL d# = 101

FOR count = 1 TO 100

PRINT can1(count); can2(count); can3(count)

INPUT a$

IF a$ = "s" THEN STOP

NEXT count
```

"C:\Users\Reactor1967\vcns\work\10310407.BAS"

REM --------------Goal Reached here this program works for intended

dx

purpose.--------

REM --------------- So far N is not retreivable but r# keeps up.

REM Im trying to establish a way to increment r# controbabily and work in

REM any base. If you,ve read my previous programs I can take v# - r#, see

REM what it is divisable by, then use that as a base for coding data. That

REM means working in different bases at the same time unless I can specificly

REM control what v# - r# will be each time which is also possible by knowing

REM d#, the desired N, and what equation you are using. But instead of doing

REM that now im just seeing if I can incrment r# controbably(misspelled I do

REM that alot.)

REM Anyway this is what im doing here.

v# = 1

r# = 1

CLS

RANDOMIZE TIMER

small# = 9999999999#

large# = 0

DO

z = INT(RND * 3) + 0

d# = v# - r#

IF ((v# + (v# - r#)) - r#) + 5 >= 80 THEN z = 4

IF flag = 1 THEN z = 3

IF z = 0 THEN v# = v# + (v# - r#) + 1

IF z = 1 THEN v# = v# + (v# - r#) + 2

IF z = 2 THEN v# = v# + (v# - r#) + 3: REM also a control value don,t correct
by subtracting 2 from r# when decoding.

```
IF z = 3 THEN v# = v# + (v# - r#) + 4: REM you read this when decoding and
REM                                  derement r#
IF z = 4 THEN v# = v# + (v# - r#) + 5
IF v# - r# < small# THEN small# = v# - r#
IF v# - r# > large# THEN large# = v# - r#
REM PRINT , , z; v#; r#; v# - r#; d#; d# - (r# - sr#); r# - sr#
m# = ((v# - r#) / 5) - INT((v# - r#) / 5): m# = m# * 10: m# = INT(m#)
PRINT , , z; m#; v#; r#; v# - r#, small#; large#
sr# = r#
a$ = INKEY$
REM INPUT a$
IF a$ = "s" THEN STOP
IF z = 4 THEN r# = r# + 80
IF v# - r# >= 2 THEN r# = r# + 3: REM using base two so this is the correction
IF v# - r# < 2 THEN z = 3: REM when v# - r# less than the base don,t correct.
LOOP
```

"C:\Users\Reactor1967\vcns\work\10310359.BAS"

```
REM Im trying to establish a way to increment r# controbabily and work in
REM any base. If you,ve read my previous programs I can take v# - r#, see
REM what it is divisable by, then use that as a base for coding data. That
REM means working in different bases at the same time unless I can specificly
REM control what v# - r# will be each time which is also possible by knowing
REM d#, the desired N, and what equation you are using. But instead of doing
```

```
REM that now im just seeing if I can incrment r# controbably(misspelled I do
REM that alot.)
REM Anyway this is what im doing here.
v# = 1
r# = 1
CLS
RANDOMIZE TIMER
small# = 9999999999#
large# = 0
DO
z = INT(RND * 2) + 0
d# = v# - r#
IF ((v# + (v# - r#)) - r#) + 4 >= 80 THEN z = 3
IF flag = 1 THEN z = 2
IF z = 0 THEN v# = v# + (v# - r#) + 1
IF z = 1 THEN v# = v# + (v# - r#) + 2
IF z = 2 THEN v# = v# + (v# - r#) + 3: REM also a control value don,t correct
by subtracting 2 from r# when decoding.
IF z = 3 THEN v# = v# + (v# - r#) + 4: REM you read this when decoding and
REM                             derement r#
IF v# - r# < small# THEN small# = v# - r#
IF v# - r# > large# THEN large# = v# - r#
REM PRINT , , z; v#; r#; v# - r#; d#; d# - (r# - sr#); r# - sr#
PRINT , , z; v#; r#; v# - r#, small#; large#
sr# = r#
a$ = INKEY$
```

```
REM INPUT a$

IF a$ = "s" THEN STOP

IF z = 3 THEN r# = r# + 80

IF v# - r# >= 2 THEN r# = r# + 2: REM using base two so this is the correction

IF v# - r# < 2 THEN z = 2: REM when v# - r# less than the base don,t correct.

LOOP
```

```
"C:\Users\Reactor1967\vcns\work\10301216.BAS"

v2 = 1

v# = 1

r# = 1

CLS

DO

z = INT(RND * 2) + 0

IF z2 = 0 THEN v# = v# + (v# - r#) + 1

IF z2 = 1 THEN v# = v# + (v# - r#) + 2

r2# = v2# - (v# - r#)

IF z = 0 THEN v2# = v2# + (v2# - r2#) + 1

IF z = 1 THEN v2# = v2# + (v2# - r2#) + 2

m# = ((v# - r#) / 4) - INT((v# - r#) / 4): m# = m# * 10: m# = INT(m#)

m2# = ((v2# - r2#) / 4) - INT((v2# - r2#) / 4): m2# = m2# * 10: m2# = INT(m2#)

PRINT , z; m2#; v2#; r2#; v2# - r2#; " "; z2; m#; v#; r#; v# - r#

a$ = INKEY$
```

```
REM INPUT a$

IF a$ = "s" THEN STOP

test = (r# + 10) <= (v# + 1)

IF test = -1 THEN r# = r# + 10

IF test = 0 THEN z2 = 0

IF test = -1 THEN z2 = 1

LOOP




"C:\Users\Reactor1967\vcns\work\10301149.BAS"

vr# = 287

CLS

DO

PRINT , , vr#;

FOR count = 1 TO 50

IF (vr# / count) - INT(vr# / count) = 0 THEN PRINT ","; count;

NEXT count

PRINT " "

INPUT a$

IF a$ = "s" THEN STOP

vr# = vr# + 1

LOOP
```

"C:\Users\Reactor1967\vcns\work\10301120.BAS"

REM Knowing what ever d# is divisable by can be your base I ask myself the

REM question? Is it possible to code r# by using the distance between v - r

REM so that after decoding v I can decode r then decode v then decode r

REM ect.... If I can do this that would help my work tremendously.

d# = 9

v# = 100

RANDOMIZE TIMER

CLS

DO

z = INT(RND * 2) + 0

r# = v# - 9

IF z = 0 THEN v# = v# + (v# - r#) + 1

IF z = 1 THEN v# = v# + (v# - r#) + 2

IF z = 2 THEN v# = v# + (v# - r#) + 3

m# = ((v# - r#) / 3) - INT((v# - r#) / 3): m# = m# * 10: m# = INT(m#)

PRINT , , z; m#; v#; r#; v# - r#; d#; d# - (r# - sr#); r# - sr#

sr# = r#

a$ = INKEY$

IF a$ = "s" THEN STOP

LOOP

"C:\Users\Reactor1967\vcns\work\10301050.BAS"

```
REM BINGO!!!!!!!!!!!! YEHAAAAAAAAAAAAA!!!!!!!!!!!!!!!!!!!

REM THE SPEED OF D# CHANGES EVERYTHING HERE JESUS
CHRIST BASE 3 WITH

REM BASE 2 EQUATIONS. DAMM IM GOOD!!!!!!!!!!!!!!!!!

REM WHAT EVER D IS DIVISABLE BY CAN BE YOUR BASE. SO,
KNOWING THAT ITS

REM POSSIBLE TO CODE SEVERAL BASES AT THE SAME TIME
BASED ON D. I WAS WANTING

REM TO KNOW WHEN TO CHANGE OR INCREMENT THE SPEED OF R
OR DECREMENT IT SO

REM MAYBE I CAN USE THIS.

REM here v1# = ((v2# - r2#) / 2) + 1 for 0 or 1 of base 3

REM here v1# = ((v2# - r2#) / 2) + 2 for 2 of base 3

v# = 1

r# = 1

CLS

RANDOMIZE TIMER

DO

z = INT(RND * 3) + 0

d# = v# - r#

IF z = 0 THEN v# = v# + (v# - r#) + 1

IF z = 1 THEN v# = v# + (v# - r#) + 2

IF z = 2 THEN v# = v# + (v# - r#) + 3

m# = ((v# - r#) / 3) - INT((v# - r#) / 3): m# = m# * 10: m# = INT(m#)

PRINT , z; m#; v#; r#; v# - r#; d#; d# - (r# - sr#); r# - sr#

sr# = r#

a$ = INKEY$
```

```
IF a$ = "s" THEN STOP

DO

d# = d# + 1

IF (d# / 10) - INT(d# / 10) = 0 THEN EXIT DO

LOOP

r# = v# - d#

LOOP
```

```
"C:\Users\Reactor1967\vcns\work\10300950.BAS"

REM r# = r# + (d# * (base - 1)) seems to be the general speed of r# to aim

REM for in any base. Just get close to that speed and the chart will fly.

REM whether or not it decodes is another matter

REM speed of r# = (d# * (base - 1)) - N - 1

REM Well this did not work. I know how to calculate the speed of r# from the

REM previous v# - r#. So I need to be able to take v# - r# and calcualte

REM the speed of r# as a function of d#. Now I just need to find that

REM calculation.

v# = 1

r# = 1

d# = 0

s# = 0

CLS

RANDOMIZE TIMER

DO
```

```
z = INT(RND * 3) + 0

r# = v# - d#

IF z = 0 THEN v# = v# + ((v# - r#) * 2) + 1

IF z = 1 THEN v# = v# + ((v# - r#) * 2) + 2

IF z = 2 THEN v# = v# + ((v# - r#) * 2) + 3

m# = ((v# - r#) / 3) - INT((v# - r#) / 3): m# = m# * 10: m# = INT(m#)

PRINT , z; m#; v#; r#; v# - r#; d#; d# - (r# - s#); r# - s#

s# = r#

a$ = INKEY$

INPUT a$

IF a$ = "s" THEN STOP

d# = d# + 1

LOOP




"C:\Users\Reactor1967\vcns\work\10290517.BAS"

REM d# = ((v# - r#) - n) / base

REM v1# = v2# - (((((v2# - r2#) - n) / base) * (base - 1)) + n)

v# = 1

r# = 1

y# = 0

sr# = 1

CLS

RANDOMIZE TIMER

DO
```

sz = z

z = INT(RND * 3) + 0

s# = (v# - r#) - y#

d# = (v# - r#) - s#

r# = r# + s#

d# = v# - r#

IF z = 0 THEN v# = v# + ((v# - r#) * 2) + 1

IF z = 1 THEN v# = v# + ((v# - r#) * 2) + 2

IF z = 2 THEN v# = v# + ((v# - r#) * 2) + 3

m# = ((v# - r#) / 3) - INT((v# - r#) / 3): m# = m# * 10: m# = INT(m#)

REM 0 at v2# to 0 at v1# (y# * 2) - 2 = (r# - sr#)

REM 0 at v2# to 1 at v1# (y# * 2) - 1 = (r# - sr#)

REM 0 at v2# to 2 at v1# (y# * 2) - 0 = (r# - sr#)

REM 1 at v2# to 0 at v1# (y# * 2) - 2 = (r# - sr#)

REM 1 at v2# to 1 at v1# (y# * 2) - 1 = (r# - sr#)

REM 1 at v2# to 2 at v1# (y# * 2) - 0 = (r# - sr#)

REM 2 at v2# to 0 at v1# (y# * 2) - 2 = (r# - sr#)

REM 2 at v2# to 1 at v1# (y# * 2) - 1 = (r# - sr#)

REM 2 at v2# to 2 at v1# (y# * 2) - 0 = (r# - sr#)

test1 = (z = 0) AND (sz = 0)

test2 = (z = 0) AND (sz = 1)

test3 = (z = 0) AND (sz = 2)

test4 = (z = 1) AND (sz = 0)

test5 = (z = 1) AND (sz = 1)

test6 = (z = 1) AND (sz = 2)

test7 = (z = 2) AND (sz = 0)

```
test8 = (z = 2) AND (sz = 1)

test9 = (z = 2) AND (sz = 2)

IF test1 = -1 THEN q = 2

IF test2 = -1 THEN q = 1

IF test3 = -1 THEN q = 0

IF test4 = -1 THEN q = 2

IF test5 = -1 THEN q = 1

IF test6 = -1 THEN q = 0

IF test7 = -1 THEN q = 2

IF test8 = -1 THEN q = 1

IF test9 = -1 THEN q = 0

test = (y# * 2) - q = (r# - sr#)

IF v# < 5 THEN GOTO skip:

IF test = 0 THEN STOP

skip:

REM PRINT z; m#; v#; r#; v# - r#; d#; y#; r# - sr#; (y# * 2) - q

PRINT z; m#; v#; r#; v# - r#; d#; d# - (r# - sr#); r# - sr#

a$ = INKEY$

REM INPUT a$

IF a$ = "s" THEN STOP

IF a$ = "d" THEN EXIT DO

sr# = r#

y# = y# + 1

LOOP

Decode:

sr# = (r# - sr#)
```

```
DO

m# = ((v# - r#) / 3) - INT((v# - r#) / 3): m# = m# * 10: m# = INT(m#)

IF m# = 0 THEN z = 2

IF m# = 3 THEN z = 0

IF m# = 6 THEN z = 1

PRINT , z; m#; v#; r#; v# - r#; y#

a$ = INKEY$

REM INPUT a$

IF a$ = "s" THEN STOP

dist# = (v# - r#)

IF z = 0 THEN dist# = dist# - 1

IF z = 1 THEN dist# = dist# - 2

IF z = 2 THEN dist# = dist# - 3

dist# = ((dist# / 3) * 2)

IF z = 0 THEN dist# = dist# + 1

IF z = 1 THEN dist# = dist# + 2

IF z = 2 THEN dist# = dist# + 3

sz = z

v# = v# - dist#

r# = r# - sr#

m# = ((v# - r#) / 3) - INT((v# - r#) / 3): m# = m# * 10: m# = INT(m#)

IF m# = 0 THEN z = 2

IF m# = 3 THEN z = 0

IF m# = 6 THEN z = 1

y# = y# - 1
```

LOOP

SUB junk

test1 = (sz = 0) AND (z = 0)

test2 = (sz = 0) AND (z = 1)

test3 = (sz = 0) AND (z = 2)

test4 = (sz = 1) AND (z = 0)

test5 = (sz = 1) AND (z = 1)

test6 = (sz = 1) AND (z = 2)

test7 = (sz = 2) AND (z = 0)

test8 = (sz = 2) AND (z = 1)

test9 = (sz = 2) AND (z = 2)

IF test1 = -1 THEN q = 2

IF test2 = -1 THEN q = 1

IF test3 = -1 THEN q = 0

IF test4 = -1 THEN q = 2

IF test5 = -1 THEN q = 1

IF test6 = -1 THEN q = 0

IF test7 = -1 THEN q = 2

IF test8 = -1 THEN q = 1

IF test9 = -1 THEN q = 0

sr# = (y# * 2) - q

r# = r# - sr#

y# = y# - 1

END SUB

```
"C:\Users\Reactor1967\vcns\work\10290311.BAS"

REM d# - (r# - sr#) = 6 if the previous v# is 1

REM d# - (r# - sr#) = 7 if the previous v# is 0

REM use this info to decode. decode two v#'s down to find N

REM then configure previous speed of r# by d# and new info.

v# = 100

d# = 0

RANDOMIZE TIMER

CLS

DO

z = INT(RND * 2) + 0

DO

d# = d# + 1

test = (d# / 4) - INT(d# / 4) = 0

IF test = -1 THEN EXIT DO

LOOP

r# = v# - d#

REM IF (d# - (r# - sr#)) = 6 THEN z = 0

REM IF (d# - (r# - sr#)) = 7 THEN z = 1

IF z = 0 THEN v# = v# + (v# - r#) + 1

IF z = 1 THEN v# = v# + (v# - r#) + 2

x# = ((v# - r#) / 4) - INT((v# - r#) / 4): x# = x# * 10: x# = INT(x#)

PRINT , z; x#; v#; r#; v# - r#; d#; d# - (r# - sr#); r# - sr#
```

```
sr# = r#

a$ = INKEY$

INPUT a$

IF a$ = "s" THEN STOP

LOOP




"C:\Users\Reactor1967\vcns\work\10290057.BAS"

v# = 1

r# = 1

x# = 0

a1# = -1

a2# = 0

CLS

DO

sz = z

z = INT(RND * 2) + 0

d# = (v# - r#)

IF z = 0 THEN v# = v# + (v# - r#) + 2

IF z = 1 THEN v# = v# + (v# - r#) + 4

REM m# = ((v# - r#) / .75) - INT((v# - r#) / .75): m# = m# * 10: m# = INT(m#)

PRINT z; v#; r#; v# - r#; d#; d# - sd#; d# - (r# - sr#); r# - sr#; (r# - sr#) - srs#;
"|"; a1#; a2#

srs# = (r# - sr#)

sd# = d#
```

```
sr# = r#

a$ = INKEY$

REM INPUT a$

IF a$ = "s" THEN STOP

REM x# = 0

a1# = a1# + 2

a2# = a2# + 2

DO

s# = (v# - r#) - x#

test1 = x# - s# = a1#

test2 = x# - s# = a2#

IF test1 = -1 THEN EXIT DO

IF test2 = -1 THEN EXIT DO

a$ = INKEY$: IF a$ = "s" THEN STOP

x# = x# + 1

LOOP

r# = v# - x#

LOOP
```

"C:\Users\Reactor1967\vcns\work\10282337.BAS"

REM This is decodable but it can not keep containment.

v# = 1

r# = 1

d# = 0

```
s# = 0

z = 1

CLS

DO

sz = z

z = INT(RND * 2) + 0

d2# = v# - r#

IF z = 0 THEN v# = v# + (v# - r#) + 1

IF z = 1 THEN v# = v# + (v# - r#) + 2

PRINT , z; v#; r#; v# - r#; d2#; d# - (r# - sr#); r# - sr#; s#

a$ = INKEY$

INPUT a$

IF a$ = "s" THEN STOP

d# = (v# - r#) - s#

test = (d# / 2) - INT(d# / 2) = .5

IF test = -1 THEN s# = s# + 8

r# = r# + s#

LOOP
```

"C:\Users\Reactor1967\vcns\work\10282310.BAS"

REM One for the book v# - r# = (d# * base) + N after coding.

REM Programer and inventor Notes......

REM This program attempts to use the d# which is v# - r# before coding

REM to control the speed of r# thus also being able to use the speed

REM of r# to aid in decoding. This reading N each time a decode is completed

REM using the patten of N's to decode the speed of r#

DIM can(3): REM This is a data buffer

b = 0: REM count for data buffer

v# = 1: REM our fist vector equals v#

d# = 0: REM The most inportanted variable in this program study it well.

REM   This variable is the gas paddel and the break of this program

REM   Using this variable to control the speed of r# and v# before

REM   N is coded.

RANDOMIZE TIMER

CLS

DO

REM ------------------------------- Reading random binary pattern

z = INT(RND * 2) + 0

REM IF sd# = d# THEN z = 2

REM ------------------------------- Calculating Next v# with z

IF z = 0 THEN v# = v# + ((v# - r#) * 2) + 1

IF z = 1 THEN v# = v# + ((v# - r#) * 2) + 2

IF z = 2 THEN v# = v# + ((v# - r#) * 2) + 3

REM ------------------------------- Using v# to show N can be retrieved

m# = ((v# - r#) / 3) - INT((v# - r#) / 3): m# = m# * 10: m# = INT(m#)

m2# = (v# / 3) - INT(v# / 3): m2# = m2# * 10: m2# = INT(m2#)

PRINT , z; m2#; m#; v#; r#; v# - r#; d#; d# - (r# - sr#); r# - sr#

REM ------------------------------- Buffer storing z

b = b + 1

can(b) = z

```
IF b = 3 THEN b = 0

REM ------------------------------ Reading buffer for specific data pattern

lb = 0

FOR count = 1 TO 3

test = can(count) = 0

IF test = -1 THEN lb = lb + 1

NEXT count

REM IF lb = 3 THEN EXIT DO: REM If we find specific data pattern in buffer
exit loop

REM ------------------------------ Stop program and exit loop routine

a$ = INKEY$

INPUT a$

IF a$ = "d" THEN EXIT DO

IF a$ = "s" THEN STOP

REM ------------------------------ Saving last R# so speed of r# can be figured

sd# = d#

sr# = r#

REM ------------------------------ Calculating speed of r# so it is divisable by 3

DO

s# = (v# - r#) - d#

test = (s# / 3) - INT(s# / 3) = 0

IF test = -1 THEN EXIT DO

d# = d# + 1

a$ = INKEY$

IF a$ = "s" THEN STOP

LOOP
```

```
REM ------------------------------ Getting New r# here.

r# = v# - d#

LOOP

REM ------------------------------ Decoding Data

sr# = (r# - sr#)

REM DO

REM ------------------------------ Getting z1

m# = ((v# - r#) / 3) - INT((v# - r#) / 3): m# = m# * 10: m# = INT(m#)

IF m# = 0 THEN z1 = 2

IF m# = 6 THEN z1 = 1

IF m# = 3 THEN z1 = 0

REM ------------------------------ Pringing first line before decode

PRINT , z1; m#; v#; r#; v# - r#; d#; sr#

REM ------------------------------ Getting first decode

dist# = v# - r#

IF z1 = 0 THEN dist# = dist# - 1

IF z1 = 1 THEN dist# = dist# - 2

IF z1 = 2 THEN dist# = dist# - 3

dist# = dist# / 3

dist# = dist# * 2

IF z1 = 0 THEN dist# = dist# + 1

IF z1 = 1 THEN dist# = dist# + 2

IF z1 = 2 THEN dist# = dist# + 3

r# = r# - sr#

v# = v# - dist#

REM ------------------------------ Getting z2
```

```
m# = ((v# - r#) / 3) - INT((v# - r#) / 3): m# = m# * 10: m# = INT(m#)

IF m# = 0 THEN z2 = 2

IF m# = 6 THEN z2 = 1

IF m# = 3 THEN z2 = 0

REM ------------------------------- Printing second line before decode 2

PRINT , z2; m#; v#; r#; v# - r#

REM ------------------------------- Storing v# and r# will need this again

store1# = v#: store2# = r#: REM Use this variables when repeating this loop

REM ------------------------------- Getting seconding decode

dist# = v# - r#

IF z2 = 0 THEN dist# = dist# - 1

IF z2 = 1 THEN dist# = dist# - 2

IF z2 = 2 THEN dist# = dist# - 3

dist# = dist# / 3

dist# = dist# * 2

IF z2 = 0 THEN dist# = dist# + 1

IF z2 = 1 THEN dist# = dist# + 2

IF z2 = 2 THEN dist# = dist# + 3

v# = v# - dist#

REM ------------------------------- Getting z3

m# = ((v# - r#) / 3) - INT((v# - r#) / 3): m# = m# * 10: m# = INT(m#)

IF m# = 0 THEN z3 = 2

IF m# = 6 THEN z3 = 1

IF m# = 3 THEN z3 = 0

REM ------------------------------- Printing 3 line after 2nd decode

PRINT , , z3; v#
```

REM -------------------------------- Reading z1 and z2 and z3 for speed of r# calculation

test1 = (z1 = 0) AND (z2 = 0) AND (z3 = 0): IF test1 = -1 THEN sr# = sr# - 0

test2 = (z1 = 0) AND (z2 = 0) AND (z3 = 1): IF test2 = -1 THEN sr# = sr# - 4

test3 = (z1 = 0) AND (z2 = 0) AND (z3 = 2)

test4 = (z1 = 0) AND (z2 = 1) AND (z3 = 0)

test5 = (z1 = 0) AND (z2 = 1) AND (z3 = 1)

test6 = (z1 = 0) AND (z2 = 1) AND (z3 = 2)

test7 = (z1 = 0) AND (z2 = 2) AND (z3 = 0)

test8 = (z1 = 0) AND (z2 = 2) AND (z3 = 1)

test9 = (z1 = 0) AND (z2 = 2) AND (z3 = 2)

test10 = (z1 = 1) AND (z2 = 0) AND (z3 = 0)

test11 = (z1 = 1) AND (z2 = 0) AND (z3 = 1)

test12 = (z1 = 1) AND (z2 = 0) AND (z3 = 2)

test13 = (z1 = 1) AND (z2 = 1) AND (z3 = 0)

test14 = (z1 = 1) AND (z2 = 1) AND (z3 = 1)

test15 = (z1 = 1) AND (z2 = 1) AND (z3 = 2)

test16 = (z1 = 1) AND (z2 = 2) AND (z3 = 0)

test17 = (z1 = 1) AND (z2 = 2) AND (z3 = 1)

test18 = (z1 = 1) AND (z2 = 2) AND (z3 = 2)

test19 = (z1 = 2) AND (z2 = 0) AND (z3 = 0)

test20 = (z1 = 2) AND (z2 = 0) AND (z3 = 1)

test21 = (z1 = 2) AND (z2 = 0) AND (z3 = 2)

test22 = (z1 = 2) AND (z2 = 1) AND (z3 = 0)

test23 = (z1 = 2) AND (z2 = 1) AND (z3 = 1)

test24 = (z1 = 2) AND (z2 = 1) AND (z3 = 2)

test25 = (z1 = 2) AND (z2 = 2) AND (z3 = 0)

test26 = (z1 = 2) AND (z2 = 2) AND (z3 = 1)

test27 = (z1 = 2) AND (z2 = 2) AND (z2 = 2)

"C:\Users\Reactor1967\vcns\work\10281957.BAS"

REM Programer and inventor Notes......

REM This program attempts to use the d# which is v# - r# before coding

REM to control the speed of r# thus also being able to use the speed

REM of r# to aid in decoding. This reading N each time a decode is completed

REM using the patten of N's to decode the speed of r#

DIM can(3): REM This is a data buffer

b = 0: REM count for data buffer

v# = 1: REM our fist vector equals v#

d# = 0: REM The most inportaned variable in this program study it well.

REM   This variable is the gas paddel and the break of this program

REM   Using this variable to control the speed of r# and v# before

REM   N is coded.

RANDOMIZE TIMER

CLS

DO

REM ------------------------------ Reading random binary pattern

z = INT(RND * 2) + 0

IF sd# = d# THEN z = 2

REM ------------------------------ Calculating Next v# with z

```
IF z = 0 THEN v# = v# + ((v# - r#) * 2) + 1

IF z = 1 THEN v# = v# + ((v# - r#) * 2) + 2

IF z = 2 THEN v# = v# + ((v# - r#) * 2) + 3

REM ------------------------------ Using v# to show N can be retrieved

m# = ((v# - r#) / 3) - INT((v# - r#) / 3): m# = m# * 10: m# = INT(m#)

m2# = (v# / 3) - INT(v# / 3): m2# = m2# * 10: m2# = INT(m2#)

PRINT , z; m2#; m#; v#; r#; v# - r#; d#; d# - (r# - sr#); r# - sr#

REM ------------------------------ Buffer storing z

b = b + 1

can(b) = z

IF b = 3 THEN b = 0

REM ------------------------------ Reading buffer for specific data pattern

lb = 0

FOR count = 1 TO 3

test = can(count) = 0

IF test = -1 THEN lb = lb + 1

NEXT count

REM IF lb = 3 THEN EXIT DO: REM If we find specific data pattern in buffer
exit loop

REM ------------------------------ Stop program and exit loop routine

a$ = INKEY$

INPUT a$

IF a$ = "d" THEN EXIT DO

IF a$ = "s" THEN STOP

REM ------------------------------ Saving last R# so speed of r# can be figured

sd# = d#
```

```
sr# = r#

REM ------------------------------ Calculating speed of r# so it is divisable by 3

DO

s# = (v# - r#) - d#

test = (s# / 3) - INT(s# / 3) = 0

IF test = -1 THEN EXIT DO

d# = d# + 1

a$ = INKEY$

IF a$ = "s" THEN STOP

LOOP

REM ------------------------------ Getting New r# here.

r# = v# - d#

LOOP

REM ------------------------------ Decoding Data

sr# = (r# - sr#)

REM DO

REM ------------------------------ Getting z1

m# = ((v# - r#) / 3) - INT((v# - r#) / 3): m# = m# * 10: m# = INT(m#)

IF m# = 0 THEN z1 = 2

IF m# = 6 THEN z1 = 1

IF m# = 3 THEN z1 = 0

REM ------------------------------ Pringing first line before decode

PRINT , z1; m#; v#; r#; v# - r#; d#; sr#

REM ------------------------------ Getting first decode

dist# = v# - r#

IF z1 = 0 THEN dist# = dist# - 1
```

```
IF z1 = 1 THEN dist# = dist# - 2

IF z1 = 2 THEN dist# = dist# - 3

dist# = dist# / 3

dist# = dist# * 2

IF z1 = 0 THEN dist# = dist# + 1

IF z1 = 1 THEN dist# = dist# + 2

IF z1 = 2 THEN dist# = dist# + 3

r# = r# - sr#

v# = v# - dist#

REM ------------------------------ Getting z2

m# = ((v# - r#) / 3) - INT((v# - r#) / 3): m# = m# * 10: m# = INT(m#)

IF m# = 0 THEN z2 = 2

IF m# = 6 THEN z2 = 1

IF m# = 3 THEN z2 = 0

REM ------------------------------ Printing second line before decode 2

PRINT , z2; m#; v#; r#; v# - r#

REM ------------------------------ Storing v# and r# will need this again

store1# = v#: store2# = r#: REM Use this variables when repeating this loop

REM ------------------------------ Getting seconding decode

dist# = v# - r#

IF z2 = 0 THEN dist# = dist# - 1

IF z2 = 1 THEN dist# = dist# - 2

IF z2 = 2 THEN dist# = dist# - 3

dist# = dist# / 3

dist# = dist# * 2

IF z2 = 0 THEN dist# = dist# + 1
```

IF z2 = 1 THEN dist# = dist# + 2

IF z2 = 2 THEN dist# = dist# + 3

v# = v# - dist#

REM ------------------------------- Getting z3

m# = ((v# - r#) / 3) - INT((v# - r#) / 3): m# = m# * 10: m# = INT(m#)

IF m# = 0 THEN z3 = 2

IF m# = 6 THEN z3 = 1

IF m# = 3 THEN z3 = 0

REM ------------------------------- Printing 3 line after 2nd decode

PRINT , , z3; v#

REM ------------------------------- Reading z1 and z2 and z3 for speed of r# calculation

test1 = (z1 = 0) AND (z2 = 0) AND (z3 = 0): IF test1 = -1 THEN sr# = sr# - 0

test2 = (z1 = 0) AND (z2 = 0) AND (z3 = 1): IF test2 = -1 THEN sr# = sr# - 4

test3 = (z1 = 0) AND (z2 = 0) AND (z3 = 2)

test4 = (z1 = 0) AND (z2 = 1) AND (z3 = 0)

test5 = (z1 = 0) AND (z2 = 1) AND (z3 = 1)

test6 = (z1 = 0) AND (z2 = 1) AND (z3 = 2)

test7 = (z1 = 0) AND (z2 = 2) AND (z3 = 0)

test8 = (z1 = 0) AND (z2 = 2) AND (z3 = 1)

test9 = (z1 = 0) AND (z2 = 2) AND (z3 = 2)

test10 = (z1 = 1) AND (z2 = 0) AND (z3 = 0)

test11 = (z1 = 1) AND (z2 = 0) AND (z3 = 1)

test12 = (z1 = 1) AND (z2 = 0) AND (z3 = 2)

test13 = (z1 = 1) AND (z2 = 1) AND (z3 = 0)

test14 = (z1 = 1) AND (z2 = 1) AND (z3 = 1)

test15 = (z1 = 1) AND (z2 = 1) AND (z3 = 2)

test16 = (z1 = 1) AND (z2 = 2) AND (z3 = 0)

test17 = (z1 = 1) AND (z2 = 2) AND (z3 = 1)

test18 = (z1 = 1) AND (z2 = 2) AND (z3 = 2)

test19 = (z1 = 2) AND (z2 = 0) AND (z3 = 0)

test20 = (z1 = 2) AND (z2 = 0) AND (z3 = 1)

test21 = (z1 = 2) AND (z2 = 0) AND (z3 = 2)

test22 = (z1 = 2) AND (z2 = 1) AND (z3 = 0)

test23 = (z1 = 2) AND (z2 = 1) AND (z3 = 1)

test24 = (z1 = 2) AND (z2 = 1) AND (z3 = 2)

test25 = (z1 = 2) AND (z2 = 2) AND (z3 = 0)

test26 = (z1 = 2) AND (z2 = 2) AND (z3 = 1)

test27 = (z1 = 2) AND (z2 = 2) AND (z2 = 2)

"C:\Users\Reactor1967\vcns\work\10281058.BAS"

REM 2 or 7 = 0

REM 0 or 5 = 1

v# = 100

r# = 1

CLS

RANDOMIZE TIMER

y# = 0

DO

z = INT(RND * 2) + 0

```
d# = v# - r#

IF z = 0 THEN v# = v# + (v# - r#) + 1

IF z = 1 THEN v# = v# + (v# - r#) + 2

m# = ((v# - r#) / 4) - INT((v# - r#) / 4): m# = m# * 10: m# = INT(m#)

PRINT , z; m#; v#; r#; v# - r#; d#; y#; sr#

a$ = INKEY$

IF a$ = "s" THEN STOP

IF a$ = "d" THEN EXIT DO

y# = y# + 1

d# = 0

DO

sr# = d# - y#

test = sr# + d# = (v# - r#)

IF test = -1 THEN EXIT DO

REM a$ = INKEY$

REM IF a$ = "s" THEN STOP

d# = d# + 1

LOOP UNTIL d# > (v# - r#)

r# = r# + sr#

LOOP

DO

m# = ((v# - r#) / 4) - INT((v# - r#) / 4): m# = m# * 10: m# = INT(m#)

IF m# = 0 THEN z = 1

IF m# = 5 THEN z = 1

IF m# = 2 THEN z = 0

IF m# = 7 THEN z = 0
```

```
PRINT , z; m#; v#; r#; v# - r#; d#; y#; sr#

a$ = INKEY$

IF a$ = "s" THEN STOP

dist# = (v# - r#)

IF (dist# / 2) - INT(dist# / 2) = .5 THEN dist# = dist# - 1

dist# = dist# / 2

dist# = dist# + 1

v# = v# - dist#

r# = r# - sr#

m# = ((v# - r#) / 4) - INT((v# - r#) / 4): m# = m# * 10: m# = INT(m#)

IF m# = 0 THEN z = 1

IF m# = 5 THEN z = 1

IF m# = 2 THEN z = 0

IF m# = 7 THEN z = 0

IF (dist# / 2) - INT(dist# / 2) = .5 THEN dist# = dist# - 1

dist# = dist# / 2

dist# = dist# + 1

IF z = 0 THEN dist# = dist# - 1

IF z = 1 THEN dist# = dist# - 2

y# = y# - 1

sr# = (v# - r#) - dist# - y#

PRINT , z; m#; v#; r#; v# - r#; d#; y#; sr#

STOP

LOOP
```

"C:\Users\Reactor1967\vcns\work\10280429.BAS"

REM notice the speed of r# is the (base * d#) => than d#. This is what is

REM needed for the speed of r# to keep up with d# for the program to maintain

REM its integerity.

v# = 1

r# = 1

CLS

RANDOMIZE TIMER

DO

z = INT(RND * 3) + 0

d# = v# - r#

IF z = 0 THEN v# = v# + ((v# - r#) * 2) + 1

IF z = 1 THEN v# = v# + ((v# - r#) * 2) + 2

IF z = 2 THEN v# = v# + ((v# - r#) * 2) + 3

PRINT , z; v#; r#; v# - r#; d#; d# - (r# - sr#); r# - sr#

sr# = r#

a$ = INKEY$

INPUT a$

IF a$ = "s" THEN STOP

dist# = INT((v# - r#) / 3)

r# = v# - dist#

LOOP

```
"C:\Users\Reactor1967\vcns\work\10270618.BAS"

v# = 1

r# = 1

sr# = 1

x# = 0

CLS

RANDOMIZE TIMER

DO

z = INT(RND * 2) + 0

d# = v# - r#

IF z = 0 THEN v# = v# + (v# - r#) + 2

IF z = 1 THEN v# = v# + (v# - r#) + 4

PRINT z; v#; r#; v# - r#; (v# - r#) - svr#; d#; d# - sd#; d# - (r# - sr#); r# - sr#;
(r# - sr#) - rsr#

sd# = d#

rsr# = (r# - sr#)

svr# = (v# - r#)

a$ = INKEY$

REM INPUT a$

IF a$ = "s" THEN STOP

y# = d# - (r# - sr#)

sr# = r#

DO

s# = (v# - r#) - x#

IF x# - s# = y# + 2 THEN EXIT DO
```

```
x# = x# + 1

REM a$ = INKEY$

REM IF a$ = "s" THEN STOP

LOOP

r# = v# - x#

LOOP
```

```
"C:\Users\Reactor1967\vcns\work\10270009.BAS"

v# = 100

d# = 0

CLS

DO

z = INT(RND * 2) + 0

DO

d# = d# + 1

m# = (d# / 4) - INT(d# / 4)

m# = m# * 10

m# = INT(m#)

IF m# = 5 THEN EXIT DO

LOOP

r# = v# - d#

IF z = 0 THEN v# = v# + (v# - r#) + 2

IF z = 1 THEN v# = v# + (v# - r#) + 4

m2# = ((v# - r#) / 4) - INT((v# - r#) / 4): m2# = m2# * 10: m2# = INT(m2#)

x# = x# + 1
```

```
PRINT , z; m2#; v#; r#; v# - r#; d#; d# - (r# - sr#); r# - sr#

sr# = r#

a$ = INKEY$

IF a$ = "s" THEN STOP

LOOP
```

"C:\Users\Reactor1967\vcns\work\10262047.BAS"

```
REM playing with v# - r#;d#;and speed of r#

vr# = 2703

REM x# = INT((vr#) / 2) - 20

REM IF (x# / 2) - INT(x# / 2) = 0 THEN x# = x# + 1

x# = 0

CLS

DO

DO

s# = (v# - r#) - x#

o# = 0: REM This controls if v# - r# gets larger or smaller.

test = (((x# + o#) - s#) > (d# - y#)) AND ((x# / 2) - INT(x# / 2) = .5)

REM test = (x# - s#) = ((d# - y#) + 2)

IF test = -1 THEN EXIT DO

a$ = INKEY$

IF a$ = "s" THEN STOP

x# = x# + 1

LOOP
```

```
PRINT , , vr#; x#; x# - s#; s#

INPUT a$

IF a$ = "s" THEN STOP

x# = x# + 1

LOOP
```

```
vr# = 27212

x# = 0

y# = 408: REM play with this variable make it anything that is even.

CLS

DO

s# = vr# - x#

PRINT , , vr#; x#; x# - s#; s#

a$ = INKEY$

IF a$ = "s" THEN STOP

IF x# - s# = y# THEN STOP

x# = x# + 1

LOOP
```

```
vr# = 1092184
```

```
REM d# has to be odd

REM d# - sr# needs to be even

x# = 1

CLS

DO

s# = vr# - x#

PRINT , , vr#; x#; x# - s#; s#

a$ = INKEY$

REM INPUT a$

IF a$ = "s" THEN STOP

x# = x# + 2

LOOP
```

```
"C:\Users\Reactor1967\vcns\work\10261823.BAS"

REM v1# = (((v2# - r2#) / base) * (base - 1)) + (N - 1):coding up

REM NEW DISCOVERY!!! IT IS POSSIBLE FOR V - R TO STAY
POSITIVE AS LONG

REM AS D STAYS RIGID TOO.

v# = 1

r# = 1

CLS

RANDOMIZE TIMER

DO

z = INT(RND * 2) + 0
```

```
d# = v# - r#

IF z = 0 THEN v# = v# + (v# - r#) + 2

IF z = 1 THEN v# = v# + (v# - r#) + 4

m# = ((v# - r#) / 4) - INT((v# - r#) / 4): m# = m# * 10: m# = INT(m#)

PRINT z; m#; v#; r#; v# - r#; d#; d# - (r# - sr#); r# - sr#

lb# = d# - (r# - sr#)

y# = (r# - sr#)

sr# = r#

a$ = INKEY$

REM INPUT a$

IF a$ = "s" THEN STOP

x# = 0

x# = INT((v# - r#) / 2) - 20

DO

s# = (v# - r#) - x#

o# = 0: REM This controls if v# - r# gets larger or smaller.

test = (((x# + o#) - s#) > (d# - y#)) AND ((x# / 2) - INT(x# / 2) = .5) AND ((x#
- s#) = (lb# + 2))

IF test = -1 THEN EXIT DO

REM a$ = INKEY$

REM IF a$ = "s" THEN STOP

x# = x# + 1

LOOP

r# = v# - x#

LOOP
```

```
"C:\Users\Reactor1967\vcns\work\10250107.BAS"

REM vr1# and vr2# are (v# - r#) after coding

REM d# = v# - r# before coding

REM s1# and s2# are speed of r#

REM d# - s1# and d# - s2# is the distance between d# and speed of r#

REM this program show the possibilities for speed of r# and for d#

vr1# = 237

vr2# = 238

CLS

d# = 0

DO

s1# = vr1# - d#

s2# = vr2# - d#

PRINT , , vr1#; d#; d# - s1#; s1#

PRINT , , vr2#; d#; d# - s2#; s2#

INPUT a$

IF a$ = "s" THEN STOP

d# = d# + 1

LOOP
```

```
"C:\Users\Reactor1967\vcns\work\10250040.BAS"

v# = 1
```

```
r# = 1
sr# = 0
d# = 0
CLS
z = 1
DO
z = INT(RND * 2) + 0
d# = v# - r#
IF z = 0 THEN v# = v# + (v# - r#) + 1
IF z = 1 THEN v# = v# + (v# - r#) + 2
m# = ((v# - r#) / 4) - INT((v# - r#) / 4): m# = m# * 10: m# = INT(m#)
PRINT , z; m#; v#; r#; v# - r#; d#; d# - sd#; r# - sr#
a$ = INKEY$
REM INPUT a$
IF a$ = "s" THEN STOP
sd# = d#
s# = r# - sr#
sr# = r#
REM x# = 0
x# = INT((v# - r#) / 2) - 25
DO
IF x# - ((v# - r#) - x#) > (d# - s#) THEN EXIT DO
a$ = INKEY$
IF a$ = "s" THEN STOP
x# = x# + 1
LOOP
```

```
r# = v# - x#

LOOP




"C:\Users\Reactor1967\vcns\work\10242342.BAS"

d# = 0

sr# = 0

v# = 1

r# = 1

CLS

RANDOMIZE TIMER

z = 0

DO

d# = v# - r#

IF z = 0 THEN v# = v# + (v# - r#) + 1

IF z = 1 THEN v# = v# + (v# - r#) + 2

m# = ((v# - r#) / 4) - INT((v# - r#) / 4): m# = m# * 10: m# = INT(m#)

PRINT z; m#; v#; r#; v# - r#; d#; d# - sr#; sr#

a$ = INKEY$

IF a$ = "s" THEN STOP

REM ----------------------------

REM this sub takes v - r then tell it the desired result for d minus speed of

REM r and it gives it to you provided it can do that for the v - r provided.

REM CLS

a# = v# - r#
```

sz = z

z = INT(RND * 2) + 0

REM ---

test1 = (sz = 0) AND (z = 0) AND (d# - sr# = -1): REM at -1 0 to 0 = -1

IF test1 = -1 THEN b# = -1

test2 = (sz = 0) AND (z = 0) AND (d# - sr# = 0): REM at 0 0 to 0 = 1

IF test2 = -1 THEN b# = 1

test3 = (sz = 0) AND (z = 0) AND (d# - sr# = 1): REM at 1 0 to 0 = 3

IF test3 = -1 THEN b# = 3

test4 = (sz = 0) AND (z = 0) AND (d# - sr# = 2): REM at 2 0 to 0 = 5

IF test4 = -1 THEN b# = 5

test5 = (sz = 0) AND (z = 0) AND (d# - sr# = 3): REM at 3 0 to 0 = -1

IF test5 = -1 THEN b# = -1

test6 = (sz = 0) AND (z = 0) AND (d# - sr# = 4): REM at 4 0 to 0 = 1

IF test6 = -1 THEN b# = 1

test7 = (sz = 0) AND (z = 0) AND (d# - sr# = 5): REM at 5 0 to 0 = 3

IF test7 = -1 THEN b# = 3

test8 = (sz = 0) AND (z = 0) AND (d# - sr# = 6): REM at 6 0 to 0 = 5

IF test8 = -1 THEN b# = 5

REM ---

test9 = (sz = 0) AND (z = 1) AND (d# - sr# = -1): REM at -1 0 to 1 = -1

IF test9 = -1 THEN b# = -1

test10 = (sz = 0) AND (z = 1) AND (d# - sr# = 0): REM at 0 0 to 1 = 1

IF test10 = -1 THEN b# = 1

test11 = (sz = 0) AND (z = 1) AND (d# - sr# = 1): REM at 1 0 to 1 = 3

IF test11 = -1 THEN b# = 3

dli

test12 = (sz = 0) AND (z = 1) AND (d# - sr# = 2): REM at 2 0 to 1 = 5

IF test12 = -1 THEN b# = 5

test13 = (sz = 0) AND (z = 1) AND (d# - sr# = 3): REM at 3 0 to 1 = -1

IF test13 = -1 THEN b# = -1

test14 = (sz = 0) AND (z = 1) AND (d# - sr# = 4): REM at 4 0 to 1 = 1

IF test14 = -1 THEN b# = 1

test15 = (sz = 0) AND (z = 1) AND (d# - sr# = 5): REM at 5 0 to 1 = 3

IF test15 = -1 THEN b# = 3

test16 = (sz = 0) AND (z = 1) AND (d# - sr# = 6): REM at 6 0 to 1 = 5

IF test16 = -1 THEN b# = 5

REM ---

test17 = (sz = 1) AND (z = 0) AND (d# - sr# = -1): REM at -1 1 to 0 = 0

IF test17 = -1 THEN b# = 0

test18 = (sz = 1) AND (z = 0) AND (d# - sr# = 0): REM at 0 1 to 0 = 2

IF test18 = -1 THEN b# = 2

test19 = (sz = 1) AND (z = 0) AND (d# - sr# = 1): REM at 1 1 to 0 = 4

IF test19 = -1 THEN b# = 4

test20 = (sz = 1) AND (z = 0) AND (d# - sr# = 2): REM at 2 1 to 0 = 6

IF test20 = -1 THEN b# = 6

test21 = (sz = 1) AND (z = 0) AND (d# - sr# = 3): REM at 3 1 to 0 = 0

IF test21 = -1 THEN b# = 0

test22 = (sz = 1) AND (z = 0) AND (d# - sr# = 4): REM at 4 1 to 0 = 2

IF test22 = -1 THEN b# = 2

test23 = (sz = 1) AND (z = 0) AND (d# - sr# = 5): REM at 5 1 to 0 = 4

IF test23 = -1 THEN b# = 4

test24 = (sz = 1) AND (z = 0) AND (d# - sr# = 6): REM at 6 1 to 0 = 6

IF test24 = -1 THEN b# = 6

REM ---

test25 = (sz = 1) AND (z = 1) AND (d# - sr# = -1): REM at -1 1 to 1 = 0

IF test25 = -1 THEN b# = 0

test26 = (sz = 1) AND (z = 1) AND (d# - sr# = 0): REM at 0 1 to 1 = 2

IF test26 = -1 THEN b# = 2

test27 = (sz = 1) AND (z = 1) AND (d# - sr# = 1): REM at 1 1 to 1 = 4

IF test27 = -1 THEN b# = 4

test28 = (sz = 1) AND (z = 1) AND (d# - sr# = 2): REM at 2 1 to 1 = 6

IF test28 = -1 THEN b# = 6

test29 = (sz = 1) AND (z = 1) AND (d# - sr# = 3): REM at 3 1 to 1 = 0

IF test29 = -1 THEN b# = 0

test30 = (sz = 1) AND (z = 1) AND (d# - sr# = 4): REM at 4 1 to 1 = 2

IF test30 = -1 THEN b# = 2

test31 = (sz = 1) AND (z = 1) AND (d# - sr# = 5): REM at 5 1 to 1 = 4

IF test31 = -1 THEN b# = 4

test32 = (sz = 1) AND (z = 1) AND (d# - sr# = 6): REM at 6 1 to 1 = 6

IF test32 = -11 THEN b# = 6

REM ---

REM d# = 0

d# = INT((v# - r#) / 2) - 10

DO

s# = a# - d#

IF d# - s# = b# THEN EXIT DO

REM a$ = INKEY$

REM IF a$ = "s" THEN STOP

```
d# = d# + 1

LOOP UNTIL d# >= a#

r# = r# + s#

sr# = s#

LOOP
```

```
"C:\Users\Reactor1967\vcns\work\10242025.BAS"

REM im starting to name my files month/day/time in military instead of

REM month/day/year. The date of my files will show the year.

v# = 1

r# = 1

CLS

RANDOMIZE TIMER

sr# = 0


REM at d# - speed of r# for N1 equals d# - speed of r# at N2

REM at -1 0 to 0 = -1

REM at 0 0 to 0 = 1

REM at 1 0 to 0 = 3

REM at 2 0 to 0 = 5

REM at 3 0 to 0 = -1

REM at 4 0 to 0 = 1

REM at 5 0 to 0 = 3

REM at 6 0 to 0 = 5
```

REM ----------------

REM at -1 1 to 1 = 0

REM at 0 1 to 1 = 2

REM at 1 1 to 1 = 4

REM at 2 1 to 1 = 6

REM at 3 1 to 1 = 0

REM at 4 1 to 1 = 2

REM at 5 1 to 1 = 4

REM at 6 1 to 1 = 6

REM ----------------

REM at -1 0 to 1 = -1

REM at 0 0 to 1 = 1

REM at 1 0 to 1 = 3

REM at 2 0 to 1 = 5

REM at 3 0 to 1 = -1

REM at 4 0 to 1 = 1

REM at 5 0 to 1 = 3

REM at 6 0 to 1 = 5

REM ----------------

REM at -1 1 to 0 = 0

REM at 0 1 to 0 = 2

REM at 1 1 to 0 = 4

REM at 2 1 to 0 = 6

REM at 3 1 to 0 = 0

REM at 4 1 to 0 = 2

REM at 5 1 to 0 = 4

```
REM at  6 1 to 0 = 6

REM ----------------

DO

sz = z

z = INT(RND * 2) + 0

d# = v# - r#

IF z = 0 THEN v# = v# + (v# - r#) + 1

IF z = 1 THEN v# = v# + (v# - r#) + 2

PRINT z; v#; r#; v# - r#; "|"; z; d#; d# - sr#; sr#

a$ = INKEY$

REM ---------------------------- disabled remove rem comments to enable

REM test = (sz = 1) AND (z = 0)

REM IF test = -1 THEN INPUT a$

REM ----------------------------

IF a$ = "s" THEN STOP

IF d# - sr# >= 3 THEN sr# = sr# + 4

r# = r# + sr#

LOOP
```

"C:\Users\Reactor1967\vcns\work\10232320.BAS"

```
REM Inventor Notes: What I need to do is figure out all the distances

REM that I can code base 4 with then code binary and base 4 at the same

REM time. When I decode I can look at v - r and tell if I can use base

REM 4 with it or not. If not I use binary to tell when or when not
```

dlvi

```
REM to increment/decrement r#. If the distance v - r is a base 4 distance
REM I find my N. If it is a 3 or a 4 then I have my binary data.
d# = 1
DIM can1(100)
DIM can2(100)
DIM can3(100)
CLS
DO
PRINT d#;
can1(d#) = d#
FOR count = 1 TO 5
test = (d# / count) - INT(d# / count) = 0
IF test = -1 THEN PRINT count;
IF test = -1 THEN can3(d#) = (d# * count) + count
IF test = -1 THEN can2(d#) = (d# * count) + 1
IF test = -1 THEN x# = count
NEXT count
PRINT (d# * 2) + 1; (d# * 2) + (x# - 1);
PRINT " "
d# = d# + 1
a$ = INKEY$
INPUT a$
IF a$ = "s" THEN STOP
LOOP UNTIL d# = 101
FOR count = 1 TO 100
PRINT can1(count); can2(count); can3(count)
```

```
INPUT a$

IF a$ = "s" THEN STOP

NEXT count




"C:\Users\Reactor1967\vcns\work\1207427.BAS"

REM Essentially to make this work you need to know that if your going

REM from a 0 to a 1 then r# = r# + (0 to a 1). Increase r# first then

REM code. When decodeing you see you have a one and you decode to a 0

REM then r# = r# - (1 to a 0). There needs to be specific values

REM for increasing r# from a 0 to a 0, 0 to a 0, 1 to a 0, and 1 to a 1.

REM The distance between v# and r# needs to stay the same before coding

REM and increasing r# to make this work or at least have it going in a

REM specific pattern in relation to the binary code.

v# = 1

r# = 1

CLS

DO

z = INT(RND * 2) + 0

v# = v# + (v# - r#) + 2

PRINT , , "1"; v#; r#; v# - r#

a$ = INKEY$

IF a$ = "s" THEN STOP

r# = r# + 2

LOOP
```

"C:\Users\Reactor1967\vcns\work\01030139.BAS"

REM Here you keep up with your speed of r#. Look at your N to tell what

REM to subtract from your speed of r#. Then decode r# to your previous

REM r# to decode v# again. You find N by looking at both your v# and your

REM r#. This seems to be a good start of something but Not fully tested

REM as of yet.

redo:

RANDOMIZE TIMER

v# = 1

r# = 1

sr# = 0

CLS

DO

z = INT(RND * 2) + 0

IF z = 0 THEN v# = v# + (v# - r# + 1)

IF z = 1 THEN v# = v# + (v# - r# + 2)

m# = (v# / 4) - INT(v# / 4): m# = m# * 10: m# = INT(m#)

PRINT , z; m#; v#; r#; v# - r#; sr#; r# - rsr#

rsr# = r#

a$ = INKEY$

REM INPUT a$

IF a$ = "s" THEN STOP

REM --

```
test = (v# = 224) AND (z = 0) AND ((r# / 2) - INT(r# / 2) = .5)

IF test = -1 THEN PRINT , , , "COOL!!!!!!!!!!!!!!!!!!!!!!!!!!!!!!!!!!!!!"

IF test = -1 THEN SLEEP 2

test2 = (v# = 224) AND (z = 1) AND ((r# / 2) - INT(r# / 2) = .5)

IF test2 = -1 THEN BEEP

IF test2 = -1 THEN STOP

REM --------------------------------------------------------------

test = (v# = 224) AND (z = 1) AND ((r# / 2) - INT(r# / 2) = 0)

IF test = -1 THEN PRINT , , , "COOL!!!!!!!!!!!!!!!!!!!!!!!!!!!!!!!!!!!!!"

IF test = -1 THEN SLEEP 2

test2 = (v# = 224) AND (z = 0) AND ((r# / 2) - INT(r# / 2) = 0)

IF test2 = -1 THEN BEEP

IF test2 = -1 THEN STOP

REM --------------------------------------------------------------

IF v# > 224 THEN GOTO redo:

r# = r# + sr#

IF z = 0 THEN sr# = sr# + 1

IF z = 1 THEN sr# = sr# + 2

LOOP




"C:\Users\Reactor1967\vcns\work\102403A.BAS"

REM BINGO!!! BINGO!!! BINGO!!!!

REM HOW TO KEEP TRACK OF YOUR SPEED OF R#

REM STEP 1. Keep track of your speed of r# and d# as you code.
```

dlx

REM step 2. When you finish coding take note of

REM   A. (Speed of r#) - d# or d# - (speed of r#) How ever you wish.

REM   B. Decode to your previous v#. Use speed of r# to decode to

REM   previous r#. Take Note of N2 and N1.

REM   1. You can use sysmetery of v# or sysmetery of (v# - r#)

REM   How ever you wish.

REM   C. Find your previous d# take note of it.

REM   D. Knowing ((N2 & N1) and ((speed of r#) - d#)) use chart to

REM   calculate of what previous (speed of r#) - d# should be.

REM   E. Knowing this distance take your prevous d# and find your

REM   speed of r# with your new found data.

REM THIS PROGRAM DOES NOT FULLY IMPLEMENT THE FUNCTION DESCRIBED ABOVE.

REM BUT IT SEEMS IT CAN BE IMPLEMENTED. To do that you have to

REM 1. Take N1 and N2. Take d# - speed of r#. Know what your next

REM   d# - speed of r# should be.

REM 2. Work to get your next d# - speed of r#

REM 3. So that when you decode you can take your speed of r#

REM 4. Find N2 and N1. Know what your previous d# - speed of r# should be.

REM 5. Find your previous d# and find your previous speed of r# from d#.

REM d# - speed of r# is a variable to be used to determine how the

REM speed of r# changes for N. Use it to find a new math equation.

v# = 1

r# = 1

CLS

RANDOMIZE TIMER

```
sr# = 0

x# = 0

DO

FOR count = 1 TO 2

z = INT(RND * 2) + 0

d# = v# - r#

IF z = 0 THEN v# = v# + (v# - r#) + 1

IF z = 1 THEN v# = v# + (v# - r#) + 2

PRINT , , z; v#; r#; v# - r#; d#; d# - sr#; sr#: REM Enabled

REM PRINT , , z; sr#; d# - sr#: REM Disabled

A$ = INKEY$

INPUT A$

IF A$ = "s" THEN STOP

IF d# - sr# >= 3 THEN sr# = sr# + 4

r# = r# + sr#

NEXT count

PRINT "------------------------------------------------------------"

LOOP




"C:\Users\Reactor1967\vcns\work\102303I.BAS"

vr# = 1

d# = 0

r# = 0

z = 0: REM 1 is neutral here for N.
```

```
RANDOMIZE TIMER

CLS

DO

d# = INT(vr# / 2)

DO

d# = d# + 1

IF (vr# - d#) >= r# THEN EXIT DO

a$ = INKEY$

IF a$ = "s" THEN STOP

LOOP

r# = vr# - d#

PRINT , , z; vr#; d#; r#

INPUT a$

IF a$ = "s" THEN STOP

z = INT(RND * 2) + 0

vr# = (d# * 2) + z + 1

LOOP
```

"C:\Users\Reactor1967\vcns\work\102303H.BAS"

```
DIM can(16000)

DIM can2(16000)

REM Here again I prove my method that you can use the speed of r# to code

REM and decode but I did not use the binary on decode to tell what to do

REM with my speed of r#. Here I mainly wanted to do some testing. So, next
```

REM project is to study the speed of v#. Learn how the speed of v# increases

REM by the binary that it codes. Learn to increase the speed of v# based

REM on the binary then try to apply that to the speed of r#. Because, its

REM only by study keeping r# up with v# on velocity that all this can become

REM possible. R needs to accelerate with v# and deaccelerate with v#. But

REM that acceleration deacceleration needs to be based on N if possible.

REM Actually d# can be used to control the speed of r# or the speed of r#

REM can be used to control d#. Or, a system can be put in place for both.

REM But knowing that the speed of r# must be relative to the speed of v#.

redo:

lb# = 1

v# = 1

r# = 1

sr# = 0

x# = 0

z = 1

can2(lb#) = z

lb# = lb# + 1

CLS

RANDOMIZE TIMER

DO

z = INT(RND * 2) + 0

can2(lb#) = z

d# = v# - r#

IF z = 0 THEN v# = v# + (v# - r#) + 1

IF z = 1 THEN v# = v# + (v# - r#) + 2

```
m# = (v# / 4) - INT(v# / 4): m# = m# * 10: m# = INT(m#)

PRINT , z; m#; v#; r#; v# - r#; d#; d# - sr#; sr#; sr# - x#

can(lb#) = sr#: IF lb# = 16000 THEN EXIT DO

a$ = INKEY$

REM INPUT a$

IF a$ = "s" THEN STOP

IF a$ = "d" THEN EXIT DO

x# = sr#

lb# = lb# + 1

IF d# - sr# >= 3 THEN sr# = sr# + 4

r# = r# + sr#

LOOP

DO

m# = (v# / 4) - INT(v# / 4): m# = m# * 10: m# = INT(m#)

test1 = (m# = 0) OR (m# = 5)

test2 = (m# = 2) OR (m# = 7)

IF test1 = -1 THEN z = 0

IF test2 = -1 THEN z = 1

d# = v# - r#: IF (d# / 2) - INT(d# / 2) = .5 THEN d# = d# - 1: d# = d# / 2

d# = d# + 1: IF z = 0 THEN d# = d# - 1: IF z = 1 THEN d# = d# - 2

PRINT , z; v#; r#; v# - r#; d#; sr#

test = (z = can2(lb#))

IF test = 0 THEN STOP

IF v# <= 1 THEN EXIT DO

a$ = INKEY$

REM INPUT a$
```

```
IF a$ = "s" THEN STOP

dist# = v# - r#

IF (dist# / 2) - INT(dist# / 2) = .5 THEN dist# = dist# - 1

dist# = dist# / 2

dist# = dist# + 1

v# = v# - dist#

r# = r# - sr#

IF r# < 1 THEN r# = 1

lb# = lb# - 1

sr# = can(lb#)

LOOP

a$ = INKEY$

IF a$ = "s" THEN STOP

GOTO redo:
```

"C:\Users\Reactor1967\vcns\work\102303G.BAS"

```
REM Try to look at sz to z to determine how much speed of r# is

REM This program codes and decodes and does it rather well but I need

REM a more solid method for knowing how much the speed of r# increases

REM or decreases by just looking at N on the decode. When I can use N

REM to control the speed of r# I can make this work.

redo:

v# = 5

r# = 1
```

```
CLS

RANDOMIZE TIMER

x# = 0

z = 0

sr# = 0

PRINT "CODING !!!!!!!!!!!!"

SLEEP 1

DO

sz = z

z = INT(RND * 2) + 0

d# = v# - r#

test1 = (sz = 0) AND (z = 0)

test2 = (sz = 1) AND (z = 0)

test3 = (sz = 0) AND (z = 1)

test4 = (sz = 1) AND (z = 1)

IF test1 = -1 THEN sr# = sr# + 1

IF test2 = -1 THEN sr# = sr# + 1

IF test3 = -1 THEN sr# = sr# + 1

IF test4 = -1 THEN sr# = sr# + 2

r# = r# + sr#

IF z = 0 THEN v# = v# + (v# - r#) + 1

IF z = 1 THEN v# = v# + (v# - r#) + 2

m# = ((v# - r#) / 4) - INT((v# - r#) / 4): m# = m# * 10: m# = INT(m#)

PRINT z; m#; v#; r#; v# - r#; d#; sr#

SLEEP 1

a$ = INKEY$
```

```
REM INPUT a$

IF a$ = "s" THEN STOP

IF a$ = "d" THEN EXIT DO

IF v# >= 555555555555555# THEN EXIT DO

LOOP

PRINT "DECODING !!!!!!!!!!!!!!!!!!!!!"

SLEEP 1

DO

m1# = ((v# - r#) / 4) - INT((v# - r#) / 4): m1# = m1# * 10: m1# = INT(m1#)

test1 = (m1# = 0) OR (m1# = 5)

test2 = (m1# = 2) OR (m1# = 7)

IF test1 = -1 THEN z1 = 1

IF test2 = -1 THEN z1 = 0

PRINT z1; v#; r#; v# - r#; sr#

SLEEP 1

REM IF v# <= 1 THEN SYSTEM

IF v# <= 1 THEN EXIT DO

a$ = INKEY$

REM INPUT a$

IF a$ = "s" THEN STOP

dist# = (v# - r#)

IF (dist# / 2) - INT(dist# / 2) = .5 THEN dist# = dist# - 1

dist# = dist# / 2

dist# = dist# + 1

v# = v# - dist#

gh# = r#
```

```
r# = r# - sr#

IF gh# = 1 THEN r# = 1

m2# = ((v# - r#) / 4) - INT((v# - r#) / 4): m2# = m2# * 10: m2# = INT(m2#)

test1 = (m2# = 0) OR (m2# = 5)

test2 = (m2# = 2) OR (m2# = 7)

IF test1 = -1 THEN z2 = 1

IF test2 = -1 THEN z2 = 0

test1 = (z1 = 0) AND (z2 = 0)

test2 = (z1 = 1) AND (z2 = 0)

test3 = (z1 = 0) AND (z2 = 1)

test4 = (z1 = 1) AND (z2 = 1)

IF test1 = -1 THEN sr# = sr# - 1

IF test2 = -1 THEN sr# = sr# - 1

IF test3 = -1 THEN sr# = sr# - 1

IF test4 = -1 THEN sr# = sr# - 2

IF sr# < 1 THEN sr# = 0

LOOP

a$ = INKEY$

IF a$ = "s" THEN STOP

GOTO redo




"C:\Users\Reactor1967\vcns\work\102303F.BAS"

REM Try to look at sz to z to determine how much speed of r# is

v# = 1
```

```
r# = 1

CLS

RANDOMIZE TIMER


sr# = 0

DO

sz = z: REM not implement here will do it in next program.

z = INT(RND * 2) + 0

d# = v# - r#

IF z = 0 THEN v# = v# + (v# - r#) + 1

IF z = 1 THEN v# = v# + (v# - r#) + 2

m# = ((v# - r#) / 4) - INT((v# - r#) / 4): m# = m# * 10: m# = INT(m#)

PRINT , , z; m#; v#; r#; v# - r#; d#; sr#; r# - x#

x# = r#

r# = r# + sr#

IF z = 1 THEN sr# = sr# + 2

IF z = 0 THEN sr# = sr# + 1

A$ = INKEY$

INPUT A$

IF A$ = "s" THEN STOP

LOOP
```

"C:\Users\Reactor1967\vcns\work\102303E.BAS"

REM Makeing sr# a function of the speed of v# is most dependable as far

```
REM as predicting how much d will increase or decrease. Also v - r is also
REM very predictable in increasing and decreasing. But again you need to
REM know your previous N.
v# = 1
r# = 1
sr# = 0
CLS
DO
sz = z
z = INT(RND * 2) + 0
d# = v# - r#
sv# = v#
IF z = 0 THEN v# = v# + (v# - r# + 1)
IF z = 1 THEN v# = v# + (v# - r# + 2)
m# = ((v# - r#) / 4) - INT((v# - r#) / 4): m# = m# * 10: m# = INT(m#)
PRINT , z; m#; v#; r#; v# - r#; (v# - r#) - d#; d#; sr#
REM sr# = r#:rem Dont use this when using d# - sr# >=3 then sr# = sr# + 4
a$ = INKEY$
REM INPUT a$
IF a$ = "s" THEN STOP
REM --------------------------------- Enabled (Only use one routine at a time)
IF d# - sr# >= 3 THEN sr# = sr# + 4: REM I guess I need to be play with this.
r# = r# + sr#
REM -------------------------------------- disabled
REM r# = r# + ((v# - sv#) - 2)
REM -------------------------------------- disabled
```

```
REM test1 = (z = 0) AND (sz = 0)

REM test2 = (z = 0) AND (sz = 1)

REM test3 = (z = 1) AND (sz = 1)

REM test4 = (z = 1) AND (sz = 0)

REM IF test1 = -1 THEN r# = r# + ((v# - sv#) - 2)

REM IF test2 = -1 THEN r# = r# + ((v# - sv#) - 1)

REM IF test3 = -1 THEN r# = r# + ((v# - sv#) - 2)

REM IF test4 = -1 THEN r# = r# + ((v# - sv#) - 3)

REM --------------------------------------------

LOOP




"C:\Users\Reactor1967\vcns\work\102303D.BAS"

REM This was very interesting but no go. I found I could keep v - r constant

REM while coding data so that (v - r) - d = speed of r. But it was to

REM clean and no way I could find to determine N.

v# = 2023

r# = 1

CLS

RANDOMIZE TIMER

DO

z = INT(RND * 2) + 0

d# = 0

vr# = (v# - r#)

DO
```

```
s# = (vr# - d#)

x# = (d# * 2)

IF z = 0 THEN x# = x# + 2

IF z = 1 THEN x# = x# + 4

test = (x# - d#) = s#

IF test = -1 THEN EXIT DO

a$ = INKEY$: IF a$ = "s" THEN STOP

d# = d# + 1

IF d# >= vr# THEN STOP

LOOP

r# = v# - d#

l# = v# - r#

IF z = 0 THEN v# = v# + (v# - r#) + 2

IF z = 1 THEN v# = v# + (v# - r#) + 4

m# = (r# / 4) - INT(r# / 4): m# = m# * 10: m# = INT(m#)

m2# = (v# / 4) - INT(v# / 4): m2# = m2# * 10: m2# = INT(m2#)

PRINT , z; m2#; m#; v#; r#; v# - r#; (v# - r#) - l#; " | "; l#; r# - sr#

sr# = r#

INPUT a$

IF a$ = "s" THEN STOP

LOOP
```

"C:\Users\Reactor1967\vcns\work\102303C.BAS"

REM here im trying to simulate (v# - r#) - d# equal to the speed of r#

```
REM It seems to work. Now to write a program to test it.

RANDOMIZE TIMER

CLS

DO

vr# = INT(RND * 20000) + 100

z = INT(RND * 2) + 0

d# = 0

DO

x# = (d# * 2) + z + 1

s# = vr# - d#

test = (x# - d#) = x#

IF test = -1 THEN EXIT DO

d# = d# + 1

IF d# >= vr# THEN STOP

a$ = INKEY$: IF a$ = "s" THEN STOP

LOOP

PRINT , , vr#; vr# - d#; s#

INPUT a$

IF a$ = "s" THEN STOP

LOOP
```

"C:\Users\Reactor1967\vcns\work\102303B.BAS"

vr# = 0

```
d# = 0

CLS

count = 0

DO

count = count + 1

vr# = vr# + 1

PRINT , , vr#; d#; (vr# - d#); (d# * 2) + 1; (d# * 2) + 2

a$ = INKEY$

INPUT a$

IF a$ = "s" THEN STOP

IF count = 2 THEN d# = d# + 1

IF count = 2 THEN count = 0

LOOP




"C:\Users\Reactor1967\vcns\work\102303A.BAS"

DIM can1(10000)

DIM can2(10000)

v# = 10

CLS

FOR count = 1 TO 10000

can1(count) = 0

can2(coount) = 0

NEXT count

RANDOMIZE TIMER
```

```
count = 0

DO

z = INT(RND * 2) + 0

FOR count = 1 TO 10000

test1 = (z = 0) AND (can1(count) = 0)

test2 = (z = 1) AND (can2(count) = 0)

IF test1 = -1 THEN GOTO skip:

IF test2 = -1 THEN GOTO skip:

NEXT count

skip:

IF count >= 10000 THEN STOP

r# = v# - count

IF z = 0 THEN can1(count) = 1

IF z = 1 THEN can2(count) = 1

IF z = 0 THEN v# = v# + (v# - r#) + 1

IF z = 1 THEN v# = v# + (v# - r#) + 2

m# = ((v# - r#) / 4) - INT((v# - r#) / 4): m# = m# * 10: m# = INT(m#)

PRINT , z; m#; v#; r#; v# - r#; count; r# - sr#

sr# = r#

a$ = INKEY$

REM INPUT a$

IF a$ = "s" THEN STOP

LOOP
```

REM here is a program that created the chart listed in vrchart102203.

REM what I can see here is that the d# needs to be used twice each time

REM the first time for 0 and the second time for 1. But that may throw off

REM the speed of r# so that I may need the speed of r# to vary a bit to

REM find N. As you know I keep the speed of r# constanst so it may need to

REM be constance for 0 but different and constant for 1.

DIM can(1000)

lb1# = 1

lb2# = 2

vr# = 1

CLS

DO

d# = vr#

IF (d# / 2) - INT(d# / 2) = .5 THEN d# = d# - 1

d# = d# / 2

d# = d# + 1

test# = (vr# / 4) - INT(vr# / 4): test# = test# * 10: test# = INT(test#)

test1 = (test# = 0) OR (test# = 5)

test2 = (test# = 2) OR (test# = 7)

IF test1 = -1 THEN z = 1

IF test2 = -1 THEN z = 0

IF test1 = -1 THEN d# = d# - 2

IF test2 = -1 THEN d# = d# - 1

DO

s# = vr# - d#

```
test = (s# / 4) - INT(s# / 4) = 0

IF test = -1 THEN EXIT DO

d# = d# + 1

a$ = INKEY$

IF a$ = "s" THEN STOP

LOOP

PRINT , , vr#; d#; " = "; (d# * 2) + 1; s#; (d# * 2) + 2

can(lb1#) = vr#: lb1# = lb1# + 1

can(lb2#) = (d# * 2) + 1: lb2# = lb2# + 1

can(lb2#) = (d# * 2) + 2: lb2# = lb2# + 1

vr# = can(lb1#)

IF lb2# > 1000 THEN SYSTEM

INPUT a$

IF a$ = "s" THEN STOP

LOOP
```

"C:\Users\Reactor1967\vcns\work\102203C.BAS"

REM here is a program that created the chart listed in vrchart102203.

REM what I can see here is that the d# needs to be used twice each time

REM the first time for 0 and the second time for 1. But that may throw off

REM the speed of r# so that I may need the speed of r# to vary a bit to

REM find N. As you know I keep the speed of r# constanst so it may need to

REM be constance for 0 but different and constant for 1.

DIM can(1000)

```
lb1# = 1

lb2# = 2

vr# = 1

CLS

DO

d# = vr#

IF (d# / 2) - INT(d# / 2) = .5 THEN d# = d# - 1

d# = d# / 2

d# = d# + 1

test# = (vr# / 4) - INT(vr# / 4): test# = test# * 10: test# = INT(test#)

test1 = (test# = 0) OR (test# = 5)

test2 = (test# = 2) OR (test# = 7)

IF test1 = -1 THEN z = 1

IF test2 = -1 THEN z = 0

IF test1 = -1 THEN d# = d# - 2

IF test2 = -1 THEN d# = d# - 1

DO

s# = vr# - d#

test = (s# / 4) - INT(s# / 4) = 0

IF test = -1 THEN EXIT DO

d# = d# + 1

a$ = INKEY$

IF a$ = "s" THEN STOP

LOOP

PRINT , , vr#; d#; " = "; (d# * 2) + 1; s#; (d# * 2) + 2

can(lb1#) = vr#: lb1# = lb1# + 1
```

```
can(lb2#) = (d# * 2) + 1: lb2# = lb2# + 1

can(lb2#) = (d# * 2) + 2: lb2# = lb2# + 1

vr# = can(lb1#)

IF lb2# > 1000 THEN SYSTEM

INPUT a$

IF a$ = "s" THEN STOP

LOOP
```

"C:\Users\Reactor1967\vcns\work\102203B.BAS"

```
REM This program plots v# - r#, d#, and speed of r#. Thats all this is for.

REM Lesson learned here. PUT IN BOOK NEW LEARNED STUFF.

REM V2# - R2# FROM V1# - R# FOR 0 AND FOR 1 WILL HAVE THE
SAME

REM D# AND SPEED OF R# SO THATREM MAKES IT POSSIBLE TO

REM CHART V# - R# IF YOU ALWAYS KNOW THE PREVIOUS V1# -
R1#.

REM So maybe take each v# - r# and calculate a d# for it so that the

REM speed of r# is divisible by 4 then take that d# and get to v# - r#s

REM for both 0 and 1.

vr# = 0

y# = vr#

x# = D#

D# = 0

s# = 0
```

```
DO

svr# = vr#

REM z = INT(RND * 2) + 0

z = 0

vr# = (D# * 2) + z + 1

s# = svr# - D#

PRINT , , z; vr#; D#; s#

z = 1

vr# = (D# * 2) + z + 1

s# = svr# - D#

PRINT , , z; vr#; D#; s#

a$ = INKEY$

IF a$ = "s" THEN STOP

PRINT "----------------------------------------------------"

DO

D# = D# + 1

s# = vr# - D#

test = (s# / 4) - INT(s# / 4) = 0

IF test = -1 THEN EXIT DO

a$ = INKEY$

IF a$ = "s" THEN STOP

LOOP

LOOP
```

"C:\Users\Reactor1967\vcns\work\102203A.BAS"

```
REM This program plots v# - r#, d#, and speed of r#. Thats all this is for.
vr# = 0
d# = 0
s# = 0
DO
svr# = vr#
z = INT(RND * 2) + 0
vr# = (d# * 2) + z + 1
s# = svr# - d#
PRINT , , z; vr#; d#; s#
a$ = INKEY$
IF a$ = "s" THEN STOP
DO
d# = d# + 1
s# = vr# - d#
test = (s# / 4) - INT(s# / 4) = 0
IF test = -1 THEN EXIT DO
a$ = INKEY$
IF a$ = "s" THEN STOP
LOOP
LOOP
```

"C:\Users\Reactor1967\vcns\work\102103C.BAS"

```
vr# = 1

sr# = 0

CLS

DO

test = (vr# / 2) - INT(vr# / 2) = .5

d# = vr#

IF test = -1 THEN d# = d# - 1

d# = d# / 2

d# = d# + 1

m# = (vr# / 4) - INT(vr# / 4): m# = m# * 10: m# = INT(m#)

test1 = (m# = 2) OR (m# = 7)

test2 = (m# = 0) OR (m# = 5)

IF test1 = -1 THEN z = 0

IF test2 = -1 THEN z = 1

IF z = 0 THEN d# = d# - 1

IF z = 1 THEN d# = d# - 2

PRINT , z; m#; vr#; d#; (vr# - d#); (d# * 2) + 1; (d# * 2) + 2

a$ = INKEY$

REM INPUT a$

IF a$ = "s" THEN STOP

vr# = vr# + 1

LOOP
```